A PRACTICAL INTRODUCTION TO THE SIMULATION OF MOLECULAR SYSTEMS

This book provides a practical introduction to the range of different techniques available for the simulation of molecular systems. The text includes a library of program modules written in FORTRAN 90 with which the simulations discussed in the text were performed.

Molecular simulation and modeling methods are undergoing rapid development. They are increasingly important tools for fundamental and applied research in academia and industry in such diverse fields as drug design and materials science. Each chapter describes a general class of methods or algorithms and then illustrates their use with example programs, written using the module library. Topics covered include energy functions, optimization of geometry and reaction-path-location techniques, normal mode analysis, molecular dynamics and Monte Carlo simulations and free-energy calculations. The examples and the module library all make use of molecular mechanical energy functions, but almost all of the techniques outlined in the book can be used with other energy functions, such as those based on quantum mechanical methods.

This book will be of interest to advanced undergraduates, graduate students and researchers who use molecular simulation techiques, particularly in theoretical and computational chemistry, biophysics and computational molecular physics.

A PRACTICAL INTRODUCTION TO THE SIMULATION OF MOLECULAR SYSTEMS

MARTIN J. FIELD

Laboratoire de Dynamique Moléculaire, Grenoble

CAMBRIDGE
UNIVERSITY PRESS

PUBLISHED BY THE PRESS SYNDICATE OF THE UNIVERSITY OF CAMBRIDGE
The Pitt Building, Trumpington Street, Cambridge, United Kingdom

CAMBRIDGE UNIVERSITY PRESS
The Edinburgh Building, Cambridge CB2 2RU, UK http://www.cup.cam.ac.uk
40 West 20th Street, New York, NY 10011-4211, USA http://www.cup.org
10 Stamford Road, Oakleigh, Melbourne 3166, Australia

First published 1999

Printed in the United Kingdom at the University Press, Cambridge

Typeset in Times Roman 11/14pt

A catalogue record for this book is available from the British Library

Library of Congress Cataloging in Publication data
Field, Martin (Martin J.)
A practical introduction to the simulation of molecular systems /
Martin J. Field
p.   cm.
Includes bibliographical references and index.
ISBN 0 521 58129 X (hardbound)
1. Molecules – Models – Computer simulation. I. Title.
QD480.F5   1999
541.2′2′0113–dc21   98-37540 CIP

ISBN 0521 58129 X hardback

Contents

Preface   *Page* ix

1 Preliminaries   **1**
   1.1 Introduction   1
   1.2 The DYNAMO module library   2
   1.3 Miscellaneous modules   4
   1.4 Example programs   6
   1.5 Notation and units   7

2 The coordinate file   **12**
   2.1 Introduction   12
   2.2 The structure of the coordinate file   14
   2.3 The atoms and sequence modules   16
   2.4 Reading and writing coordinate files   19
   2.5 Example 1   21
       Exercises   24

3 Operations on coordinates   **25**
   3.1 Introduction   25
   3.2 Connectivity   25
   3.3 Internal coordinates   29
   3.4 Example 2   32
   3.5 Miscellaneous transformations   34
   3.6 Superimposing structures   37
   3.7 Example 3   40
       Exercises   41

4 The energy function   **43**
   4.1 Introduction   43

4.2   The Born–Oppenheimer approximation   43
4.3   Strategies for obtaining energies on a potential
         energy surface   45
4.4   Typical empirical energy functions   47
        4.4.1   Bonding terms   47
        4.4.2   Non-bonding terms   52
        4.4.3   Force-field parametrization   59
4.5   Derivatives of the energy function   60
4.6   The energy modules   62
        Exercises   66

5   **Setting up the molecular mechanics system**   **67**
5.1   Introduction   67
5.2   The molecular mechanics definition file   67
5.3   Processing the MM file   79
5.4   Constructing the system definition   80
5.5   Examples 4 and 5   83
        Exercises   85

6   **Finding stationary points and reaction paths on potential
    energy surfaces**   **86**
6.1   Introduction   86
6.2   Exploring potential energy surfaces   86
6.3   Locating minima   91
6.4   Example 6   93
6.5   Locating saddle points   95
6.6   Example 7   99
6.7   Following reaction paths   100
6.8   Example 8   104
6.9   Determining complete reaction paths   105
6.10 Example 9   109
        Exercises   110

7   **Normal mode analysis**   **112**
7.1   Introduction   112
7.2   Calculation of the normal modes   112
7.3   Rotational and translational modes   118
7.4   Generating normal mode trajectories   122
7.5   Example 10   124
7.6   Calculation of thermodynamic quantities   125

| | 7.7 | Example 11 | 131 |
| | | Exercises | 134 |

8 | **Molecular dynamics simulations I** | | **135**
	8.1	Introduction	135
	8.2	Molecular dynamics	135
	8.3	Example 12	145
	8.4	Trajectory analysis	147
	8.5	Example 13	151
	8.6	Simulated annealing	155
	8.7	Example 14	157
		Exercises	163

9 | **More on non-bonding interactions** | | **164**
	9.1	Introduction	164
	9.2	Cutoff methods for the calculation of non-bonding interactions	164
	9.3	Example 15	175
	9.4	Including an environment	178
	9.5	Minimum image periodic boundary conditions	182
	9.6	Example 16	185
	9.7	Ewald summation techniques	187
	9.8	Fast methods for the evaluation of non-bonding interactions	193
		Exercise	195

10 | **Molecular dynamics simulations II** | | **196**
	10.1	Introduction	196
	10.2	Analysis of molecular dynamics trajectories	196
	10.3	Example 17	205
	10.4	Temperature and pressure control in molecular dynamics simulations	207
	10.5	Example 18	217
	10.6	Calculating free energies: umbrella sampling	219
	10.7	Examples 19 and 20	227
		Exercises	232

11 | **Monte Carlo simulations** | | **234**
	11.1	Introduction	234
	11.2	The Metropolis Monte Carlo method	234
	11.3	Monte Carlo simulations of molecules	239

11.4	Example 21	249
11.5	Calculating free energies: statistical perturbation theory	253
11.6	Example 22	259
	Exercises	265
12	**Miscellaneous topics**	**268**
12.1	Introduction	268
12.2	Z-matrics	268
12.3	Example 23	272
12.4	Constructing solvent boxes and Example 24	272
12.5	Solvating molecules	278
12.6	Example 25	279
12.7	Constraints	282
12.8	Parametrizing force fields	285
12.9	Other topics	289
	Exercises	290
Appendix	**The DYNAMO module library**	**291**
Bibliography		294
Author Index		315
Subject Index		318

Preface

The reason that I have written this book is simple. It is the book that I would have liked to have had when I was learning how to carry out simulations of complex molecular systems. There was certainly no lack of information about the theory behind the simulations but this was widely dispersed in the literature and I often discovered it only long after I needed it. Equally frustrating, the programs to which I had access were often poorly documented, sometimes not at all, and so they were difficult to use unless the people who had written them were available and preferably in the office next door! The situation has improved somewhat since then (the 1980s) with the publication of some excellent monographs but these are primarily directed at simple systems, such as liquids or Lennard-Jones fluids, and do not address many of the problems that are specific to larger molecules.

My goal has been to provide a practical introduction to the simulation of molecules using molecular mechanical potentials. After reading the book, readers should have a reasonably complete understanding of how such simulations are performed, how the programs that perform them work and, most importantly, how the example programs presented in the text can be tailored to perform other types of calculation. The book is an *introduction* aimed at advanced undergraduates, graduate students and confirmed researchers who are newcomers to the field. It does not purport to cover comprehensively the entire range of molecular simulation techniques, a task that would be difficult in 300 or so pages. Instead, I have tried to highlight some of the basic tasks that can be done with molecular simulations and to indicate some of the many exciting developments which are occurring in this rapidly evolving field. I have chosen the references which I have put in carefully as I did not want to burden the text with too much information. Inevitably such a choice is subjective and I apologize in advance to those workers whose work or part of whose work I did not explicitly acknowledge.

There are many people who directly or indirectly have helped to make this book possible and whom I would like to thank. They are: my early teachers in the field of computational chemistry, Nicholas Handy at Cambridge and Ian Hillier at Manchester; Martin Karplus and all the members of his group at Harvard (too numerous to mention!) during the period 1985–9 who introduced me to molecular dynamics simulations and molecular mechanics calculations; Bernie Brooks and Rich Pastor, at the NIH and FDA, respectively, whose lively discussion and help greatly improved my understanding of the simulations I was doing; and all the members of my laboratory at the IBS, past and present, Patricia Amara, Dominique Bicout, Celine Bret, Laurent David, Lars Hemmingsen, Konrad Hinsen, David Jourand, Flavien Proust, Olivier Roche and Aline Thomas. Finally, special thanks go to Patricia Amara and to Dick Wade at the IBS for comments on the manuscript, to Simon Capelin and the staff of CUP for their guidance with the production of the book, to the Commissariat `a l'Energie Atomique and the Centre National de la Recherche Scientifique for financial support and to my wife, Laurence, and to my sons, Mathieu and Jeremy, for their patience.

Martin J. Field
Grenoble

1

Preliminaries

1.1 Introduction

The aim of this book is to provide a practical introduction to performing simulations of molecular systems. To do this, the necessary background about the theory behind various types of simulation is covered and a library of program modules called DYNAMO is provided, which can be used to perform the simulations. The style of the book is pragmatic. Each chapter, in general, contains some theory about a particular topic together with a description of relevant program modules and examples of their use. Suggestions for further work (or exercises) are given at the end.

By the end of the book, readers should have a good idea of how to perform various types of simulation as well as some of the difficulties that are involved. The module library should also provide a reasonably convenient starting point for those wanting to write their own programs to study the systems they are interested in. The fact that users have to write their own programs to do their simulations has advantages and disadvantages. The major advantage is flexibility. Many molecular modeling programs come in a single package and so, inevitably, can provide only a limited range of options. In contrast, a comprehensive library of procedures can be employed to do whatever the user wants and all data in the program will be available for analysis. The drawback is, of course, that the programs have to be written – a task that many readers may not be familiar with or have little inclination to do themselves. However, those who fall into the latter category are urged to read on. The modules have been designed to be easy to use and should be accessible to everyone even if they have only a minimum amount of computing experience.

This chapter explains some essential background information about the programming style in which the modules and the example programs are

1

written. Details of how to obtain the source code of the library for implementation on specific machines are left to the appendix.

1.2 The DYNAMO module library

The module library and the example programs provided with this book are
all written in the programming language FORTRAN 90. The reasons for this
choice were threefold. First, FORTRAN 90 is a powerful and modern programming language. Unlike many modern languages, it is not *object-oriented*, but
it has, among its features, a compact syntax for specifying operations on
arrays and it allows modular code to be written. Second, FORTRAN is still a
very widely used programming language for scientific applications. Third, the
author has done most of his programming in FORTRAN!

The modules and their procedures, which are coded in FORTRAN 90, have
been written so as to be as clear as possible, even if this has been at the
expense of efficiency in some cases. It has also meant that the number of
options available has been kept limited so as not to obscure too much of the
flow of control in the code. The great majority of the modules were written
specifically for this book or were adapted from procedures used by the author
in his programs.

One of the major improvements of FORTRAN 90 over FORTRAN 77 and a
feature that we shall employ extensively is the notion of a *module*. In FORTRAN
77, data (most often in *common blocks*) and procedures (*functions* and *subroutines*) were kept separated. In contrast, FORTRAN 90 allows data and
procedures that are related to be grouped together into coherent units using
modules. Items in a module can be classified as being *private* or *public*. Public
data and procedures are accessible to other modules or programs whereas
private data and procedures can be accessed only from within the module.
This is advantageous because it provides a measure of protection for a module's variables and it means that the unnecessary details of the implementation
of specific algorithms or data structures can be hidden from the remaining
parts of the program. For this reason, only the public procedures and variables will be presented when a module is described in the rest of the book.

The shorthand that will be used for listing the (public) contents of a module
is as follows

```
MODULE EXAMPLE
  ! . Parameter declarations.
  ... ... ...
  ! . Scalar variable declarations.
  ... ... ...
  ! . Array variable declarations.
```

```
... ... ...
! . Type variable declarations.
... ... ...
! . Function and subroutine declarations.
CONTAINS
    SUBROUTINE EXAMPLE_SUBROUTINE1 ( ... arguments ... )
        ... argument declarations ...
    END SUBROUTINE EXAMPLE_SUBROUTINE1
    ... ... ...
END MODULE EXAMPLE
```

The public parameters are declared first, followed by any variables (usually in the order scalar, array and type). The procedure declarations come last after the CONTAINS statement. For the function and subroutine declarations, only the name of the procedure, its calling sequence and the details of its arguments are given. Although not strictly necessary in FORTRAN 90, all parameter and variable declarations list explicitly the nature of the variable (i.e. whether it is CHARACTER, INTEGER, LOGICAL or REAL) and, if it is an array, its dimension. All argument declarations have the INTENT statement which indicates whether the argument is for input only, for output only or both for input and for output.

This is all the information that is needed concerning the declaration of the public parts of a module. However, there are several extra technical points that can be made about the programming used in the modules (and the example programs).

- All arrays are allocated dynamically. That is, they are declared either as ALLOCATABLE or as pointers and then allocated to have the correct size when needed. This makes the module programming more complicated but greatly increases the modules' flexibility because they do not need to be recompiled for larger systems. The use of allocatable arrays has been preferred, in general, over the use of pointers because keeping track of the latter can be difficult.
- In the modules and the example programs, all floating point (REAL) numbers comprise 64 bits although in the listings that appear in the book this is omitted for the sake of clarity. The integer and logical types are left to assume their default values for the particular machines being used (usually 32 bits but sometimes 64 bits).
- Standard FORTRAN 90 has been used throughout. This means that the programs should run (and produce the same results!) on any machine for which a FORTRAN 90 compiler exists. The newer revision of FORTRAN 90, FORTRAN 95, introduces a number of minor, albeit useful, changes but these have not been used in the programming in this book.
- Both function and subroutine procedures are used. A function is employed only in cases in which all of the arguments are input arguments (and so remain

unchanged by the procedure), when there is only one output result and when variables of other modules or public variables of the function's own module are left unchanged. A function may, however, change private variables of its own module or generate an error.

- Most module variables are explicitly initialized when declared. This has been done so that the modules are ready to use immediately without the necessity of first calling special initialization procedures.

1.3 Miscellaneous modules

Most of the modules in the DYNAMO library are scientific modules concerned with the manipulation and simulation of molecular systems, which will be discussed in detail in the remainder of the book. There are, in addition, important modules that deal with a miscellany of other tasks such as input and output and the declaration of physical constants. These will be described only briefly because we shall not encounter the majority of them again. They may be divided into four categories.

The first category contains utility modules that take care of such functions as file management and parsing.

DEFINITIONS contains parameter definitions used by the modules and the example programs. In particular, it contains *all* the machine-dependent parameters that are used by the library. The most important of these is the parameter DP, which is short for *default precision*. It defines the model used for the floating point numbers and is used whenever a real number is declared in a module or program.

FILES deals with file management. In general, users are supposed to manage their own files, but FILES contains a subroutine that will locate the next available FORTRAN *stream* or *unit* number that is available.

IO_UNITS defines as the parameters INPUT and OUTPUT the unit numbers of the input and output streams (i.e. where the program is to read from and where it is to write to).

PARSING contains procedures that read and process data from a formatted file. It is employed by all modules that read formatted files and means that all input involving such files in the book is 'free format' (fixed format input is *not* used!).

STATUS contains the subroutine that is called whenever an error results from the execution of a module or program. This procedure, ERROR, prints out an error message and then, to avoid complications, terminates the execution of the program immediately.

STRING contains procedures that manipulate character strings.

TIME contains a procedure that prints the current date and time. FORTRAN 90,
   unlike the revision FORTRAN 95, has no intrinsic procedures that deal
   with CPU time and so this module does not contain any.

Of these modules, the beginning user is likely to need to use only
DEFINITIONS and IO_UNITS, the former for the parameter DP and the latter
for the parameters INPUT and OUTPUT.

   Modules of the second category store parameter data about various chemical, physical and mathematical constants. There are two.

CONSTANTS contains fundamental mathematical, physical and chemical constants.
ELEMENTS contains data about each element. This includes the element sym-
   bol, its atomic mass and a value for its radius (or 'effective' size).

In both these modules the data are defined in terms of parameters. In other
words, the information can be accessed, but it cannot be changed.

   The third category of modules deals with standard numerical mathematical
tasks. They are the following.

BAKER_OPTIMIZATION is a module that finds a stationary point of a multi-
   dimensional function. The algorithm it uses will be described in more
   detail in section 6.5.
CONJUGATE_GRADIENT has a subroutine for the optimization of a multidimensional function using a conjugate gradient algorithm.
DIAGONALIZATION contains procedures that find the eigenvalues and eigenvectors of real symmetric matrices.
LINEAR_ALGEBRA has procedures for performing standard operations in linear algebra, such as the normalization of vectors.
RANDOM is the module that contains procedures for generating random numbers drawn from uniform and Gaussian distributions. This is the only
   module for which it is advisable to call the module's initialization
   procedure explicitly before use because it sets the value of the seed
   for the random-number generator.
SORT contains procedures for sorting.
SPECIAL_FUNCTIONS has a procedure for computing the complementary
   error function.
STATISTICS contains procedures for performing various types of statistical
   analyses on data. It is explained in more detail in section 10.2.

   The fourth category of modules contains a single example. It is the module
DYNAMO whose main task is to declare all the remaining modules in the library

using USE statements, one for each module. This means that it is only necessary to define this module in an example program to make all the other
modules in the library accessible. There is also a single procedure,
DYNAMO_HEADER, which prints out a title indicating the version number of
the module library. The use of this module is described further below.

1.4 Example programs

In most of the chapters, example programs that illustrate the various modules
in the library are given. All the programs have a standard format and should
provide models for users wanting to write their own. The format is

```
PROGRAM EXAMPLE

! . Module declarations.
USE DYNAMO

IMPLICIT NONE

! . Parameter and variable declarations.
... ... ...

! . Initialization.
... initialization commands ...

! . Program commands.
... ... ...

! . Termination.
... termination commands ...

CONTAINS

    ... Program functions and procedures ...

END PROGRAM EXAMPLE
```

The order of statements is as follows.

1. The program starts with the PROGRAM statement followed by the name of the
   program. In the examples found in this book, all the programs have the name
   EXAMPLE followed by an integer.
2. The next statement declares that the module DYNAMO should be used by the
   program. As described in the previous section, this module lists all the modules
   that are in the program library. It simplifies the use of the module library
   because otherwise each module in the library that was required by the program
   would have to be specified in a separate USE statement. In this way only a single

statement is needed. The USE facility is extremely important because it means that the compiler can check the type characteristics of the module variables and the syntax of calls to module procedures. These checks help to eliminate lots of errors before the program is run.

3. The next statement, which should always be used if local variables are being defined, is IMPLICIT NONE. This compels the types of all variables to be declared explicitly.

4. Following the IMPLICIT NONE statement are declaration statements for all the parameters and variables that occur in the program.

5. After all the declarations come the executable statements. These can be divided into three parts.

   (a) An initialization section that can include statements that initialize module variables and write out information about the date and time, the version number of the module library and, perhaps, a title to indicate what the program is doing.

   (b) A section that contains the statements that perform the task for which the program has been written.

   (c) A termination section that includes statements that tidy up any dynamically allocated storage space, close files and write out miscellaneous information such as, for example, the date and time.

6. The executable commands are followed by the CONTAINS keyword and any local functions or subroutines that are used in the program's executable statements.

7. The text of the program terminates with the END PROGRAM statement.

Some simplifications are made in the presentation of the example programs for clarity. Only a few of the examples use local procedures and so the block following the CONTAINS statement seldom occurs. Also, to reduce the space required by the program listings, some of the declaration statements, including the IMPLICIT NONE keyword, and the initialization and termination commands are omitted. In case of confusion, readers are advised to look at the full texts of the example programs in the source code distribution which do, of course, include these statements.

To use a program once it is written it is necessary to compile it and link it with the modules from the module library. How this is done is machine dependent but where to find this information will be left until the appendix.

1.5 Notation and units

Finally, a few general points about the notation and units used in this book and the program library will be made. In the text, all program listings, module definitions and DYNAMO library procedures and variables have been represented by using characters in typewriter style, e.g. NATOMS.

FORTRAN statements are always written in upper case, although this is not necessary in the FORTRAN 90 standard, and only comment statements (preceded by a !) have lower case letters. For other symbols, normal typed letters are used for scalar quantities while bold face italic letters are employed for vectors and bold face roman for matrices. Lower case letters have generally been taken to represent the properties of individual atoms whereas upper case letters represent the properties of a group of atoms or, more usually, the entire system. Lower case Roman subscripts usually refer to atoms, upper case Roman subscripts to entire structures and Greek subscripts to other quantities, such as the Cartesian components of a vector. The more common symbols are listed in tables 1.1 and 1.2. The units of most of the quantities either employed or calculated by the module library are specified in table 1.3.

Table 1.1 *Symbols that denote quantities for atoms or for the entire system.*

Symbol	Description
Atomic quantities	
α_i	The isotropic dipole polarizability for atom i.
$\mathbf{a}_i \; (\equiv \ddot{\mathbf{r}}_i)$	The acceleration of atom i.
$\mathbf{f}_i$	The force on atom i.
$\mathbf{g}_i$	The first derivatives of the potential energy with respect to the coordinates of atom i.
$\mathbf{h}_{ij}$	The second derivatives of the potential energy with respect to the coordinates of atoms i and j.
m_i	The mass of atom i.
q_i	The charge for atom i.
$\mathbf{r}_i$	The vector of Cartesian coordinates, (x_i, y_i, z_i), for atom i.
r_{ij}	The distance between two atoms, i and j.
$\mathbf{s}_i$	The vector of Cartesian fractional coordinates for atom i.
$\mathbf{v}_i \; (\equiv \dot{\mathbf{r}}_i)$	The velocity of atom i.
w_i	A weighting factor for atom i.
x_i	The x Cartesian coordinate of atom i.
y_i	The y Cartesian coordinate of atom i.
z_i	The z Cartesian coordinate of atom i.
System quantities	
$\boldsymbol{\mu}$	The dipole-moment vector for the system.
A	The $3N$-dimensional vector of atom accelerations.
D	A $3N$-dimensional coordinate displacement vector.
F	The $3N$-dimensional vector of atom forces.
G	The $3N$-dimensional vector for first derivatives.
G_{RMS}	The root mean square (RMS) gradient for a system.
H	The $(3N \times 3N)$-dimensional matrix of second derivatives of a system.
M	The $3N \times 3N$ diagonal atomic mass matrix.
R	The $3N$-dimensional vector of atom coordinates.
R_{c}	The center of charge, geometry or mass of a system.
S	The $3N$-dimensional vector of atom fractional coordinates.
V	The $3N$-dimensional vector of atom velocities.

Table 1.2 *Miscellaneous symbols.*

Symbol	Description
Matrix and vector operations	
$\dot{a}$	The first time derivative of a vector.
$\ddot{a}$	The second time derivative of a vector.
$\hat{a}$	A normalized vector, a/a, where $a = |a|$.
a^{T}	The transpose of a vector a.
$\|\mathbf{A}\|$	The determinant of a matrix $\mathbf{A}$.
$a^{\mathrm{T}}b$	A dot or scalar product of two vectors.
ab^{T}	An outer product of two vectors.
$a \wedge b$	A cross or vector product.
Other symbols	
A	The Helmholtz free energy.
ϵ	The dielectric constant for a system.
ϵ_0	The permittivity of the vacuum.
E	The total energy of a system.
G	The Gibbs free energy.
H	The enthalpy.
$\mathcal{H}$	The classical Hamiltonian for a system.
$\mathcal{K}$	The kinetic energy for a system.
k_{B}	Boltzmann's constant.
L	The length of a side of a cubic box.
N	The number of atoms in the system.
N_{df}	The number of degrees of freedom in the system.
N_{r}	The number of residues in the system.
P	The pressure.
$\mathcal{P}$	The instantaneous pressure.
R	The molar gas constant.
T	The temperature.
$\mathcal{T}$	The instantaneous temperature.
S	The entropy.
t	The time.
U	The internal energy.
$\mathbf{U}$	A 3×3 proper or improper rotation matrix.
$\mathcal{V}$	The potential energy of a system.
V	The volume of a system.

Table 1.3 *The units employed by the* DYNAMO *library.*

Quantity	Units
Angle (input and output)	Degrees (°)
Angle (internally)	Radians
Charge	Elementary charge ($e \simeq 1.602 \times 10^{-19}$ C)
Energy	kJ mol^{-1}
Frequency	Wavenumbers (cm^{-1}) and ps^{-1}
Length	Ångström units (1 Å $\equiv 10^{-10}$ m)
Mass	Atomic mass units (1 a.m.u. $\simeq 1.661 \times 10^{-27}$ kg)
Pressure	Atmospheres (1 atm $\equiv 1.013250 \times 10^{5}$ Pa)
Temperature	Kelvins (K)
Time	Picoseconds (1 ps $\equiv 10^{-12}$ s)
Volume	Å^{3}

2

The coordinate file

2.1 Introduction

To perform any sort of modeling of a system it is first necessary to define its composition, i.e. the number and the type of particles that it contains. For calculations on very small systems a simple list of atoms would be sufficient. This turns out to be very uninformative as the system gets bigger, however, and some sort of classification or partitioning scheme becomes necessary. An example should make this clear. Consider the simulation of a protein, say haemoglobin, in water which is illustrated schematically in figure 2.1. The protein has four chains, two α and two β, each of which comprises about 150 amino acids and a haem group. The bath of solvent in which the protein is immersed will consist of several thousand molecules in addition to ionic species. It is evident that it will be much easier to manipulate various parts of the system if they are named than to have a contiguous string of anonymous atoms.

Here, the following classification is adopted. First of all it is supposed that the system can be divided up into *subsystems*, each of which will have a unique name. In the haemoglobin case, it would be reasonable to have five subsystems, one for each of the protein chains and one for the solvent molecules. Each subsystem in turn consists of an ordered sequence of *residues*. Residues have non-unique names and it is their order within the subsystems that determines their identity. For the protein example, there would be residues corresponding to the appropriate amino acids at their appropriate positions in the protein chains, to the haem groups and to the water molecules. The third and final level of subdivision is to define the atoms that occur in each residue. All the atoms in a residue are given unique names. So, for example, a water residue could have atoms named O, H1 and H2. It should be noted that not all residues of the same type need have the same number of atoms because nominally identical residues can be

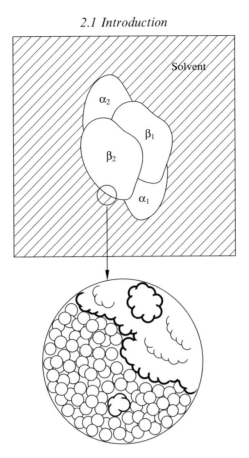

Figure 2.1. A schematic diagram of a protein, haemoglobin, in water.

slightly different. A common example would be if a group in a residue had a varying protonation state.

To summarize, an atom in a system can be identified by the name of the subsystem to which it belongs, the name and the number of the residue within the subsystem to which it belongs and its own name. It is worth emphasizing that the orders of the atoms within each residue or of the subsystems in each system are not important (because atoms and subsystems have unique names), in contrast to the case of residues, in which the order provides critical information.

Another essential element for most molecular simulation studies is a set of coordinates which define the positions of the particles in space. Given the nature of the system's atoms and their coordinates the *molecular structure* of the system is known and it is possible to deduce information about the system's physical properties and its chemistry. The generation of sets of coordinates for particular systems is the major goal of a number of important experimental techniques, including X-ray crystallography and nuclear magnetic resonance spectroscopy, and there are data banks that act as repositories for the coordinate sets of many types of molecular system obtained by such methods.

There are several alternative coordinate systems that it is possible to use to define the atom positions. For the most part in this book, *Cartesian coordinates* are employed. These give the absolute position of an atom in three-dimensional space in terms of its x, y and z coordinates. Other schemes include *crystallographic coordinates* in which the atom positions are given in a coordinate system that is based upon the symmetry of the crystal and *internal coordinates* that define the position of an atom by its position relative to a number of other atoms (usually three).

2.2 The structure of the coordinate file

Probably the simplest way to define the composition of a system and the coordinates of its particles is to use a *coordinate file*. Such a file gives the identity of the particles in the system and their classification, the particles' coordinates and, perhaps, a miscellany of other information. There are many different coordinate-file formats in use but here a unique, relatively simple, format is used, which has been adapted for the programs presented in this book. It is to be noted that we shall assume in almost all our discussions that a complete coordinate file for a system exists. Sometimes, of course, the coordinates of some atoms may be missing or even their identities may be unknown (this is often the case with experimentally derived coordinates).

We illustrate the format of the coordinate file with a small molecule, *N*-methyl-alanyl-acetamide, which is shown in figure 2.2. It is also sometimes referred to as blocked alanine (bALA) or as the alanine dipeptide. bALA is often used in modeling studies because it is relatively small but still sufficiently complex that it displays interesting behaviour. It has two peptide-like bonds that are the bonds which link consecutive amino acids in a protein together and so it is often employed as a model of protein systems when the alanyl moiety is in its L form.

The coordinate file is

```
!
! Coordinates for blocked alanine (the ''dipeptide'') in Angstroms.
!
    22    3    1 ! Number of atoms, residues and subsystems.
!=================================================================================
Subsystem    1  Blocked_Alanine
      3 ! Number of residues.
```

```
!=================================================================
Residue     1  Acetyl
     6 ! Number of atoms.
     1   CT        6      6.9340000020     4.4601000008     0.0000000000
     2   HT1       1      7.3141000006     3.9362999998    -0.9099999993
     3   HT2       1      7.3153999984     3.9354000017     0.9088999988
     4   HT3       1      7.3121000020     5.5110999979     0.0000000000
     5   C         6      5.4236000005     4.4898999982     0.0000000000
     6   O         8      4.7946000001     5.5678999999     0.0000000000
!-----------------------------------------------------------------
Residue     2  Alanyl
    10 ! Number of atoms.
     7   N         7      4.7771999992     3.2719000007     0.0000000000
     8   H         1      5.3083999976     2.4299000007     0.0000000000
     9   CA        6      3.3395000003     3.1332000026     0.0000000000
    10   HA        1      2.9463999977     3.6002000017     0.9547000006
    11   C         6      2.9402999980     1.6358999984     0.0000000000
    12   O         8      3.7873000024     0.7186999989     0.0000000000
    13   CB        6      2.6569999995     3.8236999977    -1.1855999994
    14   HB1       1      1.5575000011     3.8895999998    -1.0070000018
    15   HB2       1      3.0452000016     4.8664999979    -1.3025999983
    16   HB3       1      2.8417000019     3.2640999983    -2.1326000015
!-----------------------------------------------------------------
Residue     3  N_Methyl
     6 ! Number of atoms.
    17   N         7      1.5951999995     1.3490000024     0.0000000000
    18   H         1      0.9338999995     2.0855000019     0.0000000000
    19   CT        6      1.1209000017     0.0000000000     0.0000000000
    20   HT1       1      0.0000000000     0.0000000000     0.0000000000
    21   HT2       1      1.4953999989    -0.5448999980     0.9084999990
    22   HT3       1      1.4949999992    -0.5446000022    -0.9088000002
!-----------------------------------------------------------------
```

The system defined in the coordinate file is comprised of only one subsystem that contains the complete bALA molecule. The molecule itself is divided into three residues – terminal acetyl and *N*-methyl groups and a central alanyl residue.

The following points about the format are important.

- An exclamation mark (!) denotes a comment and any characters after it are ignored. Note that in the example lots of comments are used for clarity.
- The first non-comment line contains the numbers of atoms, residues and subsystems. The subsystems, their residues and their atoms are then listed sequentially in the appropriate order. The lines which define the subsystem and the residues contain the name of the subsystem or residue and its number. The lines immediately afterwards contain the number of residues in the subsystem or the number of atoms in the residue.

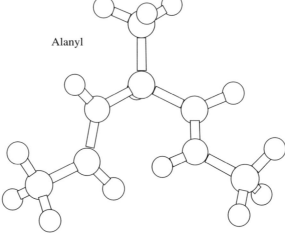

Alanyl

Acetyl   *N*–Methyl

Figure 2.2. *N*-Methyl-alanyl-acetamide.

- The lines listing the details of the atoms contain the sequential number of the atom in the file, the name of the atom, its atomic number (i.e. its element type) and its x, y and z coordinates. The units of length are ångström units (Å).
- The subsystem, residue and atom names are restricted to be contiguous strings of alphanumeric characters – no spaces are allowed.
- The file as a whole is in *free format* which means that the characters do not need to be aligned in particular columns.

2.3 The atoms and sequence modules

It is straightforward to design data structures that can store information about the composition of the system. This book employs two modules, SEQUENCE and ATOMS. The former contains the information about the subsystems and the residues while the second deals with the atoms. They are kept separate in this way because for many calculations only the atom information is needed without the accompanying sequence data. The contents of both modules are listed in full because they are the first modules that we shall describe in detail.

The definition of the atoms module is

```
MODULE ATOMS

! . Parameter declarations.
INTEGER, PARAMETER :: ATOM_NAME_LENGTH = 8
```

```
! . Scalar data.
INTEGER :: NATOMS = 0

! . Character array data.
CHARACTER ( LEN = ATOM_NAME_LENGTH ), ALLOCATABLE, DIMENSION(:) :: ATMNAM

! . Integer array data.
INTEGER, ALLOCATABLE, DIMENSION(:) :: ATMNUM

! . Real array data.
REAL, ALLOCATABLE, DIMENSION(:)    :: ATMMAS
REAL, ALLOCATABLE, DIMENSION(:,:) :: ATMCRD

! . The module procedures.
CONTAINS

   SUBROUTINE ATOMS_ALLOCATE ( N )
      INTEGER, INTENT(IN) :: N
   END SUBROUTINE ATOMS_ALLOCATE

   SUBROUTINE ATOMS_FORMULA
   END SUBROUTINE ATOMS_FORMULA

   SUBROUTINE ATOMS_INITIALIZE
   END SUBROUTINE ATOMS_INITIALIZE

   SUBROUTINE ATOMS_SUMMARY
   END SUBROUTINE ATOMS_SUMMARY

END MODULE ATOMS
```

The module has the following features.

- The number of atoms is given by the scalar variable NATOMS. It is explicitly initialized to zero and so will have this value at the start of a program.
- The atom names, which can be a maximum of ATOM_NAME_LENGTH characters long, are contained in the character array ATMNAM.
- The arrays ATMNUM and ATMMAS contain the atomic numbers and the masses of the atoms, respectively. The masses are the standard masses for the element type of the atom and are normally given in atomic mass units (a.m.u.). If specific isotopic or other special values for the masses of particular atoms are required then these values must be set by the user.
- The array ATMCRD is a two-dimensional array containing the x, y and z Cartesian coordinates for the atoms. In use the array is dimensioned ATMCRD(1:3,1 :NATOMS). If any coordinates happen to be undefined the corresponding elements in the array are filled with the number 999999.0, which is the value of the parameter UNDEFINED in the module CONSTANTS.

- The subroutine ATOMS_ALLOCATE assigns the number of atoms in the module NATOMS to be the value of its scalar integer argument, N, and allocates all the variable arrays in the module to the correct sizes.
- The subroutine ATOMS_INITIALIZE sets the number of atoms in the data structure to zero and deallocates all the module arrays.
- The subroutines ATOMS_FORMULA and ATOMS_SUMMARY print out information about the atoms of the system defined in the module.

All the arrays ATMCRD, ATMMAS, ATMNAM and ATMNUM are initially unallocated and so any attempt to access any of their elements will result in a program error. They will be automatically allocated by other modules when this is required (for example, when a coordinate file is read). It is strongly recommended that users leave the allocation of these arrays to those modules. Individual array elements can be accessed or even changed but their explicit allocation and deallocation should be avoided.

The sequence module has a similar structure to the previous module. It contains

```
MODULE SEQUENCE

! . Parameter declarations.
INTEGER, PARAMETER ::   RESIDUE_NAME_LENGTH = 32, &
                     SUBSYSTEM_NAME_LENGTH = 32

! . Scalar data.
INTEGER :: NRESID = 0, NSUBSYS = 0

! . Character array data.
CHARACTER ( LEN =  RESIDUE_NAME_LENGTH ), ALLOCATABLE, DIMENSION(:) :: RESNAM
CHARACTER ( LEN = SUBSYSTEM_NAME_LENGTH ), ALLOCATABLE, DIMENSION(:) :: SUBNAM

! . Integer array data.
INTEGER, ALLOCATABLE, DIMENSION(:) :: RESIND, SUBIND

! . The module procedures.
CONTAINS

   SUBROUTINE SEQUENCE_ALLOCATE ( NRES, NSUB )
      INTEGER, INTENT(IN) :: NRES, NSUB
   END SUBROUTINE SEQUENCE_ALLOCATE

   SUBROUTINE SEQUENCE_INITIALIZE
   END SUBROUTINE SEQUENCE_INITIALIZE

   SUBROUTINE SEQUENCE_PRINT
   END SUBROUTINE SEQUENCE_PRINT

   SUBROUTINE SEQUENCE_SUMMARY
   END SUBROUTINE SEQUENCE_SUMMARY

END MODULE SEQUENCE
```

The various items in the module are the following.

- NRESID and NSUBSYS give the numbers of residues and subsystems, respectively, in the system. They are initialized to zero at the start of the program.
- The character arrays, RESNAM and SUBNAM, contain the residue and subsystem names. The maximum lengths of these names are given by the parameters RESIDUE_NAME_LENGTH and SUBSYSTEM_NAME_LENGTH, respectively.
- The integer arrays, RESIND and SUBIND, are index arrays that indicate which atoms are in a residue and which residues are in a subsystem. For example, RESIND will normally have NRESID + 1 elements and the element RESIND(i+1) gives the number of the last atom in the *i*th residue. Thus, the first atom in the *i*th residue is given by RESIND(i)+1 and the total number of atoms in the residue by RESIND(i+1) - RESIND(i). The array SUBIND has the same structure except that it deals with residues rather than atoms. Therefore, it has NSUBSYS + 1 elements and the *i*th subsystem will have SUBIND(i+1) - SUBIND(i) residues.
- The subroutine SEQUENCE_ALLOCATE sets the numbers of residues and subsystems to the values passed in the arguments NRES and NSUB, respectively, and allocates the module's variable arrays to the appropriate sizes.
- The subroutine SEQUENCE_INITIALIZE initializes the sequence data structure variables. It sets the numbers of residues and subsystems to zero and deallocates the variable arrays.
- The subroutines SEQUENCE_PRINT and SEQUENCE_SUMMARY print miscellaneous information about the sequence data structure.

Like the arrays in the module ATOMS, the arrays in SEQUENCE are initially unallocated and their allocation is handled automatically by other modules. In contrast to the variables in ATOMS, however, it is probable that the variables in SEQUENCE need be accessed much less frequently by the user.

The only subroutines that readers are likely to need in both modules initially are those that print information about the modules, i.e. the subroutines ATOMS_FORMULA, ATOMS_SUMMARY, SEQUENCE_PRINT and SEQUENCE_SUMMARY.

2.4 Reading and writing coordinate files

In the previous sections some necessary background has been presented. In this section, the first module that readers are likely to want to use in performing their simulations is introduced. It is COORDINATE_IO, which contains procedures that read and write coordinate files of the type described in section 2.2.

The module's definition is

```
MODULE COORDINATE_IO

CONTAINS

   SUBROUTINE COORDINATES_DEFINE ( FILE )
      CHARACTER ( LEN = * ), INTENT(IN), OPTIONAL :: FILE
   END SUBROUTINE COORDINATES_DEFINE

   SUBROUTINE COORDINATES_READ ( FILE, DATA, SELECTION )
      CHARACTER ( LEN = * ), INTENT(IN), OPTIONAL :: FILE
      LOGICAL, DIMENSION(1:NATOMS),     INTENT(IN),  OPTIONAL :: SELECTION
      REAL,    DIMENSION(1:3,1:NATOMS), INTENT(OUT), OPTIONAL :: DATA
   END SUBROUTINE COORDINATES_READ

   SUBROUTINE COORDINATES_WRITE ( FILE, DATA, SELECTION )
      CHARACTER ( LEN = * ), INTENT(IN), OPTIONAL :: FILE
      LOGICAL, DIMENSION(1:NATOMS),     INTENT(IN), OPTIONAL :: SELECTION
      REAL,    DIMENSION(1:3,1:NATOMS), INTENT(IN), OPTIONAL :: DATA
   END SUBROUTINE COORDINATES_WRITE

END MODULE COORDINATE_IO
```

There are three subroutines in the module, two of which read a coordinate file and one of which writes a coordinate file. Each subroutine has an optional scalar character argument that gives the name of the coordinate file for the read or write operation. If the argument is not present then the coordinates are read from the default input stream (for the read subroutines) or written to the default output stream (for the write subroutine). The subroutines COORDINATES_READ and COORDINATES_WRITE have two more optional arguments. The first of these is DATA, which is a two-dimensional array that contains the coordinate data that are to be written by COORDINATES_WRITE or that will be filled with the coordinate data that have been read by COORDINATES_READ. The array ATMCRD is employed for these operations if this argument is absent. For the subroutine COORDINATES_DEFINE there is no choice – ATMCRD is always employed. The last optional argument, SELECTION, is an input logical array of length NATOMS which specifies which atoms' data are to be read or written. Thus, if an element of SELECTION for a particular atom is .TRUE., the atom's data are read or written whereas, if the element is .FALSE., the atom's data are ignored. If this argument is not present, the default is to read or write the data for all atoms in the system.

Of the two read subroutines, COORDINATES_DEFINE can be used if no atom or sequence information has been defined previously. It will read the coordinate file, check that the syntax is correct and create and fill the atom and sequence data structures. This is the subroutine that must be called first in a program. If it is used afterwards then all existing atom and sequence information is destroyed and redefined when the new coordinate file is read. In contrast, the subroutine COORDINATES_READ should only be called once the atom and sequence data structures exist (for example, after a previous call to COORDINATES_DEFINE). This subroutine reads the coordinate file and checks that the composition of the system in it is consistent with the data already defined in the modules ATOMS and SEQUENCE. If the data match, the x, y and z coordinate data are read from the file and put in the array DATA, if this argument is present, or in the array ATMCRD otherwise. The write subroutine is simpler and leaves all the relevant data structures unchanged. It writes out a coordinate file using the x, y and z coordinate data in the array DATA, if this is present, or those in the array ATMCRD if it is not.

All three subroutines will give error messages if there is a problem with the file (for example, if it cannot be opened) while the read subroutines will also generate errors if there are mistakes in the syntax of the coordinate file. In addition, COORDINATES_READ will fail if the sequence defined in the input file does not agree with that already defined in the ATOM and SEQUENCE data structures.

2.5 Example 1

The first example program is very simple. It reads the coordinate file for a small molecule and writes a summary of the sequence and atom information. The values of the atomic coordinates are stored in a temporary, locally defined array and another set of coordinates is read in from a second file for the same system. Finally, the coordinates are subtracted and the differences between the two sets are written out to the default output stream.

The program is

```
PROGRAM EXAMPLE1

... Declaration Statements ...

! . Local array declarations.
REAL, ALLOCATABLE, DIMENSION(:,:) :: TMPCRD

! . Define the atom and sequence data structures.
CALL COORDINATES_DEFINE ( ''bALA1.crd'' )

! . Write out information about the atom data.
CALL ATOMS_FORMULA
CALL ATOMS_SUMMARY
```

```
! . Write out information about the sequence data.
CALL SEQUENCE_PRINT
CALL SEQUENCE_SUMMARY

! . Allocate the temporary array.
ALLOCATE ( TMPCRD(1:3,1:NATOMS) )

! . Read in the second set of coordinates.
CALL COORDINATES_READ ( ''bALA2.crd'', TMPCRD )

! . Write out the differences to the output stream.
CALL COORDINATES_WRITE ( DATA = ( ATMCRD - TMPCRD ) )

! . Deallocate the temporary array.
DEALLOCATE ( TMPCRD )


END PROGRAM EXAMPLE1
```

The power of the FORTRAN 90 array syntax is illustrated by the use of the temporary coordinate array TMPCRD. The array is defined to be ALLOCATABLE in the declarations and it is allocated to have the correct size after the call to COORDINATES_DEFINE once the number of atoms in the system is known. It is passed as an argument to the subroutine COORDINATES_READ and it forms part of the argument to COORDINATES_WRITE, in which the differences between the coordinates are calculated using the simple statement ATMCRD − TMPCRD. At the end of the program it is deallocated explicitly when the differences have been printed. The deallocation is not strictly necessary because the array will be automatically freed once the program stops but it is a good idea for the user to get into the habit of deallocating arrays explicitly when they are no longer needed.

The output from the program is self-explanatory but it is given here for completeness. It is

```
Sequence read from bALA1.crd
------------------------------- Atom Formula ---------------------------------
 H      12   C      6    N      2   O      2
------------------------------------------------------------------------------


--------------------------- Summary of Atom Data -----------------------------
Number of Atoms      =            22  Number of Hydrogens =              12
Number of Heavy Atoms =           10  Number of Unknowns  =               0
Total Mass (a.m.u.)  =         144.2
------------------------------------------------------------------------------
```

```
----------------------------- System Sequence ------------------------------
Subsystem 1 BLOCKED_ALANINE:
    ACETYL                                      ALANYL
    N_METHYL
-------------------------------------------------------------------------------

----------------------- Summary of the System Sequence -----------------------
Subsystem 1 BLOCKED_ALANINE:
  ACETYL                         1  ALANYL                                1
  N_METHYL                       1
-------------------------------------------------------------------------------

Data read from bALA2.crd

Data written to the output stream.

!==============================================================================
    22     3     1 ! # of atoms, residues and subsystems.
!==============================================================================
Subsystem    1  BLOCKED_ALANINE
    3 ! # of residues.
!==============================================================================
Residue     1 ACETYL
     6 ! # of atoms.
     1    CT          6        -0.1748900000    -0.2724300000    -0.2537000000
     2    HT1         1        -0.0855300000    -0.8222900000     0.5654700000
     3    HT2         1        -0.1204100000    -0.7706900000    -1.1408500000
     4    HT3         1        -0.3259800000     0.6545000000    -0.2587000000
     5    C           6        -0.2115300000    -0.1023100000    -0.1345900000
     6    O           8        -0.3490800000    -0.0973500000    -0.0742200000
!------------------------------------------------------------------------------
Residue     2 ALANYL
    10 ! # of atoms.
     7    N           7        -0.0751800000     0.0533600000    -0.1033900000
     8    H           1         0.0014800000     0.0705700000    -0.1472700000
     9    CA          6        -0.0908200000     0.1663100000    -0.0466200000
    10    HA          1        -0.1205500000     0.4891800000     0.0754700000
    11    C           6        -0.2309600000     0.1558400000    -0.4414700000
    12    O           8        -0.8512800000     0.2148600000    -1.4942200000
    13    CB          6         0.0564500000    -0.0775700000     0.2580100000
    14    HB1         1         0.1597000000    -0.0672600000     0.5246800000
    15    HB2         1         0.0836000000    -0.3848500000     0.1573200000
    16    HB3         1         0.0582500000     0.0623000000     0.3286000000
!------------------------------------------------------------------------------
Residue     3 N_METHYL
     6 ! # of atoms.
    17    N           7         0.3670700000     0.1126000000     0.4228800000
    18    H           1         0.8152500000     0.0186900000     1.2408100000
    19    CT          6         0.2854000000     0.1334100000     0.1805300000
    20    HT1         1        -0.7827000000     0.3639300000     0.1446100000
    21    HT2         1         0.8192900000     0.0757400000     1.0173900000
    22    HT3         1         0.7724100000     0.0234400000    -0.8207100000
!==============================================================================
```

Exercises

2.1 The module ATOMS and the coordinate file classify each atom by its name and its atomic number. In many simulation studies, *composite particles* (often called *extended* or *united atoms*) representing more than one atom are used. A typical example is the use of a single particle to represent aliphatic and aromatic CH and aliphatic CH_2 and CH_3 groups. How might such particles be handled in the coordinate file and what modifications to the data structures would be needed to account for them?

2.2 The coordinate file format described in section 2.2 is unique to this book. Many other coordinate file formats exist, some of which are more widely used than others. A common one, although it is not necessarily so convenient for modeling studies, is the format of the *Protein Data Bank* (PDB). Choosing this format (or any other) write a module, equivalent to that of COORDINATE_IO, to read and write the appropriate files. The atoms and sequence modules should remain unchanged. For convenience the subroutine interfaces should be as simple, if possible the same, as those of the subroutines in COORDINATE_IO. Note that a module, called PDB_IO, that satisfies these criteria and reads PDB format files is available with the other modules that are described in this book.

3

Operations on coordinates

3.1 Introduction

The coordinates of the particles in a system can give essential information about its structure. The aim of this chapter is to describe the various ways in which coordinates can be analysed and manipulated to provide this information. Because numerous analyses can be performed on a set of coordinates, only a sampling of some of the more common ones will be covered here.

3.2 Connectivity

One of the most important properties of a chemical system is the number and type of *bonds* between its atoms that it possesses. A rigorous determination of the bonding pattern would require an analysis of the electron density of a system or of its geometry in combination with a database of chemical information about the bonding behaviours of each element in particular molecular environments. Once the bonding arrangement is known, it is straightforward to determine other aspects of the connectivity of a system, such as its *bond angles* and its *torsion* or *dihedral angles*.

In this section the bonding pattern of a system is not determined using a database, although this will be done in a later chapter. Instead a less rigorous approach that will provide a list of likely bonds is adopted. It is based upon a simple search of the distances between atoms.

If r_i is the three-dimensional vector containing the x, y and z Cartesian coordinates of atom i, then the distance, r_{ij}, between two atoms, i and j, can be written as

$$r_{ij} = |r_i - r_j|$$
$$= \sqrt{(x_i - x_j)^2 + (y_i - y_j)^2 + (z_i - z_j)^2} \qquad (3.1)$$

25

The algorithm for finding bonds works by calculating the distances between two atoms in a system and then checking to see whether they are within a certain *bonding distance* apart. The bonding distance is determined as the sum of radii that are typical 'bonding' radii for each element and a *buffer distance*, which is an empirical parameter that is set by the user. The bonding radii are also empirical parameters that have been derived from the covalent, van der Waals and ionic radii for the elements and have been shown to give reasonable results in this application.

The simplest implementation of the algorithm would calculate the distances between all possible pairs of atoms and perform the distance comparison for each pair. This is sufficient for systems with small numbers of particles but as the number increases this method becomes prohibitively expensive. To see this, assume that the system contains N atoms. The number of possible pairs of atoms in the system can be then be calculated with the following procedure. The first atom can pair with all the atoms from atom 2 to atom N giving $N - 1$ pairs. The second atom can pair with atom 1 and with the atoms from atom 3 to atom N. However, the pair with atom 1 has already been counted so only the pairs with atoms 3 to N are new ones. The procedure can be continued for all the atoms up until atom N, which contributes no new pairs. The total number of pairs, N_{pair}, is thus

$$N_{pair} = (N - 1) + (N - 2) + \cdots + 1 + 0$$

$$= \sum_{i=1}^{N} (N - i)$$

$$= \frac{1}{2} N(N - 1) \tag{3.2}$$

It is apparent that the number of pairs is approximately equal to the square of the number of particles. To denote this it is common to use the notation $N_{pair} \simeq O(N^2)$, which means that the number of pairs is of the order of the square of the number of particles. The reason for the expense of the search for large N is now apparent, for the number of distances to be calculated increases with the square of N. So, for a small system of 100 atoms the number of pairs is $O(10^4)$, but for a larger system of 10 000 atoms the number of pairs is $O(10^8)$, which is 10^4 times bigger.

The problem of how the computational effort needed for an algorithm scales with increasing size is ubiquitous in all areas of computational science, not only for molecular modeling, and it is one that we shall meet repeatedly. The aim when designing any algorithm is to develop one with as low a scaling behaviour as possible and ideally with linear behaviour, $O(N)$, or less.

There are several possible approaches that can be used to reduce the computational load for the bond-finding algorithm. In principle, it is possible to design algorithms that do scale linearly with the system's size for this particular problem but they are complicated and so we adopt a simpler approach. Rather than determining which atoms are bonded together, we locate first the residues which are within a bonding distance. Only if residues are bonded is the search over atoms performed.

To see that this method can be quicker the steps it contains are detailed below. The number of residues in the system is denoted by N_r ($N_r \ll N$) and the average number of atoms per residue as $\bar{N}$. Note that whereas N_r, like N, increases with the size of the system, $\bar{N}$ is independent of it. The algorithm is as follows.

1. A *bounding box* is determined for each residue. The bounding box is defined as the smallest rectangular box which will enclose every atom in the residue and its calculation requires $O(\bar{N})$ operations for each residue or $O(N_r\bar{N})$ operations in total.
2. A search is performed over each residue pair to see whether their bounding boxes are within bonding range. The cost of this step is $O(N_r^2)$.
3. Only if two residues are within range is a search over all possible atom pairs between the residues done. For a large system it is easy to see that the average number of residues with which any one residue interacts is constant. The number does not depend upon the system's size but only upon the average size of the residues (or, equivalently, their number per unit volume). Thus, the number of residue pairs within range scales linearly with the number of residues, i.e. as $O(N_r)$, so the total cost of the search over atoms is $O(N_r\bar{N}^2)$.

The scaling of this algorithm also goes as the square of the system's size (for the worst step), although, in this case, it is the number of residues which is the measure of the system's size rather than the number of atoms. Because, in general, there will be many fewer residues than there are atoms in a system the use of the *prescreening* step by searching over residues will make the algorithm more efficient for larger systems. It is to be noted though that for smaller systems (such as bALA) this algorithm is likely to be less efficient because there is an *overhead* cost associated with the initial search over interacting residues. This illustrates another important principle of algorithm design – that for differently sized systems different algorithms will be optimal. This is shown schematically in figure 3.1. Although a particular algorithm, algorithm 1, may exhibit worse scaling properties with system size than does another algorithm, algorithm 2, the extra overhead costs associated with algorithm 2 make it, in fact, less efficient for smaller systems. It

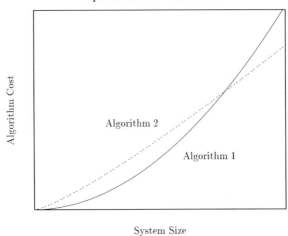

Figure 3.1. The scaling properties of two algorithms, 1 and 2, versus the system's size.

is only when a certain *critical system size* is reached that algorithm 2 will, in fact, become faster.

Once the list of bonds is known, the generation of lists of all possible bond angles and dihedral angles that are compatible with the bond list is relatively straightforward. It is possible to devise algorithms whose costs, for well-behaved systems, scale nearly linearly with the size of the system.

Procedures that generate lists of bonds, bond angles and dihedral angles for a system are implemented in the module CONNECTIVITY. The module contains three subroutines, each of which deals with the generation of a different list. Its definition is

```
MODULE CONNECTIVITY

CONTAINS

   SUBROUTINE CONNECTIVITY_ANGLES ( ANGLES, BONDS )
      INTEGER, DIMENSION(:,:), POINTER    :: ANGLES
      INTEGER, DIMENSION(:,:), INTENT(IN) :: BONDS
   END SUBROUTINE CONNECTIVITY_ANGLES

   SUBROUTINE CONNECTIVITY_BONDS ( BONDS, BUFFER )
      INTEGER, DIMENSION(:,:), POINTER    :: BONDS
      REAL,                    INTENT(IN) :: BUFFER
   END SUBROUTINE CONNECTIVITY_BONDS

   SUBROUTINE CONNECTIVITY_DIHEDRAL ( DIHEDRALS, BONDS, ANGLES )
      INTEGER, DIMENSION(:,:), POINTER    :: DIHEDRALS
      INTEGER, DIMENSION(:,:), INTENT(IN) :: ANGLES, BONDS
   END SUBROUTINE CONNECTIVITY_DIHEDRALS

END MODULE CONNECTIVITY
```

CONNECTIVITY_BONDS is the key subroutine in the module. It implements the bond-search algorithm discussed above. There are two arguments, both of which are essential. The argument BONDS is a pointer that should be undefined or not associated with a target on entry to the subroutine but will, on exit, be defined and will contain the list of bonds that has been found. Its dimensions will be BONDS(1:2,1:NBONDS), where NBONDS is the number of bonds for the system. The first and second atoms of a particular bond, I, are contained in the elements BONDS(1,I) and BONDS(2,I), respectively. The second argument is an input argument that specifies the buffer distance which is to be added to the sum of the elements' radii when a check is made to see whether two atoms are bonded. CONNECTIVITY_BONDS uses the modules ATOMS and SEQUENCE, although it leaves both of these unchanged. In particular, it uses the coordinates that are defined in the array ATMCRD to determine the connectivity.

CONNECTIVITY_ANGLES and CONNECTIVITY_DIHEDRALS are the subroutines that generate the angle and dihedral lists, respectively. Like CONNECTIVITY_BONDS, the arguments that will contain the lists of angles and dihedrals generated by each subroutine are pointers that should be undefined or not associated with a target on entry. They will be defined and filled by the subroutines themselves. The ANGLES argument for the CONNECTIVITY_ANGLES subroutine will have the dimensions ANGLES(1:3,1:NANGLES) on exit and the DIHEDRALS argument dimensions of DIHEDRALS(1:4,1:NDIHEDRALS), where NANGLES and NDIHEDRALS are the numbers of angles and dihedrals that are generated. The remaining arguments for each subroutine (BONDS for CONNECTIVITY_ANGLES and BONDS and ANGLES for CONNECTIVITY_DIHEDRALS) are lists of bonds and angles that have previously been determined. Both CONNECTIVITY_ANGLES and CONNECTIVITY_DIHEDRALS use data from the module ATOMS but leave it unchanged.

There is one very important point about these subroutines and that is that they create lists with a particular format. Both CONNECTIVITY_ANGLES and CONNECTIVITY_DIHEDRALS will not work properly unless they have bond and angle lists of the proper format, so it is advisable with these two subroutines to use arrays that themselves have been generated by the subroutines in the connectivity module.

3.3 Internal coordinates

The lists of bonds, angles and dihedrals are a start in the analysis of a system's chemical structure but they define only its overall connectivity, which will be the same for all systems in the same chemical state. To investigate differences between the structures of systems with identical connectivities it is important to calculate the values of their internal coordinates. The three principal internal coordinates are bond lengths, bond angles and dihedral angles, which are illustrated in figure 3.2.

The distance between two atoms has already been defined in equation (3.1). The angle, θ_{ijk}, subtended by three atoms i, j and k is calculated from

$$\theta_{ijk} = \arccos\left(\hat{\boldsymbol{r}}_{ij}^{\mathrm{T}}\hat{\boldsymbol{r}}_{kj}\right) \tag{3.3}$$

where the difference vectors are defined as

$$\boldsymbol{r}_{ij} = \boldsymbol{r}_i - \boldsymbol{r}_j \tag{3.4}$$

and the hat over a vector indicates that it is normalized, i.e. $\hat{\boldsymbol{r}} = \boldsymbol{r}/r$.

The definition of a dihedral angle, ϕ_{ijkl}, is slightly more ambiguous because several definitions exist. The one used here is

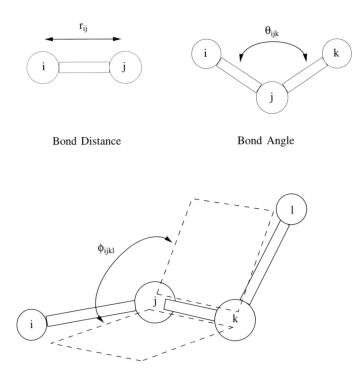

Figure 3.2. The three principal types of internal coordinate.

$$\phi_{ijkl} = \pm \arccos \left(\hat{a}^{\mathrm{T}} \hat{b} \right) \tag{3.5}$$

where the vectors a and b are defined as

$$a = r_{ij} - (r_{ij}^{\mathrm{T}} \hat{r}_{kj}) r_{kj} \tag{3.6}$$

$$b = r_{lk} - (r_{lk}^{\mathrm{T}} \hat{r}_{kj}) r_{kj} \tag{3.7}$$

The sign of the dihedral (i.e. whether to use $+$ or $-$ in equation (3.5)) is the same as the sign of the scalar quantity, $-r_{ij}^{\mathrm{T}}(r_{kj} \wedge r_{lk})$. A dihedral angle of $0°$ indicates that all the atoms are in the same plane in a *cis* conformation whereas angles of $180°$ or $-180°$ mean that the atoms are coplanar in a *trans* conformation.

The module GEOMETRY contains procedures that will calculate bond lengths, bond angles and dihedral angles from lists that specify the atoms involved in the internal coordinates. The module's definition is

```
MODULE GEOMETRY

CONTAINS

   FUNCTION GEOMETRY_ANGLES ( COORDINATES, LIST )
      INTEGER, DIMENSION(:,:), INTENT(IN) :: LIST
      REAL,    DIMENSION(:,:), INTENT(IN) :: COORDINATES
      REAL,    DIMENSION(:)               :: GEOMETRY_ANGLES
   END FUNCTION GEOMETRY_ANGLES

   FUNCTION GEOMETRY_DIHEDRALS ( COORDINATES, LIST )
      INTEGER, DIMENSION(:,:), INTENT(IN) :: LIST
      REAL,    DIMENSION(:,:), INTENT(IN) :: COORDINATES
      REAL,    DIMENSION(:)               :: GEOMETRY_DIHEDRALS
   END FUNCTION GEOMETRY_DIHEDRALS

   FUNCTION GEOMETRY_DISTANCES ( COORDINATES, LIST )
      INTEGER, DIMENSION(:,:), INTENT(IN) :: LIST
      REAL,    DIMENSION(:,:), INTENT(IN) :: COORDINATES
      REAL,    DIMENSION(:)               :: GEOMETRY_DISTANCES
   END FUNCTION GEOMETRY_DISTANCES

END MODULE GEOMETRY
```

There are three array functions in the module, each of which calculates one type of internal coordinate. The functions all have the same calling sequence consisting of two input arguments, one of which is the coordinate array from which the internal coordinates are to be calculated and the other a list of internal coordinates. For the subroutine GEOMETRY_DISTANCES the array LIST should have the dimensions LIST(1:2,1:NBONDS), where

NBONDS is the number of distances to be calculated. LIST should be dimensioned to LIST(1:3,1:NANGLES) for GEOMETRY_ANGLES and to LIST(1:4,1:NDIHEDRALS) for GEOMETRY_DIHEDRALS, where NANGLES and NDIHEDRALS are the numbers of angles and dihedrals in each type of list. The function value on return is an array of size equal to the second dimension of the array LIST which is filled with the calculated values of the internal coordinates. In contrast to the subroutines in the module CONNECTIVITY the order in which the internal coordinates appear in the list arrays is unimportant.

3.4 Example 2

The example program in this section uses the modules CONNECTIVITY and GEOMETRY to analyse the structure of the molecule, blocked alanine. The program is

```
PROGRAM EXAMPLE2

... Declaration Statements ...

! . Local scalars.
INTEGER :: I, J
REAL :: BUFFER

! . Local array declarations.
INTEGER, DIMENSION(:,:), POINTER :: LISTA, LISTB, LISTD
REAL, ALLOCATABLE, DIMENSION(:) :: DATAA, DATAB, DATAD

! . Define the system.
CALL COORDINATES_DEFINE ( ''bALA1.crd'' )

! . Generate a series of bond connectivity lists with
! . different buffer sizes.
DO I = 0,10

   ! . Calculate BUFFER.
   BUFFER = 0.1 * REAL ( I )

   ! . Calculate the lists.
   CALL CONNECTIVITY_BONDS ( LISTB, BUFFER )

   ! . Free the bond lists.
   DEALLOCATE ( LISTB )

END DO

! . Calculate the bond connectivity lists with a good
! . buffer size.
```

```
CALL CONNECTIVITY_BONDS ( LISTB, 0.5 )

! . Calculate the angle and dihedral lists.
CALL CONNECTIVITY_ANGLES ( LISTA, LISTB )
CALL CONNECTIVITY_DIHEDRALS ( LISTD, LISTB, LISTA )

! . Allocate the arrays to hold the calculated values
! . of the internal coordinates.
ALLOCATE ( DATAA(SIZE(LISTA,2)), DATAB(SIZE(LISTB,2)), &
           DATAD(SIZE(LISTD,2)) )

! . Calculate the internal coordinates.
DATAB = GEOMETRY_DISTANCES ( ATMCRD, LISTB )
DATAA = GEOMETRY_ANGLES    ( ATMCRD, LISTA )
DATAD = GEOMETRY_DIHEDRALS ( ATMCRD, LISTD )

! . Print the calculated values.
WRITE ( OUTPUT, ''(/,A,/)'' ) ''Bond Lengths (Angstroms):''
WRITE ( OUTPUT, ''(4(4X,2I4,2X,F6.3))'' ) &
                ( ( LISTB(J,I), J = 1,2 ), DATAB(I), I = 1,SIZE(DATAB) )

WRITE ( OUTPUT, ''(/,A,/)'' ) ''Angles (Degrees):''
WRITE ( OUTPUT, ''(3(4X,3I4,2X,F6.1))'' ) &
                ( ( LISTA(J,I), J = 1,3 ), DATAA(I), I = 1,SIZE(DATAA) )

WRITE ( OUTPUT, ''(/,A,/)'' ) ''Dihedrals (Degrees):''
WRITE ( OUTPUT, ''(3(4X,4I4,2X,F6.1))'' ) &
                ( ( LISTD(J,I), J = 1,4 ), DATAD(I), I = 1,SIZE(DATAD) )

! . Deallocate all arrays.
DEALLOCATE ( DATAA, DATAB, DATAD, LISTA, LISTB, LISTD )

END PROGRAM EXAMPLE2
```

In the first part of the program, lists of bonds in the system are generated using the procedure CONNECTIVITY_BONDS with buffer sizes ranging in 0.1 Å increments from 0 to 1 Å. This is for illustrative purposes only to show how the number of bonds changes as a function of the buffer length. From the output file (which is not given here) it can be seen that the number of bonds is constant (at 21) with buffer sizes from 0.3 to 0.8 Å. With smaller buffer sizes the number of bonds is less and with larger buffers it is bigger. This is typical behaviour for systems (with well-defined coordinate sets!) that contain primarily non-metal atoms – a buffer size of 0.5 Å is usually a good compromise value.

In the second part of the program after the loop, lists of bonds, angles and dihedrals are generated, with 0.5 Å for the buffer size, and the internal coordinates are calculated using the functions from the module GEOMETRY. The values obtained are printed at the end.

Several points need to be made about the arrays. First of all the list arrays are specified by pointers as required by the connectivity subroutines while the arrays containing the values of the internal coordinates are allocatable arrays. Second, after each call to CONNECTIVITY_BONDS in the loop the list of bonds in LISTB is deallocated before it is passed again to the subroutine. This is necessary to avoid a duplication of this array by the subroutine. Finally, in the ALLOCATE statement for the data arrays and when printing the data at the end, the function SIZE is used to determine the number of elements in the list and in the data arrays. This is because the number of list elements is not returned explicitly by the connectivity subroutines, but only implicitly in the output length of the list arrays. It would, of course, have been possible to define intermediate variables, such as NBONDS = SIZE (LISTB, 2), to denote these quantities for subsequent use.

3.5 Miscellaneous transformations

The previous sections presented analyses that involve the relative positions of atoms. In many cases, however, it is useful to be able to manipulate the Cartesian coordinates themselves, either for the whole system or for a subset of it. This section presents a number of simple transformations of this type, providing a necessary preliminary to the more complex operations of the next section.

The most fundamental and probably the most useful transformations are those that rotate and translate the atom coordinates. A translation of a coordinate is effected by adding a vector that specifies the translation, $\mathbf{t}$, to the coordinate vectors of the atoms, i.e.

$$r'_i = r_i + t \tag{3.8}$$

where the prime denotes a modified coordinate. A rotation involves the multiplication of the coordinate vector of each atom by a 3×3 orthogonal matrix, $\mathbf{U}$, which specifies the rotation:

$$r'_i = \mathbf{U}r_i \tag{3.9}$$

If the matrix has a determinant of 1 ($\|\mathbf{U}\| = 1$), then the rotation is a *proper rotation* and the handedness of the coordinate system is preserved. If the matrix has a determinant of -1 ($\|\mathbf{U}\| = -1$), then the rotation is an *improper rotation* and involves a mirror reflection or an inversion and can lead to changes in the chirality of the system.

In many cases a coordinate set for a system will be centered and oriented arbitrarily. It is often useful to be able to remove these effects and to orient the system in a systematic fashion. A standard way to center a system is to translate it so that its *center of mass* is at the origin. The position vector of the center of mass of a system, R_c, is defined by

$$R_c = \frac{\sum_{i=1}^{N} w_i r_i}{\sum_{i=1}^{N} w_i} \tag{3.10}$$

where w_i are *weights* associated with each particle. Note that, for the center of mass, these will be the atomic masses, but it is possible to employ other values. For example, if the weights are all equal to 1 then the *center of geometry* is calculated, whereas if the weights are the charges on the atoms, it is the *center of charge* that is obtained. To put the center of mass at the origin it is necessary to calculate R_c for the system and then translate the coordinates of all atoms by $-R_c$ (using equation (3.8)).

To orient a system in a systematic fashion it is common to use a *principal axis transformation* that rotates a system so that the off-diagonal elements of its *inertia matrix* are zero. The inertia matrix, $\mathcal{I}$, is a 3×3 symmetric matrix that is defined as

$$\mathcal{I} = \sum_{i=1}^{N} w_i \left(r_i^2 \mathbf{I} - r_i r_i^{\mathrm{T}} \right) \tag{3.11}$$

where $\mathbf{I}$ is the 3×3 identity matrix and the superscript T denotes the transpose of a vector. Alternatively, the individual components of the matrix can be written as, for example,

$$\mathcal{I}_{xx} = \sum_{i=1}^{N} w_i \left(r_i^2 - x_i^2 \right)$$

$$\mathcal{I}_{xy} = -\sum_{i=1}^{N} w_i x_i y_i$$

$$= \mathcal{I}_{yx} \tag{3.12}$$

As before, the weights for the system can be the masses for each atom or other values depending upon the property that is being studied.

The *moments of inertia* and the *principal axes* of the system are the *eigenvalues*, $\mathcal{I}_\alpha$, and *eigenvectors*, e_α, obtained by *diagonalizing* the inertia matrix. This means that the following equation is solved:

$$\mathcal{I} e_\alpha = \mathcal{I}_\alpha e_\alpha \qquad \alpha = 1,2 \text{ or } 3 \tag{3.13}$$

The inertia matrix and the moments of inertia are important properties because they determine the overall rotational motion of a system.

The principal axis transformation is the one that makes the inertia matrix diagonal and converts all the diagonal elements of the matrix to the moments of inertia. This can be achieved by rotating the coordinates of the atoms of the system with the rotation matrix, $\mathbf{U}$, which is equal to the transpose of the matrix of eigenvectors:

$$\mathbf{U} = (e_1, e_2, e_3)^{\mathrm{T}} \tag{3.14}$$

The transformations discussed above are implemented in subroutines included in the module TRANSFORMATION. Its definition is

```
MODULE TRANSFORMATION

CONTAINS

    FUNCTION CENTER ( COORDINATES, WEIGHTS )
        REAL, DIMENSION(:,:), INTENT(IN)                :: COORDINATES
        REAL, DIMENSION(:),   INTENT(IN), OPTIONAL :: WEIGHTS
        REAL, DIMENSION(1:3)                            :: CENTER
    END FUNCTION CENTER_OF_MASS

    FUNCTION INERTIA_MATRIX ( COORDINATES, WEIGHTS )
        REAL, DIMENSION(:,:),     INTENT(IN)  :: COORDINATES
        REAL, DIMENSION(:),       INTENT(IN)  :: WEIGHTS
        REAL, DIMENSION(1:3,1:3), INTENT(OUT) :: INERTIA_MATRIX
    END FUNCTION INERTIA_MATRIX

    SUBROUTINE MOMENTS_OF_INERTIA ( COORDINATES, MOMENTS, AXES, WEIGHTS )
        REAL, DIMENSION(:,:),     INTENT(IN)                :: COORDINATES
        REAL, DIMENSION(1:3),     INTENT(OUT)               :: MOMENTS
        REAL, DIMENSION(1:3,1:3), INTENT(OUT), OPTIONAL :: AXES
        REAL, DIMENSION(:),       INTENT(IN),  OPTIONAL :: WEIGHTS
    END SUBROUTINE MOMENTS_OF_INERTIA

    SUBROUTINE ROTATE ( COORDINATES, ROTATION )
        REAL, DIMENSION(:,:),     INTENT(INOUT) :: COORDINATES
        REAL, DIMENSION(1:3,1:3), INTENT(IN)    :: ROTATION
    END SUBROUTINE ROTATE

    SUBROUTINE TRANSLATE ( COORDINATES, TRANSLATION )
        REAL, DIMENSION(:,:), INTENT(INOUT) :: COORDINATES
        REAL, DIMENSION(1:3), INTENT(IN)    :: TRANSLATION
    END SUBROUTINE TRANSLATE

    SUBROUTINE TRANSLATE_TO_CENTER ( COORDINATES, WEIGHTS )
        REAL, DIMENSION(:,:), INTENT(INOUT) :: COORDINATES
        REAL, DIMENSION(:),   INTENT(IN)    :: WEIGHTS
```

```
END SUBROUTINE TRANSLATE_TO_CENTER

SUBROUTINE TO_PRINCIPAL_AXES ( COORDINATES, WEIGHTS )
    REAL, DIMENSION(:,:), INTENT(INOUT)        :: COORDINATES
    REAL, DIMENSION(:),   INTENT(IN), OPTIONAL :: WEIGHTS
END SUBROUTINE TO_PRINCIPAL_AXES

END MODULE TRANSFORMATION
```

The rotation and translation of a set of coordinates are effected with the subroutines ROTATE and TRANSLATE. Both these subroutines take an input set of coordinates (in the array COORDINATES) and apply the appropriate transformation, either a rotation using the rotation matrix in the array ROTATION or a translation using the translation vector in the array TRANSLATION.

The function CENTER calculates a center for the system using the coordinates in the array COORDINATES and the weights in the array WEIGHTS. The latter array is optional and, if it is not present, a weight of 1 is used for each coordinate. To calculate the center of mass the array WEIGHTS should be set equal to ATMMAS from the module ATOMS. If only certain atoms are to be included in the calculation of the center then the weights for those atoms that are not needed can be set to zero. The position vector of the center which is calculated is returned as the value of the function. This function leaves the coordinates of the system unchanged.

The function INERTIA_MATRIX calculates the inertia matrix for the system. The comments about the previous function apply except that it is the inertia matrix that is returned as the function result rather than the center.

MOMENTS_OF_INERTIA is a companion subroutine to INERTIA_MATRIX. It too takes as input a coordinate array and, optionally, an array of weights, but it returns the moments of inertia in the array MOMENTS and, optionally, the axes of inertia in the array AXES. In this array the axes are stored such that AXES(1:3,i) contains the vector for the ith axis.

The two remaining subroutines in the module modify the values of the input coordinates. The first, TRANSLATE_TO_CENTER, moves the system so that it is at its center (whose type is determined by the input weights). The second, TO_PRINCIPAL_AXES, both moves the input coordinate set to its center and performs a principal axis transformation.

3.6 Superimposing structures

The transformations introduced in the previous section modified a single set of coordinates. In many instances, though, it is necessary to compare the structures defined by two or more sets of coordinates. It is possible, of course, to compare two structures by comparing the values of their internal coordinates. This gives useful information but can be cumbersome when the system and, hence, the number of internal coordinates are large. A simpler and widely used measure of the difference between two structures, I and J, is the *RMS coordinate deviation*, σ_{IJ}. It provides a quicker, albeit cruder, measure insofar as it is a single number. It is defined as

$$\sigma_{IJ} = \sqrt{\frac{\sum_{i=1}^{N} w_i\left(r_i^I - r_i^J\right)^2}{\sum_{i=1}^{N} w_i}} \tag{3.15}$$

where the superscripts I and J refer to the coordinates of the first and second structures, respectively.

A moment's consideration shows that the RMS coordinate deviation will not be a useful measure for the comparison of structures unless the coordinate sets are somehow oriented with respect to each other. It is evident that if one set has undergone a translation or a rotation with respect to the other then the RMS coordinate deviation can take any value. A possible solution is to orient the two sets of coordinates separately using a principal axis transformation and then calculate the RMS coordinate deviation between them. This can provide a satisfactory measure of comparison in some circumstances. However, a better method is to choose one set of coordinates as a reference structure and then find the transformation that superimposes the other coordinate set upon it.

There are various methods for superimposing structures. One of the original ones was developed by W. Kabsch but a more recent one that uses *quaternions* has been proposed by G. Kneller. Both algorithms work by initially translating the coordinate set of the structure to be moved so that it is has the same center as the reference set. The first structure is then rotated so that its RMS coordinate deviation from the reference structure is minimized. The expression for the rotation matrix, $\mathbf{U}$, which effects this rotation can be constructed by minimizing the equation for the RMS coordinate deviation between the two structures, σ_{IJ}, which is itself a function of the rotation matrix:

$$\sigma_{IJ}^2(\mathbf{U}) \propto \sum_{i=1}^{N} w_i\left(r_i^I - \mathbf{U}r_i^J\right)^2 \tag{3.16}$$

The difference between the two methods (which should give equivalent results) lies in how the rotation matrix is specified. In the method of Kabsch, the minimization in equation (3.16) is performed subject to the constraint that the matrix is orthogonal, i.e. that the inverse of the matrix is its transpose or $\mathbf{U}^T\mathbf{U} = \mathbf{I}$, where $\mathbf{I}$ is the identity matrix. Expressions for the elements of the rotation matrix can then be obtained using the standard method of Lagrange multipliers. In the quaternion algorithm, in constrast, the rotation matrix is expressed in terms of four parameters q_0, q_1, q_2 and q_3 as

$$\mathbf{U} = \begin{pmatrix} q_0^2 + q_1^2 - q_2^2 - q_3^2 & 2(-q_0q_3 + q_1q_2) & 2(q_0q_2 + q_1q_3) \\ 2(q_0q_3 + q_1q_2) & q_0^2 - q_1^2 + q_2^2 - q_3^2 & 2(-q_0q_1 + q_2q_3) \\ 2(-q_0q_2 + q_1q_3) & 2(q_0q_1 + q_2q_3) & q_0^2 - q_1^2 - q_2^2 + q_3^2 \end{pmatrix} \quad (3.17)$$

The matrix defined in equation (3.17) automatically defines an orthogonal matrix if the vector of quaternion parameters is normalized, i.e. if $q_0^2 + q_1^2 + q_2^2 + q_3^2 = 1$. The minimization in equation (3.16) can then be done in terms of these parameters subject to the normalization condition.

The calculation of the RMS coordinate deviation and the superposition of two coordinate sets is handled by three subroutines in the module SUPERIMPOSE. Its definition is

```
MODULE SUPERIMPOSE

CONTAINS

    SUBROUTINE RMS_DEVIATION ( SET1, SET2, WEIGHTS, RMS )
        REAL, DIMENSION(:,:), INTENT(IN) :: SET1
        REAL, DIMENSION(:,:), INTENT(IN) :: SET2
        REAL, DIMENSION(:), INTENT(IN), OPTIONAL :: WEIGHTS
        REAL, INTENT(OUT), OPTIONAL              :: RMS
    END SUBROUTINE RMS_DEVIATION

    SUBROUTINE SUPERIMPOSE_KABSCH ( SET1, SET2, WEIGHTS )
        REAL, DIMENSION(:,:), INTENT(IN)    :: SET1
        REAL, DIMENSION(:,:), INTENT(INOUT) :: SET2
        REAL, DIMENSION(:), INTENT(IN), OPTIONAL :: WEIGHTS
    END SUBROUTINE SUPERIMPOSE_KABSCH

    SUBROUTINE SUPERIMPOSE_QUATERNION ( SET1, SET2, WEIGHTS )
        REAL, DIMENSION(:,:), INTENT(IN)    :: SET1
        REAL, DIMENSION(:,:), INTENT(INOUT) :: SET2
        REAL, DIMENSION(:), INTENT(IN), OPTIONAL :: WEIGHTS
    END SUBROUTINE SUPERIMPOSE_QUATERNION

END MODULE SUPERIMPOSE
```

The first subroutine in the module, RMS_DEVIATION, calculates the RMS coordinate deviation between two structures. The arrays SET1 and SET2 contain the coordinates of the two structures and the optional array, WEIGHTS, contains the weights to use for each atom. If the weight array is absent then a value of 1 will be taken for each atom. There is a fourth argument, an optional real scalar, which, if specified, will be assigned the value of the calculated RMS coordinate deviation on output. This means that the value can be used in subsequent calculations. Note that, in any case, the value of the coordinate deviation will be automatically printed to the output file.

The subroutines SUPERIMPOSE_KABSCH and SUPERIMPOSE_QUATERNION both have three arguments. SET1 and SET2 are arrays containing the coordinates of the reference structure and the structure that is to be superimposed upon it, respectively. Only the coordinates for the second structure in SET2 are changed. The third, optional argument is an array that contains the weights for the calculation.

3.7 Example 3

The program in this section illustrates the use of the subroutines in the modules TRANSFORMATION and SUPERIMPOSE. In the first part of the example the transformation subroutines are used to manipulate a single set of coordinates. First the coordinates for the system are translated to its center of mass and then they are transformed to their principal axes. As a check the inertia matrix is calculated before and after the transformation – before the transformation the matrix will, in general, have non-zero values for all its components, but afterwards it should be diagonal.

In the second part of the program a second set of coordinates is read and superimposed upon the first set of coordinates which had previously been stored in the array TMPCRD. The RMS coordinate deviations between the structures before and after the superposition are calculated. Note that the atomic masses are used as weights for the transformation and superposition operations in both halves of the program. It is left to the reader to see what happens if other values are used.

The program is

```
PROGRAM EXAMPLE3

... Declaration Statements ...

! . Local scalar declarations.
INTEGER :: I

! . Local array declarations.
REAL,            DIMENSION(1:3,1:3) :: INERTIA
REAL, ALLOCATABLE, DIMENSION(:,:)      :: TMPCRD
```

```
!  .  Define the atom and sequence data structures.
CALL COORDINATES_DEFINE ( ''bALA1.crd'' )

!  .  Move the structure to its center of mass.
CALL TRANSLATE_TO_CENTER ( ATMCRD, ATMMAS )

!  .  Calculate and print the inertia matrix.
INERTIA = INERTIA_MATRIX ( ATMCRD, ATMMAS )
WRITE ( OUTPUT, ''(/,A)'' ) ''Inertia matrix before reorientation:''
WRITE ( OUTPUT, ''(3F20.4)'' ) ( INERTIA(I,1:3), I = 1,3 )

!  .  Transform the coordinates to their principal axes.
CALL TO_PRINCIPAL_AXES ( ATMCRD, ATMMAS )

!  .  Calculate and print the inertia matrix in the new orientation.
INERTIA = INERTIA_MATRIX ( ATMCRD, ATMMAS )
WRITE ( OUTPUT, ''(/,A)'' ) ''Inertia matrix after reorientation:''
WRITE ( OUTPUT, ''(3F20.4)'' ) ( INERTIA(I,1:3), I = 1,3 )

!  .  Allocate a temporary coordinate array.
ALLOCATE ( TMPCRD(1:3,1:NATOMS) )

!  .  Read in the second set of coordinates.
CALL COORDINATES_READ ( ''bALA2.crd'', TMPCRD )

!  .  Calculate the RMS coordinate deviation between the two structures.
CALL RMS_DEVIATION ( ATMCRD, TMPCRD, ATMMAS )

!  .  Superimpose the two structures.
CALL SUPERIMPOSE_KABSCH ( ATMCRD, TMPCRD, ATMMAS )

!  .  Recalculate the RMS coordinate deviation between the two structures.
CALL RMS_DEVIATION ( ATMCRD, TMPCRD, ATMMAS )

!  .  Deallocate the temporary coordinate array.
DEALLOCATE ( TMPCRD )

END PROGRAM EXAMPLE3
```

Exercises

3.1 A property of a molecule closely related to its moments of inertia is its *radius of gyration*, R_{gyr}, which is defined as

$$R_{gyr} = \sqrt{\frac{\sum_{i=1}^{N} w_i (r_i - R_c)^2}{\sum_{i=1}^{N} w_i}} \tag{3.18}$$

where R_c is the center of the molecule calculated with the same weights as those used for the calculation of R_{gyr}. Write a subroutine that calculates this quantity.

3.2 Many molecules possess *chiral* centers that make them optically active. In principle, knowledge of the connectivity of the atoms in the molecule together with their coordinates and masses is sufficient if the different optical isomers are to be labelled. Write a module to identify the character of an asymmetric center (i.e. whether it is in its R or in its S form) and apply it to the structure of the blocked alanine molecule appearing in the text. Is the alanyl group of this molecule in its D or in its L form?

4

The energy function

4.1 Introduction

In the last chapter we dealt with how to perform manipulations on a set of coordinates and how to compare the structures defined by two sets of coordinates. This is useful for distinguishing between two different structures but it gives little indication regarding which structure is the more probable; i.e., which structure is most likely to be found experimentally. To do this, it is necessary to be able to evaluate the intrinsic stability of a structure, which is determined by its *potential energy*. The differences between the energies of different structures, their *relative energies*, will then determine which structure is the more stable and, hence, which structure is most likely to be observed. This chapter gives some theoretical background about the calculation of the energies of molecular systems and then goes on to describe in detail the method that is to be used in the rest of the book.

4.2 The Born–Oppenheimer approximation

As far as is known, the theory of *quantum mechanics*, which was developed during the first decades of the twentieth century as a result of inadequacies in the existing *classical mechanics*, is adequate to explain all atomic and molecular phenomena. In fact, in an oft-quoted, but nevertheless pertinent, remark, Dirac, one of the founders of quantum mechanics, said in 1929:

The underlying physical laws necessary for the mathematical theory of a large part of physics and the whole of chemistry are thus completely known, and the difficulty is only that the exact application of these laws leads to equations much too complicated to be soluble. It therefore becomes desirable that approximate practical methods of applying quantum mechanics should be developed, which can lead to an explanation of the main features of complex atomic systems without too much computation.

One of the equations that Dirac was talking about and which, in principle, determines the complete behaviour of a (non-relativistic) molecular system is the *time-dependent Schrödinger equation*. It has the form

$$\hat{\mathcal{H}}\Psi = i\hbar\frac{\partial\Psi}{\partial t} \tag{4.1}$$

In this equation, $\hat{\mathcal{H}}$ is what is known as the *Hamiltonian operator* which operates on the *wavefunction* for the system denoted by Ψ. The other symbols are i, which is the imaginary number, $\hbar$, which is Planck's constant, h, divided by 2π and t, which is the time.

A little further qualitative explanation of some of these terms is necessary. The wavefunction, Ψ, is, in general, a function of the position coordinates of all the particles in the system, the time and some specifically quantum mechanical variables that determine each particle's *spin*. It is the wavefunction that gives complete information about the system and which is the goal when solving equation (4.1). The wavefunction is important because its square is the probability density distribution for the particles in the system. To make this clearer, consider a system consisting of one particle that is constrained to move in one dimension. Then the wavefunction is a function of two variables, the position coordinate, x, and the time, t (for convenience spin is ignored) – this dependence is written as $\Psi(x, t)$. The probability that the particle will be found in the range x to $x + \delta x$, where δx is a small number, at a time t is $|\Psi(x, t)|^2\delta x$. In quantum mechanics this is the best that can be done – it is possible to know only the probability that a particle will have certain values for its variables – unlike classical mechanics, in which, in principle, the exact values for each of a particle's variables can be defined.

The wavefunction is important but its behaviour is determined by the Hamiltonian operator of the system. This operator will consist of a sum of two other operators – the kinetic energy operator, $\hat{\mathcal{K}}$, and the potential energy operator, $\hat{\mathcal{V}}$. The former determines the *kinetic energy* of the system, which is the energy due to 'movement' of the particles, and the latter the energy due to interactions between the particles and with their environment.

In certain cases the system can exist in what is called a *stationary state* in which its wavefunction and, thus, its particles' probability distribution do not change with time. For stationary states the time-dependent equation (equation (4.1)) reduces to a simpler form, the *time-independent Schrödinger equation*:

$$\hat{\mathcal{H}}\Psi = E\Psi \tag{4.2}$$

where E is the energy of the stationary state which is a constant.

For atomic and molecular systems, there are essentially two types of particles – electrons and the atomic nuclei. The latter will differ in their mass and their charge depending upon the element and its isotope. These classes of particles have greatly disparate masses. As an example, the lightest nucleus – that for hydrogen which consists of a single proton – has a mass about 1836 times greater than that of an electron. These very different masses will cause the electrons and the nuclei to have very different motions and this means that, to a good approximation, their dynamics can be treated separately. This is the basis of the *Born–Oppenheimer approximation*. It leads to the following procedure. The first step is to tackle the electronic problem by solving the *electronic Schrödinger equation* for a specific set of nuclear variables. That is, the nuclear coordinates are regarded as fixed and the wavefunction that is determined gives the distribution of the electrons only. The energy for which this equation is solved is no longer a constant but is a function of the nuclear positions. The second stage is to treat the nuclear dynamics by using the energy obtained from the solution of the electronic problem as an *effective potential* for the interaction of the nuclei.

The electronic equation is

$$\hat{\mathcal{H}}_{el}\Psi_{el} = E_{el}(\mathbf{r}_1, \mathbf{r}_2, \ldots, \mathbf{r}_N)\Psi_{el} \tag{4.3}$$

where the Hamiltonian is the same as that in equation (4.2) except that the kinetic energy operator for the nuclei has been omitted and the wavefunction, Ψ_{el}, gives the distribution of electrons only. Because the nuclei have been fixed the wavefunction and the energy, E_{el}, both depend parametrically upon the nuclear coordinates and so for each different configuration or structure there will be a different electronic distribution and a different electronic energy.

The electronic energy, E_{el}, is the energy in which we will be primarily interested. Because it is a function of the positions of all the nuclei in the system it is a multidimensional function that will often have a very complex form. This function defines the *potential energy surface* for the system and it is this which determines the effective interactions between the nuclei and, hence, their dynamics.

4.3 Strategies for obtaining energies on a potential energy surface

One of the most important problems when performing molecular simulations is that of how to obtain accurate values for the electronic energy of a system as a function of its nuclear coordinates. Various strategies are possible.

The most fundamental approach is to attempt to calculate the energies from first principles by solving the electronic Schrödinger equation (equation (4.3)) for the electronic energy at each nuclear configuration. Many methods for doing this are available but three of the more commonly used types are based upon *density functional theory, molecular orbital theory* and *valence bond theory*. All three are first principles or *ab initio* methods in the sense that they attempt to solve equation (4.3) with as few assumptions as possible. Although these methods can give very accurate results in many circumstances they are expensive and so cheaper alternatives have been developed.

One way of making progress is to drop the restriction of performing first-principles calculations and seek ways of simplifying the *ab initio* methods outlined above. These so-called *semi-empirical* methods have the same overall formalism as that of the *ab initio* methods but they approximate various time-consuming parts of the calculation with simpler approaches. Of course, because approximations have been introduced the methods must be calibrated to ensure that the results they produce are meaningful. This often means that the values of various *empirical parameters* in the methods have to be chosen so that the results of the calculations agree with experimental data or with the results of accurate *ab initio* quantum mechanical calculations. Semi-empirical versions of all the *ab initio* methods mentioned above exist.

A second and even cheaper approach is to employ an entirely *empirical potential energy function*. This consists of choosing an analytic form for the function which is to represent the potential energy surface for the system and then *parametrizing* the function so that the energies that it produces agree with experimental data or with the results of accurate *ab initio* quantum mechanical calculations. Very many different types of empirical energy function have been devised. Many have been designed for the description of a single system. For example, studies of simple reactions such as $H_2 + H \rightarrow H + H_2$ often use surfaces of special forms. For studies on larger molecules more general functions have been developed and it is these that we shall use in this book.

To finish this section it should be emphasized that there is often considerable overlap between the types of method discussed above. In particular, *hybrid methods* that combine quantum mechanical and empirical potential energy function calculations are frequently found. For example, there are methods that treat the σ-electron, single-bond structure of a system with an empirical energy function and the delocalized π-electron structure with a quantum mechanical technique. Other methods that are the subject of active research use a quantum mechanical approximation to treat a small part of a system, such as the active site of an enzyme, and an empirical energy function to treat the remaining atoms. Such methods are outside the scope of the discussion in this book.

4.4 Typical empirical energy functions

This section presents the general form of the empirical energy functions that are used in molecular simulations. To start, though, a point of notation will be clarified. Several different terms are employed to denote empirical energy functions in the literature and, no doubt, inadvertently, in this book. Common terms, which all refer to the same thing, include *empirical energy function*, *potential energy function*, *empirical potential* and *force field*. The use of empirical potentials to study molecular conformations is often termed *molecular mechanics*.

Some of the earliest empirical potentials were derived by vibrational spectroscopists interested in interpreting their spectra (this was, in fact, the origin of the term 'force field'), but the type of empirical potential that is described here was developed at the end of the 1960s and the beginning of the 1970s. Two prominent proponents of this approach were S. Lifson and N. Allinger. These types of force field are appropriate for studying conformations of molecules close to their equilibrium positions but different approaches must be used to study chemical reactions.

Because the form of a empirical potential function is, to some extent, arbitrary, various forms exist. Most potential functions, however, have two categories of terms that treat the *bonding* and the *non-bonding interactions* between atoms. These will be discussed separately.

4.4.1 Bonding terms

The bonding energy terms are those that help define the bonding or *covalent* structure of the molecule, i.e. its local shape. In a typical, simple force field, the bonding or covalent energy, V_{cov}, will consist of a sum of terms for the *bond, angle, dihedral* (or *torsion*) and *out-of-plane distortion* (or *improper dihedral*) energies:

$$V_{cov} = V_{bond} + V_{angle} + V_{dihedral} + V_{improper} \qquad (4.4)$$

The bond energy is often taken to have a *harmonic* form:

$$V_{bond} = \sum_{bonds} \frac{1}{2} k_b (b - b_0)^2 \qquad (4.5)$$

where k_b is the *force constant* for the bond, b is the actual bond length in the structure between the two atoms defining the bond and b_0 is the *equilibrium distance* for the bond. The sum runs over all the bonds that have been defined in the system. Because the energy is harmonic in form (see figure 4.1) it means that the energy of the bond will increase steadily without limit as it is distorted from its equilibrium value, b_0.

Harmonic terms are sufficient for many studies, but sometimes it is important to have a form for the bond energy that permits dissociation. An example would be if a reaction were being studied. One form that does this is the *Morse potential* which is shown in figure 4.2. The Morse energy, V_{Morse}, is given by

$$V_{Morse} = \sum_{bonds} D\{\exp[-a(b - b_0)] - 1\}^2 - D \qquad (4.6)$$

where the two new parameters are D, which is the dissociation energy of the bond, and a, which determines the width of the potential well.

The angle energy term is designed to imitate how the energy of a bond angle changes when it is distorted away from its equilibrium position. Like the bond energy term it too is often taken to be harmonic:

$$V_{angle} = \sum_{angles} \frac{1}{2} k_\theta (\theta - \theta_0)^2 \qquad (4.7)$$

The extra parameters are similar to those of the bond energy – k_θ is the force constant for the angle and θ_0 is its equilibrium value. The sum runs over all

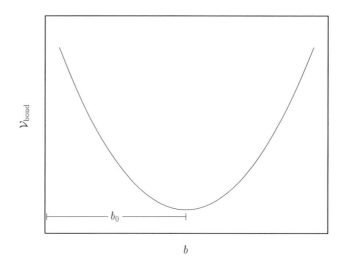

Figure 4.1. The harmonic bond energy term.

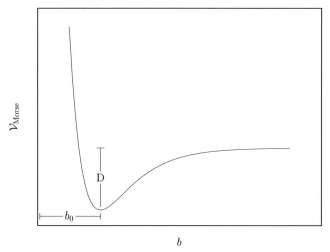

Figure 4.2. The Morse function bond energy term.

the angles in the system and each angle is defined in the same way as described in section 3.3.

The third type of bonding term is the term that describes how the energy of a molecule changes as it undergoes a rotation about one of its bonds, i.e. the dihedral or torsion energy for the system. In contrast to the bond and angle terms a harmonic form for the dihedral energy is not usually appropriate. This is because, for many dihedral angles in molecules, the whole range of angles from $0°$ to $360°$ can be accessible with not too large differences in energy. Such effects can be reproduced with a periodic function that is continuous throughout the complete range of possible angles (see figure 4.3). The dihedral energy can then be written as

$$V_{\text{dihedral}} = \sum_{\text{dihedrals}} \frac{1}{2} V_n[1 + \cos(n\phi - \delta)] \tag{4.8}$$

Once again the sum is over all the dihedrals that are defined in the system and the form of the dihedral angle is the same as that given in the previous chapter. In the formula, n is the *periodicity* of the angle (which determines how many peaks and wells there are in the potential), δ is the *phase* of the angle and V_n is the force constant. Often δ is restricted to taking the values $0°$ or $180°$, in which case it is only the sign of the cosine term in the expression that will change. It is to be noted that the periodicity of each term in the sum can change depending upon the type of dihedral and that values of n from 1 to 6 are most commonly used. It is also worth remarking that, in many force fields, multiple terms with different periodicities are used for some dihedral

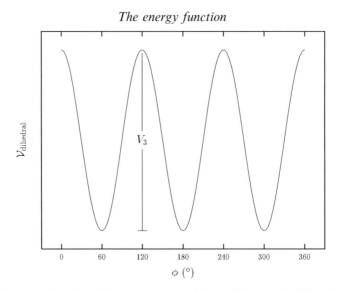

Figure 4.3. A dihedral angle energy term with a periodicity of 3.

angles. Thus, it may be that a single term with a periodicity of three, for example, is inadequate to reproduce the change in energy during a torsional motion accurately and other terms with periodicities of 1 and 2 will be added.

The fourth term in the sum in equation (4.4) is a more complicated one that describes the energy of out-of-plane motions. It is often necessary for planar groups, such as sp^2 hybridized carbons in carbonyl groups and in aromatic systems, because it is found that use of the dihedral terms described above is not sufficient to maintain the planarity of these groups during calculations. A common way to avoid this problem is to define an *improper dihedral angle*, which differs from the *proper dihedral angle* in that the atoms which define the dihedral angle, i–j–k–l, are not directly bonded to each other. The calculation of the angle, however, remains exactly the same. An example is shown in figure 4.4. With this definition of an improper dihedral angle, which we denote by ω, some force fields use the same form for the energy as that in equation (4.8), i.e.

$$\mathcal{V}_{\text{improper}} = \sum_{\text{impropers}} \frac{1}{2} V_n[1 + \cos(n\omega - \delta)] \qquad (4.9)$$

while others employ a harmonic form:

$$\mathcal{V}_{\text{improper}} = \sum_{\text{impropers}} \frac{1}{2} k_\omega (\omega - \omega_0)^2 \qquad (4.10)$$

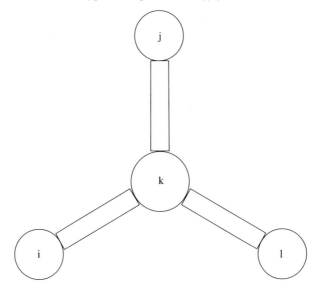

Figure 4.4. The arrangement of atoms in an improper dihedral angle.

where k_ω and ω_0 are the force constant for the energy term and the equilibrium value of the improper dihedral angle, respectively.

An alternative but less widely used way of defining the distortion due to out-of-plane motions is to define a distance rather than an angle as a measure of the distortion. One example is given in figure 4.5. The distance calculated in this way can then be used in, for example, an energy term of harmonic form.

The four terms mentioned above are the only bonding terms that we shall consider, although other types can be encountered in some force fields. In general, the extra terms are added to obtain better agreement with experimental data (especially vibrational spectra) but they increase the complexity

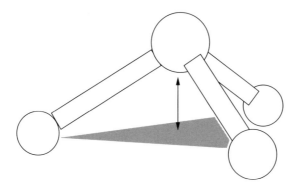

Figure 4.5. An out-of-plane distance.

of the force field and the number of parameters that need to be obtained. Extra terms that are sometimes added include bond and angle terms of the same form as equations (4.5) and (4.7) except that the terms are no longer harmonic – linear, cubic and quartic terms are possible. Other types that are sometimes seen are *cross-terms* that couple distortions in different internal coordinates. For example, a bond/angle cross-term could be proportional to $(b - b_0)(\theta - \theta_0)$.

4.4.2 Non-bonding terms

The bonding energy terms help to define the covalent energy of a molecule. The non-bonding terms describe the interactions between the atoms of different molecules or between atoms that are not directly bonded together in the same molecule. These interactions help to determine the overall conformation of a molecular system.

The non-bonding interactions are used to mimic the interactions arising from the interactions between the electronic distributions surrounding different atoms. The theory of intermolecular interactions is well developed and leads to the identification of a number of important types of interaction. At short range the interactions are primarily repulsive due to the interactions between the electron clouds and to a purely quantum mechanical effect called *exchange repulsion*, which arises when the two clouds are pushed together. At long ranges there are several important classes of interaction. The first are the *electrostatic* interactions that arise from the interaction of the charge distributions (including the nuclei) about each molecule or portion of a molecule. Second are the *dispersion* interactions which are produced by correlated fluctuations in the charge distributions of the two groups. Finally, there are *induced* or *polarization* interactions that are caused by the distortion of the charge distribution of a molecule as it interacts with neighbouring groups.

The non-bonding terms in an empirical force field attempt to reproduce all these types of interaction. Here we shall consider a non-bonding energy consisting of the sum of three terms:

$$\mathcal{V}_{nb} = \mathcal{V}_{elect} + \mathcal{V}_{LJ} + \mathcal{V}_{polar} \qquad (4.11)$$

The *electrostatic energy*, $\mathcal{V}_{elect}$, mimics the energy arising from the electrostatic interactions between two charge distributions. In a real molecule the positively charged nuclei will be surrounded by the negatively charged electron cloud. It is quite easy to calculate the electrostatic interactions between two systems that have arbitrarily shaped charge distributions. In general, though, we will not know the form of the electronic distribution unless we do a quantum mechanical calculation, which we are trying to avoid! Thus, in force fields, models of the charge distribution around a molecule must be chosen. The simplest representation is to assign a fractional charge to each atom, which represents its total nett charge (i.e. the sum of the nuclear charge and the charge in the part of the electron cloud that surrounds the atom). The electrostatic energy is calculated as

$$V_{\text{elect}} = \frac{1}{4\pi\epsilon_0\epsilon} \sum_{ij \text{ pairs}} \frac{q_i q_j}{r_{ij}} \tag{4.12}$$

where q_i and q_j are the fractional charges on atoms i and j and r_{ij} is the distance between the two particles. The terms in the prefactor are $1/(4\pi\epsilon_0)$, which is the standard term when calculating electrostatic interactions in the MKSA (metre, kilogram, second, ampere) system of units, and ϵ, which is the dielectric constant that will have the value 1 when the system is in vacuum. The sum in equation (4.12) runs over all pairs of atoms for which an electrostatic interaction is to be calculated. Note that the fractional charges on the atoms are constants and do not change during a calculation.

It is possible to define other representations of the charge distribution. For example, instead of fractional charges at the atoms' centers (i.e. on the nuclei), charges could be assigned off-center along bonds or higher moments, such as dipoles, could be used. Such representations are not usually favoured because they are more complex and more expensive than the simple point-charge model and they do not give significantly better results.

The second term in equation (4.11) is the *Lennard-Jones energy* which mimics the long-range dispersion interactions and the short-range repulsive interactions. It has the form

$$V_{\text{LJ}} = \sum_{ij \text{ pairs}} \frac{A_{ij}}{r_{ij}^{12}} - \frac{B_{ij}}{r_{ij}^{6}} \tag{4.13}$$

where A_{ij} and B_{ij} are positive constants whose values depend upon the types of the atoms, i and j, and the sum is over all pairs of atoms for which the interaction is to be calculated.

The shape of the Lennard-Jones potential is plotted in figure 4.6. The repulsive part of the curve is produced by the $1/r_{ij}^{12}$ term and the attractive part by $1/r_{ij}^{6}$. The inverse sixth power form of the attraction arises naturally from the theory of dispersion interactions. The choice of an inverse twelfth power for the repulsion is less well founded and other forms for the repulsion have been used, including other inverse powers, such as eight and ten, and an exponential form that leads to the so-called *Buckingham potential*. Most simple force fields seem to use the Lennard-Jones form.

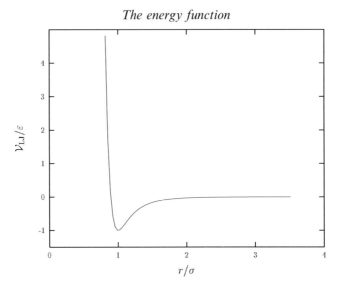

Figure 4.6. The Lennard-Jones energy for a pair of atoms.

To complete the specification of the Lennard-Jones energy it is necessary to have a recipe for determining the parameters A_{ij} and B_{ij}. These are usually defined in terms of the depth of the Lennard-Jones well, ε_{ij}, and either the distance at which the energy of the interaction is zero, s_{ij}, or the position of the bottom of the well, σ_{ij} (see figure 4.6). With these parameters the Lennard-Jones interaction between atoms i and j, V_{LJ}^{ij}, takes the form

$$V_{LJ}^{ij} = 4\varepsilon_{ij}\left[\left(\frac{s_{ij}}{r_{ij}}\right)^{12} - \left(\frac{s_{ij}}{r_{ij}}\right)^{6}\right] \tag{4.14}$$

$$= \varepsilon_{ij}\left[\left(\frac{\sigma_{ij}}{r_{ij}}\right)^{12} - 2\left(\frac{\sigma_{ij}}{r_{ij}}\right)^{6}\right] \tag{4.15}$$

where $\sigma_{ij}^{6} = 2s_{ij}^{6}$.

Although each of the parameters, ε_{ij}, s_{ij} and σ_{ij}, depends formally on two atoms, it is normal to specify a set of *combination rules* so that they can be defined from the parameters for single atoms. It is usual to use the geometrical mean as a combination rule for the well depths:

$$\varepsilon_{ij} = \sqrt{\varepsilon_{ii}\varepsilon_{jj}} \tag{4.16}$$

For the distance parameters both arithmetical and geometrical mean combination rules are common. For the σ_{ij} parameter, for example, we have either

$$\sigma_{ij} = \sqrt{\sigma_{ii}\sigma_{jj}} \qquad (4.17)$$

or

$$\sigma_{ij} = \tfrac{1}{2}\left(\sigma_{ii} + \sigma_{jj}\right) \qquad (4.18)$$

The third type of non-bonding energy term that is considered in this section is the *polarization energy*. In contrast to the previous two terms, the electrostatic and Lennard-Jones energies, this energy term is not a standard term in many force fields and it shall not be used in any of the calculations in this book. However, it is instructive to introduce it here for a number of reasons. First, polarization interactions are important in many systems. Second, the nature of the polarization-energy calculation is fundamentally different from that of the energy-term calculations that have been discussed up to now, so some interesting new principles can be introduced. Third, it is a term that will be increasingly used in future molecular modeling studies.

As mentioned above, the polarization energy arises from the fact that the charge distribution of a group or molecule is distorted by interactions with its neighbours. In the point-charge model of electrostatic interactions (equation (4.12)) the charges assigned to atoms are constants and so the charge distribution of the molecule is constant for a given nuclear configuration irrespective of its environment. To model changes in the charge distribution of a molecule a number of techniques has been developed, but the most common one is to let each atom be polarizable by giving it an *isotropic dipole polarizability*. This means that, in the presence of an electric field produced by the charge distribution of the environment, a *dipole moment* is induced on the atom that is proportional in size and parallel to the field at the atom. If the polarizability for an atom i is denoted by α_i and the field at the atom by E_i (it is a vector quantity) then the dipole induced at the atom, μ_i, is

$$\mu_i = \alpha_i E_i \qquad (4.19)$$

The polarizability model we use here is isotropic because α_i is a scalar quantity. For an *anisotropic* model it would be a 3×3 matrix, which means that the atom could be more polarizable in some directions (such as along a bond) than in others. The polarizability is called a dipole polarizability because the field induces a dipole at the atom. In more complicated versions of the theory the electric field (or its derivatives) can induce different effects in the charge distribution.

The field at each atom is produced by the sum of the fields due to the charges, E_i^q, and the induced dipoles, E_i^μ, on the other atoms. These fields have the form

$$E_i^q = \frac{1}{4\pi\epsilon_0\epsilon} \sum_{j\neq i=1}^{N} \frac{q_j \mathbf{r}_{ij}}{r_{ij}^3} \tag{4.20}$$

$$E_i^\mu = \frac{1}{4\pi\epsilon_0\epsilon} \sum_{j\neq i=1}^{N} \mathbf{T}_{ij}\boldsymbol{\mu}_j \tag{4.21}$$

where $\mathbf{T}_{ij}$ is a 3×3 matrix that is given by

$$\mathbf{T}_{ij} = -\nabla_i \frac{\mathbf{r}_{ij}}{r_{ij}^3}$$

$$= \frac{1}{r_{ij}^5}\begin{pmatrix} 3x_{ij}^2 - r_{ij}^2 & 3x_{ij}y_{ij} & 3x_{ij}z_{ij} \\ 3y_{ij}x_{ij} & 3y_{ij}^2 - r_{ij}^2 & 3y_{ij}z_{ij} \\ 3z_{ij}x_{ij} & 3z_{ij}y_{ij} & 3z_{ij}^2 - r_{ij}^2 \end{pmatrix} \tag{4.22}$$

It is now possible to substitute the equations for the fields, equations (4.20) and (4.21), into the equation for the dipole, equation (4.19). From the form for the field due to the induced dipoles, it is evident that the dipole on atom i depends upon the induced dipoles of all the other atoms. To make the dependence more explicit it is possible to combine the N equations for the dipoles on each atom into a single equation leading, after a little manipulation, to

$$\mathbf{A}\boldsymbol{\mu} = \mathbf{B} \tag{4.23}$$

$\boldsymbol{\mu}$ and $\mathbf{B}$ are both vectors of $3N$ components with the forms

$$\boldsymbol{\mu} = \begin{pmatrix} \boldsymbol{\mu}_1 \\ \boldsymbol{\mu}_2 \\ \vdots \\ \boldsymbol{\mu}_N \end{pmatrix} \qquad \mathbf{B} = \begin{pmatrix} \alpha_1 \mathbf{E}_1^q \\ \alpha_2 \mathbf{E}_2^q \\ \vdots \\ \alpha_N \mathbf{E}_N^q \end{pmatrix} \tag{4.24}$$

$\mathbf{A}$ is a $3N \times 3N$ matrix that can be taken to consist of N^2 3×3 submatrices, $\mathbf{a}_{ij}$:

$$\mathbf{A} = \begin{pmatrix} \mathbf{a}_{11} & \mathbf{a}_{12} & \cdots & \mathbf{a}_{1N} \\ \mathbf{a}_{21} & \mathbf{a}_{22} & \cdots & \mathbf{a}_{2N} \\ \vdots & \vdots & \vdots & \vdots \\ \mathbf{a}_{N1} & \mathbf{a}_{N2} & \cdots & \mathbf{a}_{NN} \end{pmatrix} \tag{4.25}$$

The diagonal matrices and the off-diagonal matrices have different forms:

$$\mathbf{a}_{ii} = \begin{pmatrix} 1 & 0 & 0 \\ 0 & 1 & 0 \\ 0 & 0 & 1 \end{pmatrix} \quad \text{and} \quad \mathbf{a}_{ij} = -\frac{1}{4\pi\epsilon_0\epsilon}\alpha_i\mathbf{T}_{ij} \tag{4.26}$$

Equation (4.23) defines a set of $3N$ linear equations that can be solved to obtain the induced dipoles, μ_i, on each atom. Once these are known the polarization energy can be computed. It is

$$V_{\text{polar}} = -\frac{1}{2}\sum_{i=1}^{N} \mu_i^T E_i^q \tag{4.27}$$

This energy arises from the sum of three contributions. There is the energy due to the interaction of the induced dipoles in the system, $V_{\mu\mu}$, the energy due to the interaction of the induced dipoles with the permanent charges, $V_{\mu q}$, and an energy (which is positive) that arises because it costs a certain amount to produce the induced dipoles, V_{induced}. Their expressions are

$$V_{\mu\mu} = -\frac{1}{2}\sum_{i=1}^{N} \mu_i^T E_i^\mu \tag{4.28}$$

$$V_{\mu q} = -\sum_{i=1}^{N} \mu_i^T E_i^q \tag{4.29}$$

$$V_{\text{induced}} = \frac{1}{2}\sum_{i=1}^{N} \mu_i^T E_i \tag{4.30}$$

It is not difficult to show that their sum is equal to equation (4.27).

The fact that there is a set of linear equations to solve (equation (4.23)) marks the major difference between the calculation of all the previous energy terms that have been mentioned and the calculation of the polarization energy. The electrostatic and Lennard-Jones energies are what are known as *pairwise additive*. Each interaction can be calculated separately and is independent of the others. For the polarization energy this is not the case in that the magnitude of the induced dipole on each atom depends upon the induced dipoles of all the other atoms and so the dipoles for all the atoms must be calculated before the energy. The polarization energy is a type of *many-body* term.

The other distinguishing feature of the calculation of the polarization energy is its expense. The calculations of the electrostatic and Lennard-Jones energies as written above both involve $O(N^2)$ operations. In contrast, the calculation of V_{polar} is much more expensive because the solution of the $3N$ linear equations, which is the most time-consuming part of the calculation, formally scales as $O(N^3)$.

For the bonding energy terms, the bond energies are pairwise additive while the remaining terms are, strictly speaking, many-body terms because they depend on either three or four atoms (for the angles and the proper and improper dihedral energies, respectively). The number of each of these four types of terms is, however, roughly proportional to the number of atoms so the expense of calculating them is only about $O(N)$.

It is, therefore, the calculation of the non-bonding energies – the electrostatic, Lennard-Jones and polarization terms – that is the most expensive part of an energy calculation. In the next few chapters the simple, $O(N^2)$ method for the calculation of the electrostatics and Lennard-Jones energies will be employed, but, as discussed in section 3.2, there are ways of reducing the cost of an $O(N^2)$ calculation to more manageable proportions. This is a very important topic whose discussion will be left until a later chapter.

There is one additional point that needs elaboration for the calculation of the electrostatic, Lennard-Jones and polarization energies. This concerns which interactions between particles are to be included in the sums for the electrostatic and Lennard-Jones energies (equations (4.12) and (4.13), respectively) or for the calculation of the fields in the case of the polarization energy (equations (4.20) and (4.21)). For particles that are far from each other there is little problem and the interactions can be calculated as described above. For particles that are bonded together or are separated by only a few bonds there are two problems. First, the non-bonding interactions between them are large because their interparticle separations are small and, second, there will also be bonding terms (bonds, angles, dihedrals, etc.) between such atoms.

This dilemma is resolved by introducing the concept of *non-bonding exclusions* (see figure 4.7). Non-bonding interactions are calculated only for particles that are not involved in direct bonding interactions. For particles that are bonded or separated by only a few bonds, the non-bonding interactions between them are not calculated and it is the bonding terms that determine their energy of interaction. This avoids the problems of having very large interaction energies and of the overcounting that would result if both types of interaction were included. If both bonding and non-bonding terms between atoms close to each other were to be calculated then the analytic forms of the interactions described above would probably need to be significantly modified.

The number and type of non-bonding exclusions used depend on the force field. It is typical to exclude interactions between atoms that are directly bonded together (the so-called *1–2 interactions*) and those which are separated by two bonds (*1–3 interactions*). The treatment of interactions between atoms separated by three bonds (*1–4 interactions*) is the most variable. In some force fields they are excluded, in others they are included and in yet other cases they will be included but either the interactions will be scaled by

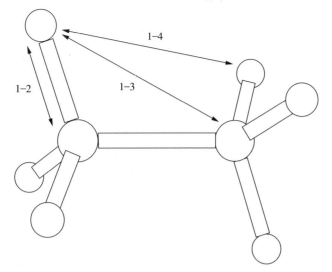

Figure 4.7. Examples of 1–2, 1–3 and 1–4 non-bonding exclusions for a hydrogen atom in an ethane molecule

some factor or special 1–4 sets of charges or Lennard-Jones parameters will be used. The reason for the different treatment of 1–4 interactions is that it is the combination of the dihedral angle bonding terms and the 1–4 electrostatic and Lennard-Jones terms that determines the barriers to rotation about bonds.

To close this section it is worth mentioning that, as was the case for the bonding energy terms, other types of non-bonding term are used in some force fields. These include alternatives to the explicit forms of the electrostatic and Lennard-Jones interactions mentioned above but also other terms. An important example included in some force fields is an energy introduced to mimic *hydrogen-bonding*. Various forms for this energy have been used but this term is often omitted from recent force fields because experience has shown that the combination of electrostatic and dispersion/repulsion inter-actions is sufficient to reproduce the hydrogen-bonding interaction to a reas-onable precision.

4.4.3 Force-field parametrization

The last sections have discussed the general form of the energy terms that are in common use in many force fields. One of the more striking aspects of the presentation is the number of different parameters that is needed. Where do they all come from and how does one go about getting new ones? This question encapsulates the problem of the *parametrization* of force fields, which is an important topic and one that we shall defer for a more detailed discussion until the end of the book. However, to paraphrase what will be said there, the parameters are obtained by a fitting procedure that involves calculating various properties for a system using the force field and comparing the calculated values with the values for these properties (or *observables*) obtained from experiment or accurate quantum mechanical calculations. The parameters are then modified and the process is repeated until satisfactory agreement has been obtained. It is relatively easy to obtain a rough guess at any parameter and, depending upon the type of application, this may be all that is required. If it is done rigorously, though, the parametrization of a force field can be a time-consuming and complicated task.

As an aside it is to be noted that the absolute energy value given by the types of energy function described here does not have any real meaning. It is only the relative energies (i.e. energy differences) between two different conformations of the same system that can be considered meaningful. It is possible to make the absolute value of the energy correspond to some physical value, such as the enthalpy of formation of the structure at a particular temperature, but extra terms have to be added to the force field and the parametrization of the energy function has to be done accordingly.

4.5 Derivatives of the energy function

Up to now the energy of a system and the form of the energy function has been the center of attention but, in many applications, it is equally or even more important to know the values of various derivatives of the energy function with respect to certain parameters. In such cases, it is crucial that the derivatives of the energy function be calculable in as efficient a manner as the energy. One useful property of the energy function given above is that its derivatives can be calculated *analytically*, which means that it is possible to derive explicit analytic formulae for the derivatives by direct differentiation of the energy function.

The alternative to analytically calculated derivatives is derivatives that are calculated *numerically*. One particularly common and reasonably effective way of calculating numerical derivatives is to use a *central-step finite-difference method*. If V is the potential energy of a system and p is the parameter with respect to which the derivative is required, the first derivative is approximated as

$$\frac{\partial V(p)}{\partial p} \sim \frac{V(p + \delta p) - V(p - \delta p)}{2\delta p} \tag{4.31}$$

where δp is a small change in the parameter value. Note that, when calculating the derivative with respect to p, only the value of p is changed; the values of all the other parameters in the function are kept constant.

In general, analytically calculated derivatives have a number of advantages over those that are numerically calculated. First, they are almost always more accurate. Analytic derivatives will be accurate to machine precision whereas the accuracy of numerical ones will depend to a large extent on the values of the finite steps (δp) taken in the numerical algorithm. Second, analytic derivatives are usually much less expensive, especially when the number of derivatives to be calculated is large.

Several different types of derivative of the energy function are useful, but by far the most important are the derivatives of the energy with respect to the positions of the atoms. The first derivatives or *gradients* are employed most extensively throughout this book, although the second derivatives are also necessary in certain applications. Because the energy function is dependent upon all the coordinates of the atoms in a system, there will be $3N$ first derivatives of the energy with respect to the coordinates, i.e.

$$\mathbf{g}_i = \frac{\partial V}{\partial \mathbf{r}_i} = \begin{pmatrix} \dfrac{\partial V}{\partial x_i} \\[2mm] \dfrac{\partial V}{\partial y_i} \\[2mm] \dfrac{\partial V}{\partial z_i} \end{pmatrix} \qquad \forall\, i = 1, \dots, N \tag{4.32}$$

The second derivatives of the energy with respect to the coordinates of two atoms, i and j, are

$$\mathbf{h}_{ij} = \frac{\partial^2 V}{\partial \mathbf{r}_i \, \partial \mathbf{r}_j} = \begin{pmatrix} \dfrac{\partial^2 V}{\partial x_i \, \partial x_j} & \dfrac{\partial^2 V}{\partial x_i \, \partial y_j} & \dfrac{\partial^2 V}{\partial x_i \, \partial z_j} \\[3mm] \dfrac{\partial^2 V}{\partial y_i \, \partial x_j} & \dfrac{\partial^2 V}{\partial y_i \, \partial y_j} & \dfrac{\partial^2 V}{\partial y_i \, \partial z_j} \\[3mm] \dfrac{\partial^2 V}{\partial z_i \, \partial x_j} & \dfrac{\partial^2 V}{\partial z_i \, \partial y_j} & \dfrac{\partial^2 V}{\partial z_i \, \partial z_j} \end{pmatrix} \tag{4.33}$$

Whereas the first derivatives can be considered to form a vector of dimension $3N$, the second derivatives of the energy with respect to the coordinates form a $3N \times 3N$ matrix. Owing to the properties of differentiation, the

second-derivative matrix is symmetric and only $3N \times (3N+1)/2$ components of the matrix will be different. This means that the second derivatives
involving the coordinates of the i and j particles will be the same, no matter in
which order the differentiation is performed; i.e. $\partial^2 V / \partial x_i \, \partial y_j = \partial^2 V / \partial y_j \, \partial x_i$. If
the particles are the same $(i = j)$ then, of course, the symmetric nature of the
second derivatives is immediately apparent. The second-derivative matrix is
often called the *Hessian*.

4.6 The energy modules

The principal module introduced in this chapter is the module
POTENTIAL_ENERGY which calculates the energy and coordinate derivatives
of the energy function for a system. The module is called after the potential
energy to distinguish it from the kinetic energy which will be required later.
The module's definition is

```
MODULE POTENTIAL_ENERGY

! . Module scalars.
REAL :: EANGLE    = 0.0, EBOND    = 0.0, &
        EDIHEDRAL = 0.0, EIMPROPER = 0.0, &
        EELECT    = 0.0, ELJ      = 0.0, &
        TOTAL_PE  = 0.0

! . Module arrays.
REAL, ALLOCATABLE, DIMENSION(:)   :: ATMHES
REAL, ALLOCATABLE, DIMENSION(:,:) :: ATMDER

CONTAINS

   SUBROUTINE ENERGY ( PRINT )
      LOGICAL, INTENT(IN), OPTIONAL :: PRINT
   END SUBROUTINE ENERGY

   SUBROUTINE GRADIENT ( PRINT )
      LOGICAL, INTENT(IN), OPTIONAL :: PRINT
   END SUBROUTINE GRADIENT

   SUBROUTINE HESSIAN ( PRINT )
      LOGICAL, INTENT(IN), OPTIONAL :: PRINT
   END SUBROUTINE HESSIAN

   SUBROUTINE ENERGY_INITIALIZE
   END SUBROUTINE ENERGY_INITIALIZE

   SUBROUTINE ENERGY_PRINT
   END SUBROUTINE ENERGY_PRINT

END MODULE POTENTIAL_ENERGY
```

Three subroutines are concerned with the calculation of the energy. The first, ENERGY, calculates the energy of the system, the second, GRADIENT, calculates the energy and its first derivatives with respect to the atomic coordinates and the third, HESSIAN, calculates the energy and both the first and second derivatives. Each of these subroutines has a single optional logical argument, PRINT, which specifies whether the energies are to be printed at the end of the calculation. If PRINT is not present then the energies are printed.

The module scalars and arrays are used to store the energies and derivatives calculated as a result of calls to each of the subroutines, ENERGY, GRADIENT and HESSIAN. The total potential energy is put into the scalar variable TOTAL_PE, the first derivatives into the array ATMDER and the second derivatives into the array ATMHES. There are also other scalar variables that store individual energy terms. These are EANGLE that stores the angle energy, EBOND the bond energy, EDIHEDRAL the dihedral energy, EIMPROPER the improper dihedral energy, EELECT the electrostatic energy and ELJ the Lennard-Jones energy.

The derivative arrays are automatically allocated to have the correct size when either of the subroutines GRADIENT or HESSIAN is called. Their dimensions are ATMDER(1:3,1:NATOMS) and ATMHES(1:(3*NATOMS*(3*NATOMS+1))/2) for the first- and second-derivative arrays, respectively. Note that when ENERGY is called both derivative arrays are deallocated and that when GRADIENT is called the second-derivative array is freed. This means that old energy and derivative results are not kept in the module and that only the minimum amount of space is used to store the new results.

Because the second-derivative matrix is symmetric, only its unique elements are kept. Two storage schemes for symmetric matrices are in common use, the so-called *lower triangle* and *upper triangle* schemes. The module POTENTIAL_ENERGY uses an upper triangle format for the elements in ATMHES, which means that, of the two indices defining an element in the matrix, the second index is always greater than or equal to the first. Thus, the ijth element of the second-derivative matrix will be stored if $j \geq i$ and it will be accessed as the $\{i + [j(j-1)]/2\}$th element in the array ATMHES. Within each dimension the elements are stored in the same order as for the first derivatives, i.e. x, y, z for atom 1, x, y, z for atom 2, etc.

The module POTENTIAL_ENERGY employs the data and the procedures from a number of other modules. The first is ATOMS because the subroutines ENERGY, GRADIENT and HESSIAN calculate the energy and derivatives using the coordinates of the system that are in the array ATMCRD. There are two modules that are used to calculate the individual energy terms.

ENERGY_COVALENT contains procedures for the evaluation of the covalent energy terms (bonds, bond angles, dihedral angles and improper angles) and ENERGY_NON_BONDING contains procedures for the evaluation of the non-bonding electrostatic and Lennard-Jones energies. Although it is not directly used by POTENTIAL_ENERGY there is another module of importance, called MM_TERMS, that contains the definitions and the parameters of all the energy terms that need to be calculated. This module is used by both the modules ENERGY_COVALENT and ENERGY_NON_BONDING. The calculation of an energy and its derivatives leaves all modules in the same state, except for the scalar and array variables in the module POTENTIAL_ENERGY described above.

The other two subroutines in the potential energy module perform miscellaneous tasks. ENERGY_INITIALIZE initializes completely the module's scalar and array variables. It should not normally be needed insofar as these variables are automatically initialized at the start of a program. The subroutine ENERGY_PRINT prints details of the energy terms to the output file. It is called each time the subroutines ENERGY, GRADIENT and HESSIAN are called unless the argument PRINT for each of these subroutines is set equal to .FALSE..

ENERGY_PRINT prints not only the total energy and the individual energy terms but also the *RMS gradient* of the system if the first derivatives of the energy have been calculated. The RMS gradient, G_{RMS}, is defined as

$$G_{RMS} = \sqrt{\frac{1}{3N} \sum_{i=1}^{N} g_i^T g_i} \tag{4.34}$$

The definition of the two modules that calculate the separate energy terms will not be described in detail because they will not be needed for the examples used in this book other than when used via the module POTENTIAL_ENERGY. It is necessary, though, to define exactly what they calculate. The module ENERGY_COVALENT computes the bonding energy terms and their coordinate derivatives for the bonds (equation (4.5)), angles (equation (4.7)), dihedral angles (equation (4.8)) and improper dihedral angles (equation (4.9)). To calculate the proper and improper dihedral angles and their derivatives the definition in section 3.3 is used.

The module ENERGY_NON_BONDING calculates the electrostatic and Lennard-Jones energies using equations (4.12) and (4.13), respectively. The version of this module that we shall be using in the next few chapters automatically includes *all* interactions between atoms in the calculation except the 1–2 and 1–3 non-bonding exclusions, which are omitted. The 1–4 exclusions are treated specially insofar as the contributions arising from these terms are not completely included but are scaled by an empirical factor. Geometrical mean combination rules are employed both for the Lennard-Jones well depths and for radii (equations (4.16) and (4.17), respectively).

Within the module ENERGY_NON_BONDING there is one subroutine that users might need to access. It is ENERGY_NON_BONDING_OPTIONS and has the following interface:

```
MODULE ENERGY_NON_BONDING

CONTAINS

    SUBROUTINE ENERGY_NON_BONDING_OPTIONS ( DIELECTRIC )
        REAL, INTENT(IN), OPTIONAL :: DIELECTRIC
    END SUBROUTINE ENERGY_NON_BONDING_OPTIONS

END MODULE ENERGY_NON_BONDING
```

The subroutine takes a single optional real argument, which allows the value of the dielectric in the calculation of the electrostatic energy to be set. Its default value is unity. The subroutine always prints out a summary of the current non-bonding energy options even if the DIELECTRIC argument has not been specified.

Both modules, ENERGY_COVALENT and ENERGY_NON_BONDING, use the data in the modules ATOMS and MM_TERMS. As mentioned above, all the parameters to use in the energy calculation, including the number of each type of term to calculate, are stored in the module MM_TERMS. How the data structures in it are set up is left as the topic of the next chapter.

We consider one other module in this section, namely that which calculates the first- and second-coordinate derivatives for a system numerically using a central-step finite-difference procedure. As mentioned above, analytic differentiation is usually preferable but numerical methods are important as checks to ensure that the derivatives are consistent with the energy. The module's definition is

```
MODULE NUMERICAL_DERIVATIVES

CONTAINS

    SUBROUTINE NUMERICAL_GRADIENT ( DELTA, GRADIENT )
        REAL,                   INTENT(IN)  :: DELTA
        REAL, DIMENSION(:,:), INTENT(OUT) :: GRADIENT
    END SUBROUTINE NUMERICAL_GRADIENT

    SUBROUTINE NUMERICAL_HESSIAN ( DELTA, HESSIAN )
        REAL,                 INTENT(IN)  :: DELTA
        REAL, DIMENSION(:), INTENT(OUT) :: HESSIAN
    END SUBROUTINE NUMERICAL_HESSIAN

END MODULE NUMERICAL_DERIVATIVES
```

There are two subroutines. NUMERICAL_GRADIENT calculates the first derivatives using a finite-difference differentiation of the energy (equation (4.31))
and NUMERICAL_HESSIAN determines the second derivatives using a finite-
difference differentiation of the analytically calculated first derivatives.
Both subroutines have two arguments. The first, DELTA, is an input real scalar
that specifies the step length to use in the finite-difference calculation. The
best value will vary depending upon the system but 10^{-4} Å is usually a good
compromise value, although, to be sure, it is often necessary to try several
values for the step size to see how the values of the derivatives change. The
second argument is the array that will be filled with the derivatives. It should
be dimensioned before entry to be the same size as the equivalent derivative
array in the module POTENTIAL_ENERGY. With an optimum step size the
numerical and analytic first derivatives should agree to within approximately
10^{-4} kJ mol^{-1} Å^{-1}. Similar accuracy can be expected for the second derivatives as well.

Exercises

4.1 Select a simple molecule, such as ethane, and list all the bonding and non-
bonding interactions that are required in order to calculate its energy using
the force field. How many different parameters are there? How does the number change if the molecule is transformed slightly (to ethanol, for example)?

4.2 As mentioned in the discussion, the calculation of the non-bonding interactions
is the most time-consuming part of an energy calculation for most systems. The
number of electrostatic and Lennard-Jones interactions is $[N(N - 1)]/2$ minus
the number of exclusions. What possible strategies are there for reducing the
time required for the calculation? A fuller discussion of this point can be found
in chapter 9.

5

Setting up the molecular mechanics system

5.1 Introduction

In the last chapter we discussed methods for calculating the potential energy
and its derivatives for a system. In particular, we saw what the general form
of many molecular mechanics force fields was like as well as the specific form
of the force field that will be used for the calculations in this book. What was
not mentioned, however, was how to define the various molecular mechanical
terms so that the energy of a system can be calculated. This will be the task of
this chapter.

5.2 The molecular mechanics definition file

A quick glance at the equations describing the molecular mechanics energy
function and its constituent terms shows that there are several quantities that
need to be defined before the energy for a system can be calculated. These
include the number of terms of each type, the atoms involved in each energy
term and the values of their parameters. To specify these terms and parameters, a new file is required, the *molecular mechanics definition file* (the MM
file), that contains the necessary information. The file acts as a database for
the various parts of the energy function and how to apply it to different
systems.

  Obviously the type and nature of the terms in the energy function and the
values of the parameters will depend upon the force field that we choose to
use. Up to now, we have not mentioned any of the specific force fields that
are available for performing molecular simulations, but now one has to be
selected if we are to perform any calculations! There are many from which to
choose (a partial list is given in the references). Some are fairly intricate and
designed for highly accurate calculations of smaller molecules (such as the
MM2 and MM3 force fields developed by N. Allinger) whereas others, such

as those devised for the simulation of biomacromolecules, are simpler and resemble the 'typical' force field that was discussed in the last chapter. For a number of reasons, primarily because it is widely used and a large variety of parameters for it have been published in the literature, the programs in this book employ the all-atom version of the OPLS force field (OPLS-AA) that has been developed by W. Jorgensen and his collaborators. This is a force field of the simpler type which has been applied to a wide range of systems, including biomacromolecules, such as proteins and nucleic acids, organic liquids and solutes in solution. It should be emphasized that, although the energies or properties for a system that are calculated with this force field may differ from those obtained using other force fields, the general principles of the calculations that we perform are completely independent of the specific choice of the force field that has been made.

The exact analytic form of the force field is repeated here for convenience. The total potential energy of the system, V, is the sum of bond, angle, torsion, improper torsion and non-bonding terms. The expressions for the bond and angle energies are the same as those in equations (4.5) and (4.7) except that the factor of one half has been omitted, i.e.

$$V_{bond} = \sum_{bonds} k_b(b - b_0)^2 \tag{5.1}$$

$$V_{angle} = \sum_{angles} k_\theta(\theta - \theta_0)^2 \tag{5.2}$$

The dihedral energy is similar to that in (4.8). For each dihedral, three terms in the Fourier expansion are included with periodicities of 1, 2 and 3. The phase of each of the terms is $0°$ for periodicities 1 and 3 and $180°$ for periodicity 2. This gives the following expression in the most general case in which none of the Fourier coefficients, V_n, is zero:

$$V_{dihedral} = \sum_{dihedrals} \frac{V_1}{2}(1 + \cos \phi) + \frac{V_2}{2}(1 - \cos 2\phi) + \frac{V_3}{2}(1 + \cos 3\phi) \tag{5.3}$$

The improper dihedral energy, $V_{improper}$, has the same form as the dihedral energy except that the dihedral angles, ϕ, are replaced by improper ones, ω.

The choice of which terms to include in the sums in the bond, angle, dihedral and improper dihedral energies depends upon the system and the type of energy being calculated. For the bond and the improper dihedral energies each bond and improper dihedral angle in the sum must be defined explicitly. We shall see how this is done later. For the bond angle and dihedral angle energies, all possible bond angles and dihedral angles are generated automatically from the list of bonds in the system using the same procedures as those in section 3.2. Those bond angle and dihedral angle terms whose force constants are non-zero are then included in the calculation of the energy.

The non-bonding energy, $\mathcal{V}_{\text{nb}}$, consists of the sum of electrostatic and Lennard-Jones terms with the form

$$\mathcal{V}_{\text{nb}} = \sum_{ij \text{ pairs}} \left\{ \frac{q_i q_j}{4\pi\epsilon_0 \epsilon r_{ij}} + 4\varepsilon_{ij} \left[\left(\frac{s_{ij}}{r_{ij}}\right)^{12} - \left(\frac{s_{ij}}{r_{ij}}\right)^{6} \right] \right\} f_{ij} \tag{5.4}$$

The variable f_{ij} is a weighting factor for the interactions. For 1–2 and 1–3 interactions it has the value 0 and so these interactions are excluded from the sum. For the 1–4 interactions the value is $\frac{1}{2}$ and for all other interactions its value is 1.

A final point about the origin of some of the parameters in the OPLS force field should be made. In their work Jorgensen and co-workers concentrated on the optimization of the non-bonding and torsional parameters. The internal bond, bond angle and improper dihedral angle parameters were taken mostly from the AMBER force field developed by P. Kollman and his group and, to a lesser extent, from the CHARMM force field developed by M. Karplus and his collaborators.

We shall illustrate the format of the MM file that is used to define the system and its OPLS force-field parameters with two examples. The first is for a model of water and two ions and the second is for the molecule bALA. The file for the ion–water system is

```
MM_Definitions OPLS_AA 1.0

! . Atom Type Definitions.
Types
! Atom Name        Atomic Number        S          Epsilon
HW                      1             0.00000      0.00000
OW                      8             3.15061      0.15210
NA                     11             1.89744      1.60714
CL                     17             4.41724      0.11779
End


! . Electrostatics and Lennard-Jones Options.
Electrostatics Scale 0.5
Lennard_Jones  Scale 0.5


! . Units specification.
Units kcal/mole


! . Residue Definitions.
Residues
```

```
Residue Chloride
! # Atoms, Bonds and Impropers.
  1  0  0
CL        CL          -1.0


Residue Sodium
! # Atoms, Bonds and Impropers.
  1  0  0
NA        NA           1.0


Residue Water
! # Atoms, Bonds and Impropers.
  3  3  0
O         OW          -0.834
H1        HW           0.417
H2        HW           0.417


O H1 ; O H2 ; H1 H2

End


! . Parameter Definitions.
Parameters


Bonds
! Atoms     FC        Equil.
HW OW     529.6       0.9572
HW-HW     38.25       1.5139
End


Angles
! Atoms     FC        Equil.
HW OW HW  34.05       104.52
HW HW OW   0.0        37.74
End


End
End
```

The general structure for the file is as follows. The first non-comment statement in the file is the keyword MM_Definitions and the last is the End keyword. After the MM_Definitions title comes the name of the force field (in this case OPLS_AA for the 'All-Atom OPLS' force field) and the version number of the module library (in this case version 1.0). Both the force field's name and the version number are verified by the module reading the file to ensure that these can be handled by the module library. The next section in the file defines the atom types and their properties. It starts with the

Types keyword and ends with End. After this section come the remaining definitions which consist of two data blocks that contain the residue and the parameter definitions and a number of single-line options to define such things as the scaling factors for the calculation of the electrostatic and Lennard-Jones energies and the units for some of the terms that are used in the file. The order of all data after the atom-type-definition section is arbitrary.

The first section in the MM file defines the atom types. Some explanation is needed here. In principle, if the energy function being used were completely accurate, the only property that would be needed in order to identify an atom for the calculation of an energy would be its element type. In practice, because the empirical energy functions in use are not sufficiently flexible in this regard, it is necessary to define different types of atom that correspond to the same element. Thus, for example, a hydrogen bound to an oxygen atom will be defined to be of a type different from a hydrogen bound to an aliphatic carbon and the two types will have different parameters associated with them. Similarly, aliphatic carbons will have different types from aromatic carbons and carbonyl carbons. In the example file given above, four types are defined, each of which corresponds to a different element. The types are HW for hydrogen in water, OW for oxygen in water, NA for a sodium cation and CL for a chloride anion. Each type is followed by an integer that gives its atomic number and two real numbers that give its Lennard-Jones parameters (the radius and well depth). The OPLS force field uses geometrical combination rules for the Lennard-Jones parameters.

The next three commands are single-line commands that define various miscellaneous options. The Electrostatics Scale and Lennard_Jones Scale commands set the factors (f_{ij} in equation (5.4)) by which the 1–4 non-bonding interactions are to be scaled. As mentioned in the last chapter, it is often necessary to reduce these values because the dihedral terms also influence the interaction between atoms which are in 1–4 positions relative to each other. The recommended values for the f_{ij} factors are $\frac{1}{2}$ both for the electrostatics and for the Lennard-Jones interactions. The third command, Units, defines the units for some of the parameters in the file. The keyword kcal/mole indicates that all the parameters listed in the file associated with an energy have units of kcal mol^{-1}. The default for use in the program library is kJ mol^{-1} and so the program will automatically make the conversion to these units when the MM file is processed. The default units for lengths and angles in the MM file are ångström units and degrees except for the angle-term force constants, for which they are radians (i.e. the units are kJ mol^{-1} radians^{-2}).

The next section in the example file defines the residues. This block starts with the keyword Residues and terminates with the keyword End. Each residue definition within the block has the same form. There is the Residue keyword followed by the name of the residue. On the next command line there are three integers giving the number of atoms, bonds and impropers in the residue. It is not necessary to specify the angles and dihedrals because these are generated automatically once the bonds in the system are known. After this line, the names of the atoms in the residue are specified, together with their atom types and their charges (in units of the electron's charge, *e*). Each specification is given on a separate line. All the atoms in the residue must be given distinct names, although, of course, they can be of similar type. The sum of the charges for each atom gives the total electronic charge on the residue. After the list of atoms in the residue, the bond and improper definitions are given. The bond definitions precede the improper definitions and each definition is given on a separate line. In the example MM file given above there are no improper definitions and only three bond definitions for water. Each bond definition consists of the names of two atoms for which there is a bond. Likewise an improper definition consists of four names. Note that a new line can be denoted either by a carriage return or by a semicolon.

There is one point that needs further elaboration. Why is it that the atomic charges are specified here rather than with, for example, the Lennard-Jones data in the atom-types section? The reason is important and is related to the *transferability* of the various parameters that are used in empirical force fields. Experience has shown that, whereas it is possible to define a single set of Lennard-Jones parameters that can be used for each atom type no matter in which environment it occurs, the same is not true for the charges. These are much more variable and their value depends critically upon the chemical environment within which the atom finds itself. The alternative to specifying the charges in the residue definitions is to define different atom types for each atom which has a different charge. This is a less elegant solution insofar as the number of atom types and hence the number of definitions needed in the MM file would be multiplied unnecessarily.

The only remaining data block in the first example MM file is the block specifying the parameters for the bonding terms in the energy function. The parameters section starts with the Parameters keyword and ends with the End keyword. There can be up to four blocks within this section, specifying the parameters for the bond, angle, dihedral and improper energy function terms. Each block starts with a keyword, Bonds, Angles, Dihedrals or

Impropers, and is terminated by the keyword End. Note that, in the example file given above, the only two blocks that are found are for the bond and angle parameters. There are no dihedral or improper terms for water (or for the ions!). The bond and angle definitions should be self-explanatory. A bond definition consists of the name of two atom types (*not* atom names) followed by the values for the force constant and the equilibrium length for the bond (k_b and b_0 in equation (4.5)). The angle definition is similar except that three atom types need to be specified.

It is to be noted in this example that a bond has been defined between the two hydrogens in the water molecule residue. Such 'unchemical' bonds (and other terms) are often included in force-field models of molecules in order to obtain better calculated properties. Because the bond is included, parameters for the bond and for the angles of type HW HW OW that are generated from it must be included. The force constant for the angle parameter is zero so, in fact, these terms are omitted, leaving only a single extra bond term between the hydrogens.

A slightly more complicated example of an MM file is the one for the molecule bALA. The file is

```
MM_Definitions OPLS_AA 1.0

! . Atom Type Definitions.
Types
! Atom Name       Atomic Number          S          Epsilon
H                      1               0.00000       0.00000
HC                     1               2.50000       0.03000
C                      6               3.75000       0.10500
CT                     6               3.50000       0.06600
N                      7               3.25000       0.17000
O                      8               2.96000       0.21000
End

! . Electrostatics and Lennard-Jones Options.
Electrostatics Scale 0.5
Lennard_Jones  Scale 0.5

! . Units specification.
Units kcal/mole

! . Residue Definitions.
Residues

Residue ACETYL
! # Atoms, Bonds and Impropers.
  6  6  1
```

```
CT        CT       -0.18
HT1       HC        0.06
HT2       HC        0.06
HT3       HC        0.06
C         C         0.50
O         O        -0.50


CT HT1 ; CT  HT2 ; CT  HT3 ; CT  C ; C  O ; C  +R


CT +R   C   O


Residue ALANYL
! # Atoms, Bonds and Impropers.
  10 11 2
N         N        -0.50
H         H         0.30
CA        CT        0.14
HA        HC        0.06
CB        CT       -0.18
HB1       HC        0.06
HB2       HC        0.06
HB3       HC        0.06
C         C         0.50
O         O        -0.50


-R  N    ; N  H   ; N  CA  ; CA  HA ; CA  CB ; CA  C
 CB HB1 ; CB HB2 ; CB HB3 ; C     O ; C    +R


-R  CA  N  H
 CA +R  C  O


Residue N_METHYL
! # Atoms, Bonds and Impropers.
  6  6  1
N         N        -0.50
H         H         0.30
CT        CT        0.02
HT1       HC        0.06
HT2       HC        0.06
HT3       HC        0.06


-R  N ; N  H ; N  CT ; CT  HT1 ; CT  HT2 ; CT  HT3


-R  CT   N   H


End


! . Parameter Definitions.
Parameters


Bonds
! Atoms       FC        Equil.
```

```
C    CT     317.0      1.522
C    O      570.0      1.229
C    N      490.0      1.335
CT   HC     340.0      1.090
CT   CT     268.0      1.529
CT   N      337.0      1.449
H    N      434.0      1.010
End
```

Angles

! Atoms			FC	Equil.
CT	C	O	80.0	120.40
CT	C	N	70.0	116.60
N	C	O	80.0	122.90
HC	CT	HC	33.0	107.80
HC	CT	N	35.0	109.50
C	CT	HC	35.0	109.50
C	CT	CT	63.0	111.10
C	CT	N	63.0	110.10
C	N	CT	50.0	121.90
C	N	H	35.0	119.80
CT	N	H	38.0	118.40
CT	CT	N	80.0	109.70
CT	CT	HC	37.5	110.70

End

Dihedrals

! Atoms				V0	V1	V2	V3
N	C	CT	CT	0.000	1.173	0.189	-1.200
N	C	CT	HC	0.000	0.000	0.000	0.000
N	C	CT	N	0.000	1.816	1.222	1.581
O	C	CT	CT	0.000	0.000	0.000	0.000
O	C	CT	HC	0.000	0.000	0.000	0.000
O	C	CT	N	0.000	0.000	0.000	0.000
CT	C	N	CT	0.000	2.800	6.089	0.000
CT	C	N	H	0.000	0.000	4.900	0.000
O	C	N	CT	0.000	0.000	6.089	0.000
O	C	N	H	0.000	0.000	4.900	0.000
C	CT	CT	HC	0.000	0.000	0.000	-0.076
HC	CT	CT	HC	0.000	0.000	0.000	0.318
HC	CT	CT	N	0.000	0.000	0.000	0.464
C	CT	N	C	0.000	-2.365	0.912	-0.850
C	CT	N	H	0.000	0.000	0.000	0.000
CT	CT	N	C	0.000	0.000	0.462	0.000
CT	CT	N	H	0.000	0.000	0.000	0.000
HC	CT	N	C	0.000	0.000	0.000	0.000
HC	CT	N	H	0.000	0.000	0.000	0.000

End

Impropers

! Atoms	V0	V1	V2	V3

```
CT N  C   O     0.000      0.000     10.500      0.000
C  CT N   H     0.000      0.000      1.000      0.000
End


End
End
```

The structure of the file is the same, although there are some new features. First of all, there are more atom types. In addition to the single types for nitrogen and oxygen, OPLS uses two types for carbon (C for any carbonyl sp^2 carbon and CT for any sp^3 carbon) and two types for hydrogen (H for a hydrogen attached to a nitrogen and HC for a hydrogen attached to an aliphatic carbon).

The first major difference in the file compared with the previous example occurs in the residue definitions. In the previous file only residues that corresponded to distinct molecules or ions were considered. For the bALA molecule there are three residues but these have bonds between them that it is essential to specify if the energy of the system is to be correctly calculated. The way this is done is to introduce into the definitions of the bonds and impropers a notation that indicates atoms of neighbouring residues. -R denotes an atom of the preceding residue in the residue sequence for the system and +R an atom of the residue that follows. When using this notation it is important to remember that only bonds to the immediately preceding and following residues can be specified and that these definitions must be matched in the bond definitions for the other residues. In other words, in the definition for the residue ACETYL there is a bond to the following residue defined as C +R. In the sequence (to be defined in the next section) the residue that follows is ALANYL and so there must be a corresponding use of -R in a bond definition for that residue. In this case it is N -R so that the carbonyl carbon of ACETYL is bonded to the nitrogen of ALANYL. It can be seen that the -R and +R types can also be present in the improper definitions. The only restriction on their use in this way is that, if they are present, they must be accompanied by a corresponding bond definition.

The second major difference in the file is the presence in the parameter section of the blocks that define the dihedral and improper energy term parameters. The dihedral definition consists of a list of four atom types followed by four real numbers. The first number is unused in the current work but the remaining numbers give the force constants V_1, V_2 and V_3 that are defined in equation (5.3). It is also possible to specify dummy atom types, X, in the definition of the dihedral, although this notation does not appear in this example. A dummy type refers to any atom type and can be used to specify both the first atom and the fourth atom in the definition, although the second and third atom types must be specified explicitly. If there are two (or more) definitions for a dihedral involving explicitly stated types and dummy types, then the more specific definition is used in preference.

The format for the improper definition is identical to the dihedral definition although there are some subtle differences. First the order of the atoms is different – the exact definitions are shown in figure 4.4. The third atom in the list is the central atom and the remaining atoms are bonded to it. The order of the atoms is important in that a change will produce a different dihedral and, hence, a different energy. The AMBER force field, from which the OPLS improper dihedrals are taken, recommends that only one improper dihedral (at most) be specified per central atom and that the first, second and fourth atoms of the improper be placed in alphabetical order of atom type or, if two or more of these atom types are the same, in order of atom number. The second difference from the dihedral terms is that the use of the dummy atom names is different. Instead of the first and fourth atoms that can be replaced by X, it is possible to replace the first atom type by X or both the first and the second by X.

There are two more data blocks that can be specified in the MM file and are needed for constructing more complicated systems. These are the Variants ... End and Links ... End blocks, both of which modify the definitions of one or more residues. A *variant* of a residue is used to alter in some way the definition of a single, existing residue. Thus, for example, it may be that a residue has been defined to have a particular protonation state. Instead of using another residue definition to define the same residue with such a minor change, a variant that just modifies the existing residue can be used. Other examples are variants to specify the N- or C-terminal modifications of amino acids in a peptide or protein chain. A *link* is used to create a bond between two residues. Such modifications are needed to cross-link residues in a polymer, for example.

The general format of a variant is best illustrated with an example. Here we take the variant which changes the residue definition for ALANYL (given above) so that its N-terminal has a NH_3^+ group. The definition is

```
Variant N_TERMINAL
! # Deletes, adds, charges, bonds and impropers.
  2  4  1  4  0

N ; H

N     N3    -0.30
```

```
H1      H3      0.33
H2      H3      0.33
H3      H3      0.33
CA      0.25

N  H1 ; N  H2 ; N  H3 ; N  CA
```

The definition starts with the `Variant` keyword and is followed by the variant's name – in this case, `N_TERMINAL`. It is also possible to have an additional keyword after the variant's name, which gives the name of the residue to which the variant can be applied. This is useful when it would be convenient to keep the overall name of the variant the same but apply it differently to various residues. Thus, in the MM definition file for proteins which is supplied with the example programs, there are three variants with the name `N_TERMINAL`. Two are specific, one for the amino acid glycine and one for the amino acid proline, because these residues require special treatment; the third one is general and can be applied to any of the other amino acid residues.

On the second line of the variant definition there are five counters defining the various modifications that need to be made to the residue definition. These are the number of atoms to delete from the original definition, the number of new atoms to add, the number of atoms in the original definition whose charges need to be changed and the number of new bond and improper definitions. Each of these changes is listed on separate lines in the same order as that in which they appeared in the previous sentence. Thus, in the example, the two N-terminal atoms are deleted and the four atoms that constitute the NH_3^+ group are added. The deleted atoms are listed by name whereas the names of the added atoms, their types and their charges must all be given. There is one atom in the original residue whose charge is to be changed. The syntax is the name of the atom in the original residue definition followed by its new charge. Last come the bond and improper definitions, which take the same format as that used in a residue definition. It is to be noted that, when a variant is applied to a residue, the operations on the residue are performed in the same order. Thus, the atoms are deleted from the original residue first (together with *all* their corresponding bond and improper terms), the new atoms are added, the charges of certain atoms changed and finally the new bond and improper terms constructed. The order of the operations is important because it is possible to apply more than one variant to the same residue in the system and in certain circumstances specifying the variants in different orders will produce different results.

The link definition is similar to a variant definition, except that, because a link is between two residues, information for two variants, one for each residue, must be listed. The general format is

```
Link LINK_NAME
! . Variant for residue 1.
...

! . Variant for residue 2.
...
```

The definition is started with the keyword Link and is followed by the name of the link modification. After the line containing the link name, the variant information that is applicable to each residue is listed in the same format as that for the variant definition. The only difference is that both variant definitions must contain at least one bond that specifies how the residues are to be linked. This is done by defining for each residue a bond between an atom of that residue and an atom with name *R (corresponding to the types +R and -R in the residue definitions), which denotes an atom of the other residue.

5.3 Processing the MM file

In the previous section the format of the MM file which contains the information about the residue and parameter definitions for the energy function was described. In this section, we introduce a module that processes the information in the file and converts it into a form that can be used to define the terms needed for the calculation of the energy of a particular system.

The module is called MM_FILE_IO and it contains a single subroutine. The module's definition is

```
MODULE MM_FILE_IO

CONTAINS

   SUBROUTINE MM_FILE_PROCESS ( FILE_OUT, FILE_IN )
      CHARACTER ( LEN = * ), INTENT(IN)           :: FILE_OUT
      CHARACTER ( LEN = * ), INTENT(IN), OPTIONAL :: FILE_IN
   END SUBROUTINE MM_FILE_PROCESS

END MODULE MM_FILE_IO
```

The subroutine, MM_FILE_PROCESS, has two arguments. Both arguments are characters and specify file names. There is an essential argument that gives the name of the *binary* file to which the information processed by the subroutine is written. The other argument is optional and gives the name of the input MM file (with the format described above) which is to be processed. If no input file name is given the MM file is read from the input stream. The subroutine reads in the MM file, from whatever source, verifies and processes the information in it and then writes out the data to the binary output file.

One additional point can be made here. In the last section two relatively simple examples of MM files were given, both of which were specific to particular systems. These were for illustrative purposes only. In practice, it would be more usual to create a file containing all the information that is likely to be required for a general class of systems, such as proteins or nucleic acids, process the file using MM_FILE_PROCESS once and then use the binary, processed file subsequently whenever it is needed.

5.4 Constructing the system definition

Once the energy-function data have been defined, they can be used to construct the data structures that contain the information about the energy-function terms that are needed for the particular systems being studied. To do this there must be a way of specifying the composition of the system. One way has already been described in chapter 2 when the system's composition was read straight from a coordinate file. This method is inappropriate for more complicated systems, however, when variants and links of certain residues may be defined. To define the composition of a system, it is necessary to write a *sequence file*.

The format of the sequence file is the simplest of any of the files encountered so far and consists of a listing of the residues in each subsystem in the complete system. For the bALA molecule the file has the form

```
Sequence
1 Subsystem BLOCKED_ALANINE
3
ACETYL ; ALANYL ; N_METHYL
End
End
```

The file starts with the keyword Sequence and is terminated by End. The second command line contains an integer denoting the number of subsystems in the system. For each subsystem there then follows a subsystem block starting with the keyword Subsystem and terminated by an End. After the Subsystem keyword comes the name of the subsystem. On the next line is an integer with the number of residues in the subsystem. This is followed by the names of these residues, in order, on separate lines. Thus, for bALA, there is one subsystem, which has three residues.

A more complicated example is illustrated by looking at the file for a small protein, crambin, which consists of a chain of 46 amino acid residues:

```
Sequence
1
Subsystem CRAMBIN

46
THR ; THR ; CYS ; CYS ; PRO ; SER ; ILE ; VAL ; ALA ; ARG
SER ; ASN ; PHE ; ASN ; VAL ; CYS ; ARG ; LEU ; PRO ; GLY
THR ; PRO ; GLU ; ALA ; ILE ; CYS ; ALA ; THR ; TYR ; THR
GLY ; CYS ; ILE ; ILE ; ILE ; PRO ; GLY ; ALA ; THR ; CYS
PRO ; GLY ; ASP ; TYR ; ALA ; ASN

Variant N_TERMINAL THR  1
Variant C_TERMINAL ASN 46

End


Link DISULPHIDE_BRIDGE CRAMBIN CYS  3 CRAMBIN CYS 40
Link DISULPHIDE_BRIDGE CRAMBIN CYS  4 CRAMBIN CYS 32
Link DISULPHIDE_BRIDGE CRAMBIN CYS 16 CRAMBIN CYS 26

End
```

The system defined by this file again consists of one subsystem only, but it illustrates the use of the variant and link modifications. The subsystem block defining the protein chain, in addition to the residue sequence, contains two lines that declare variants for particular residues. The commands start off with the keyword Variant and are followed by the name of the variant and the name and the number (in the subsystem) of the residue to which the variant is to be applied. Thus, the variants declare that an N-terminal modification is to be applied to the first residue and a C-terminal variant to the last.

The commands specifying links between residues are given outside the subsystem blocks at the end of the file. Each command line starts with the keyword Link and is followed by the name of the link modification and two residue specifications, both of which consist of a subsystem name and the name and the number (in the subsystem) of the residue. In the crambin example, there are link definitions specifying that three disulphide bridges are to be formed between the various cysteine residues in the protein chain.

With the information in the sequence file discussed above and the MM file of the previous section, it is possible to construct the appropriate data structures for an energy calculation of a system. The module used to do this is called MM_SYSTEM and it contains a single accessible subroutine:

```
MODULE MM_SYSTEM

CONTAINS

   SUBROUTINE MM_SYSTEM_CONSTRUCT ( MMFILE, SEQUENCE )
      CHARACTER ( LEN = * ), INTENT(IN)           :: MMFILE
      CHARACTER ( LEN = * ), INTENT(IN), OPTIONAL :: SEQUENCE
   END SUBROUTINE MM_SYSTEM_CONSTRUCT

END MODULE MM_SYSTEM
```

The subroutine, MM_SYSTEM_CONSTRUCT, has two arguments that, like MM_FILE_PROCESS, both specify file names. The essential argument gives the name of the binary, processed MM file. The second argument is optional and gives the name of the file in which the sequence is listed. If no sequence file name is present the sequence information is read from the default input stream.

The subroutine works by reading the sequence information from the sequence file and then collating it with the information from the processed MM file about each residue and about its parameters. Once the subroutine has finished, if there are no errors, the subroutine fills the MM_TERMS data structure with all the data needed for an energy calculation of the system.

There is one other module that will be introduced in this chapter. It is MM_SYSTEM_IO and it contains two subroutines:

```
MODULE MM_SYSTEM_IO

CONTAINS

   SUBROUTINE MM_SYSTEM_READ ( FILE )
      CHARACTER ( LEN = * ), INTENT(IN) :: FILE
   END SUBROUTINE MM_SYSTEM_READ

   SUBROUTINE MM_SYSTEM_WRITE ( FILE )
      CHARACTER ( LEN = * ), INTENT(IN) :: FILE
   END SUBROUTINE MM_SYSTEM_WRITE

END MODULE MM_SYSTEM_IO
```

The subroutines in the module are used to read and write all the information that is needed to calculate the molecular mechanics energy for a system to an external, *binary* file. All information from the ATOMS, SEQUENCE and MM_TERMS modules is on the file, except the values of the atomic coordinates. This is useful because the setting up of a system for an energy calculation can be time-consuming, especially for large systems. Thus, once a particular system has been set up, its *system file* can be saved and then read in subsequent calculations without having to repeat the setting up process. Both subroutines in the module have a single-character argument that specifies the name of the system file to be read or written.

5.5 Examples 4 and 5

To illustrate the use of the modules introduced in this chapter and the last, two examples are described. The first program constructs the system file for the bALA molecule whereas the second calculates its energy and first derivatives.

The first program is

```
PROGRAM EXAMPLE4

... Declaration Statements ...

! . Process the MM file for bALA.
CALL MM_FILE_PROCESS ( ''bALA.opls_bin'', ''bALA.opls'' )

! . Construct the system file for bALA.
CALL MM_SYSTEM_CONSTRUCT ( ''bALA.opls_bin'', ''bALA.seq'' )

! . Save the system file for later use.
CALL MM_SYSTEM_WRITE ( ''bALA.sys_bin'' )

END PROGRAM EXAMPLE4
```

This example is straightforward. It takes the MM definition file explained in section 5.2, which is in bALA.opls, processes it and writes it out to the binary file bALA.opls_bin. The system file for the bALA molecule is then constructed using the processed MM file and the sequence file for the bALA molecule which is in bALA.seq. Finally the system file is written out to the file bALA.sys_bin for later use (including in the next example).

The next program is as follows:

```
PROGRAM EXAMPLE5

... Declaration Statements ...

! . Local scalars.
REAL :: E1, E2

! . Local arrays.
REAL, ALLOCATABLE, DIMENSION(:,:) :: DEDR_ANALYTIC, DEDR_NUMERIC
```

```
! . Read in the system file.
CALL MM_SYSTEM_READ ( ''bALA.sys_bin'' )

! . Allocate space for the derivative arrays.
ALLOCATE ( DEDR_ANALYTIC(1:3,1:NATOMS), DEDR_NUMERIC(1:3,1:NATOMS) )

! . Read in a first set of coordinates.
CALL COORDINATES_READ ( ''bALA1.crd'' )

! . Calculate and print an energy.
CALL ENERGY ; E1 = TOTAL_PE

! . Read in a second set of coordinates.
CALL COORDINATES_READ ( ''bALA2.crd'' )

! . Calculate the energy and gradients for the second structure.
CALL GRADIENT ; E2 = TOTAL_PE ; DEDR_ANALYTIC = ATMDER

! . Print out the difference in the energies.
WRITE ( OUTPUT, ''(/A,F12.4)'' ) ''Energy difference = '', E1 - E2

! . Calculate the gradients numerically for the second structure.
CALL NUMERICAL_GRADIENT ( 0.0001, DEDR_NUMERIC )

! . Print out the largest difference.
WRITE ( OUTPUT, ''(/A,G20.6)'' ) ''Largest derivative difference = '', &
                    MAXVAL ( ABS ( DEDR_ANALYTIC - DEDR_NUMERIC ) )

! . Deallocate the arrays.
DEALLOCATE ( DEDR_ANALYTIC, DEDR_NUMERIC )

END PROGRAM EXAMPLE5
```

In the first step, this program reads in the system file created by the previous example and afterwards allocates two arrays, each of dimension (1:3,1:NATOMS), to store the gradients which are to be calculated analytically and numerically later. In the subsequent step a set of coordinates is read from a coordinate file. This is because all the information necessary for the calculation of an energy is stored in the system file, *except* the values of the atomic coordinates. The energy is then computed and printed. The total potential energy for the bALA molecule in the configuration stored in the coordinate file, bALA1.crd, is saved in the variable, E1, by equating it to the variable TOTAL_PE from the module POTENTIAL_ENERGY.

In the next group of commands in the program, another configuration for the molecule is read in from a different coordinate file, bALA2.crd, and its energy is also calculated and printed. The energy of the second configuration is stored in the variable, E2, and the difference in energy between the two configurations is printed. In addition to the energy, the derivatives of the energy for the second configuration are calculated and stored in one of the arrays that was allocated earlier, DEDR_ANALYTIC.

In the final part of the program, the derivatives of the energy for the second configuration are calculated numerically using a finite-difference step length of 10^{-4} Å. The results are stored in the array DEDR_NUMERIC and the largest difference between the analytic and numeric results is printed out. The program terminates with the explicit deallocation of the derivative arrays.

Exercises

5.1 Repeat the calculation of the energy using the examples in the last section but for other systems of interest. A possible example is the protein crambin, whose sequence was given in section 5.4, and for which an appropriate MM file is supplied with the program library. For other systems, readers will need to construct their own MM files.

5.2 Using the MM file for water and for the sodium and chloride ions, calculate the energies of some water–water and water–ion complexes. In each case, the geometry of the water molecules can be kept fixed (i.e. with hydrogen—oxygen bond lengths of 0.9572 Å and a bond angle of 104.52°), but the relative orientation of the molecules or molecule and ions can be altered. Devise a search procedure to investigate automatically and systematically a range of configurations for the complexes. What is the shape of the potential energy surface for the system? What are the most stable configurations for the interaction of a water molecule with the sodium cation, the chloride anion and another water molecule? The energy of interaction between the molecules is, of course, due to non-bonding terms only. What are the relative contributions of the electrostatic and Lennard-Jones terms for each of the three complexes?

6

Finding stationary points and reaction paths on potential energy surfaces

6.1 Introduction

We have already discussed how to calculate the potential energy and some of its derivatives for a single geometry of the atoms in a system. Although the calculation of an energy for one configuration or a small number of configurations may sometimes be necessary, it can give only limited information about the properties of a system. To investigate the latter more thoroughly it is necessary to identify the interesting or important regions on the system's potential energy surface and develop ways in which they can be explored. Methods to do this will be investigated in this chapter.

6.2 Exploring potential energy surfaces

The energy function that is used to calculate the energy of a molecular system is a function of the positions of all the atoms in the system and so it is a multidimensional function, which defines the system's potential energy surface. It is this surface which determines, in large part, the behaviour and the properties of the system.

A little reflection shows that the number of configurations or geometries available to a system with more than a few atoms is enormous. A simple example should make this clear. Take a diatomic molecule or, more generally, any system comprising two atoms in vacuum. The geometry of such a molecule is completely determined by specifying the distance between the two atoms and so the potential energy surface is a function of only one geometrical variable. It is easy to search the entire potential energy surface for this system. Start with a small interatomic distance, calculate the energy, increase the distance by a certain amount and then repeat the procedure. In this way we can obtain a picture similar to those in figures 4.1 and 4.6.

Consider next a three-atom system, such as a molecule of water, for which there are three independent geometrical parameters. These can be specified, for example, as the three interatomic distances or by two distances and an angle. If the atoms are collinear then there will be only two independent variables. In either case, the potential energy surface can again be explored, although substantially more calculations will be required. To see this, suppose that n values of each independent parameter are required in the search. It will be necessary to calculate n^2 energies for the linear system and n^3 energies in the general case. If n takes the value 10 – a reasonable value – then 100 and 1000 energy calculations will be needed, respectively. The extension of this argument to larger systems is obvious and shows that to search a potential energy surface in this simplistic fashion for a system with m geometrical parameters requires n^m energy calculations – a prohibitively large number except for small values of m and n.

One problem with the simplistic search scheme outlined above is that it is very wasteful in that most of the configurations for which the energy will be calculated will not be chemically reasonable. Their energies will be too high, either because some atoms will be much too close and overlap or because some atoms that should be bonded together will be too far apart. Important configurations will be those that have a low energy and so it is for these that we would like to look in any search procedure that we adopt. To make this criterion more precise, we normally search for points on the surface that are *minima*, i.e. points that have the lowest energy on a particular part of a potential energy surface. A minimum is characterized by the property that any small changes in the geometry of a system that is at a minimum will lead to an increase in energy. Minima and the regions of the potential energy surface around them correspond, roughly speaking, to the *stable states* in which a system will normally be found.

Minima are an example of *stationary points* on a potential energy surface. Stationary points are defined as points for which the first derivatives of the energy with respect to the geometrical parameters are zero. For the special case in which the Cartesian coordinates of the atoms are the geometrical parameters, we have the condition

$$G = \frac{\partial \mathcal{V}}{\mathrm{d}R} = 0 \tag{6.1}$$

To distinguish between different types of stationary points, a second condition, which uses the second derivatives of the energy with respect to the geometrical parameters, is needed. The condition states that the stationary points of a potential energy surface can be classified into different types depending upon the number of negative eigenvalues that its second derivative matrix possesses. If there are no negative eigenvalues then the point is a minimum, if there is one negative eigenvalue then the point is a *first-order saddle point* (more usually shortened to just *saddle point*) and if there are n negative eigenvalues the point is an *nth-order saddle point*.

The eigenvalues of the second-derivative matrix are determined by diagonalizing the matrix, which means that the *characteristic* or *secular equation* for the matrix is solved. This is

$$\|\mathbf{H} - \lambda\mathbf{I}\| = 0 \tag{6.2}$$

where $\mathbf{H}$ is the matrix of second derivatives, $\mathbf{I}$ is the identity matrix, λ denotes the eigenvalues and the double straight lines denote the determinant. If Cartesian coordinates are used, then the matrices $\mathbf{H}$ and $\mathbf{I}$ will each have dimension $3N \times 3N$, where N is the number of atoms, and the left-hand side of equation (6.2) will be a polynomial of order $3N$ in λ. The problem then reduces to finding the roots of this polynomial. Because the dimension of the problem is $3N$ there will be $3N$ eigenvalues. These are guaranteed to be real (either negative or positive, but not complex) because the matrix $\mathbf{H}$ is symmetric.

Associated with each eigenvalue is a vector, the eigenvector, which specifies a direction in the space of the geometrical parameters. The eigenvector, e_i, for a particular eigenvalue, λ_i, satisfies the following equation:

$$\mathbf{H}e_i = \lambda_i e_i \tag{6.3}$$

Again, due to the fact that the matrix $\mathbf{H}$ is symmetric, the eigenvectors have some useful properties. These include the fact that the eigenvectors form an *orthonormal set*, which means that the eigenvectors corresponding to different eigenvalues have zero overlap or a zero dot product, i.e.

$$e_i^T e_j = 0 \qquad i \neq j \tag{6.4}$$

Physically, if a stationary point has a negative eigenvalue it means that a small (or, to be exact, an infinitesimal) displacement of the geometry of the system along the direction defined by the eigenvector corresponding to the negative eigenvalue will lead to a reduction in the energy of the system. Equally, a small displacement along the eigenvector of a positive eigenvalue will lead to an increase in energy. Thus, a minimum has no negative eigenvalues and so all displacements increase the energy, whereas for an *nth*-order saddle point there will be n displacements that decrease the energy and $3N -$

n displacements that increase it. If an eigenvalue has a value of zero then a small displacement along its eigenvector results in no change in energy.

Minima are not the only stationary points on the surface which are of interest, although they are probably the most important. We are also often interested in searching for first-order saddle points. The reason for this is illustrated in figure 6.1, which shows a model potential energy surface. This surface has three minima and two (first-order) saddle points. As stated above, each minimum corresponds to a stable state of the system but, if there is a reaction or the system undergoes a conformational change, the system will move from one stable state and, hence, minimum to another. Looking at figure 6.1 it can be seen that there are many possible *reaction paths* that join different minima. Remembering that a system will spend most of its time in low-energy regions of a surface, it is easily seen that the paths with the lowest energy pass through the saddle points. In other words, the first-order saddle point is the point of highest energy along the reaction path of lowest energy that joins two minima. Thus, the location of saddle points is important when studying reactions and transitions between different geometrical configurations in molecular systems.

Now that we have a specific mathematical criterion for the identification of points on the surface in which we are interested it is possible to formulate

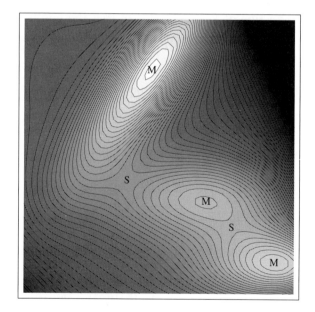

Figure 6.1 A contour plot of the Müller–Brown two-parameter model potential energy surface. The darker the shading the larger the function's value. Minima are labelled by 'M' and saddle points by 'S'.

algorithms to search for such points. In this chapter, we shall discuss a few of these that all have the common property that they are *local search algorithms*. That is they start at a given geometrical configuration of the system and then look for a stationary point that is in the vicinity of this starting geometry. This is a useful approach but it does have drawbacks because we may be interested in finding the minimum, for example, which has the lowest energy on the entire potential energy surface. The algorithms given here will not usually be useful in this case and so-called *global search algorithms* will be required. We will return to this subject later.

To see that local search algorithms will not always be useful when searching for the *global minimum* (or the near-global minimum) we can look at particular classes of systems for which the number of minima either is known or can be estimated. One such class that is widely used as a test case for optimization algorithms is the *Lennard-Jones clusters*. These are groups of a given number of identical atoms that interact solely via Lennard-Jones terms (equation (4.13)) and that act as models of clusters of rare-gas atoms, such as argon. For small numbers of atoms it is possible to find all the minima. For larger numbers of atoms it is necessary to estimate the number. The numbers of minima and the energies of the global minima for the clusters consisting of 8–19 atoms are given in table 6.1. Note that the energies have been calculated assuming that the well depths and the atom radii in the Lennard-Jones energy term both take the value of 1. It can be seen that the number of minima for the system increases very rapidly with the number of atoms. In fact the

Table 6.1 *Numbers of minima and the lowest energies of some Lennard-Jones clusters.*

Number of atoms	Number of minima	Lowest energy (reduced units)
8	8	−19.822
9	18	−24.113
10	57	−28.420
11	145	−32.765
12	366	−37.967
13	988	−44.327
14	2617	−47.845
15	6923	−52.323
16	18316	−56.816
17	48458	−61.318
18	1.28×10^5	−66.531
19	3.39×10^5	−72.660

number of minima increases, to a reasonable approximation, as the exponential of the number of atoms in the cluster, i.e. as $\exp(N)$. This should give some idea of the difficulty of finding the global minimum for a system such as a protein, which consists of 1000 atoms or more!

6.3 Locating minima

The location of a minimum on a potential energy surface is an example of a well-studied mathematical problem – that of the minimization of a multi-dimensional function. There is a large number of methods available for performing such a task, but the choice of the best method for a particular problem is determined by such factors as the nature of the function to be minimized, whether and which derivatives are available and the number of variables.

A description of the principles behind local optimization methods will not be undertaken here and the reader is referred to the references for more details. We shall classify algorithms for the optimization of the geometry of molecular systems by the type of derivatives that they use. We mention four types.

1. *No derivatives.* These algorithms make use of function (or energy) values only. An example of these algorithms would be the *simplex* method and the crude search algorithm discussed in the last section.
2. *First derivatives.* These algorithms use the values of the energy and its first derivatives. Examples include *conjugate gradient* and *steepest descent* methods.
3. *First and exact second derivatives.* These algorithms use energy values and the first and second derivatives. Examples are the *exact Newton* and *Newton–Raphson* methods.
4. *First and approximate second derivatives.* These algorithms use energy values and the exact first derivatives of the energy. They also use approximations to the second-derivative matrix, either to the full matrix or to the derivatives in a subspace of the geometrical parameters. Methods of this type include the *quasi-Newton, reduced basis-set Newton* and *truncated Newton* algorithms.

As a rule, the algorithms which use derivative information are more efficient at finding minima. For this reason and because the derivatives of the energy are readily calculated, methods that use derivatives are more commonly employed for molecular problems. Of these algorithms, different ones will prove more efficient depending upon the size of the system. For all systems, it is likely that the exact Newton algorithms would require the least number of energy evaluations, but, when the system is large, other considerations become important. The time taken for the calculation of the first and second derivatives for a system will be a few times greater than that for the calculation of the energy alone with the simple molecular mechanics energy functions we are using, but the scaling behaviour versus the number of atoms in the system will be the same (i.e. $O(N^2)$) due to the way we have chosen to calculate the non-bonding interactions up to now). If this were the only factor, then it would usually be preferable to calculate the second derivatives because, although they may take several times longer to calculate than the first derivatives, they will usually provide more than enough additional information to compensate for this extra expense.

There are two other factors that limit the use of second-derivative methods for large systems. First of all, the second-derivative matrix uses a lot of storage space. For N atoms, there are $3N(3N + 1)/2 \simeq O(N^2)$ elements, which is often much too much when there are more than a few hundred atoms in the system. Secondly, perhaps more importantly, most algorithms which use second derivatives require that the eigenvalues and eigenvectors of the matrix be found and, because the diagonalization of a matrix is an operation that scales as $O(N^3)$, this becomes difficult or impossible when there is a large number of atoms. For first-derivative methods, the storage and computational costs associated with handling the first-derivative vector both scale as $O(N)$.

To summarize, for systems with small numbers of atoms, exact or quasi-Newton methods can be used and are likely to be the most efficient. The only methods practicable for systems with large numbers of atoms are those that use the first derivatives, such as conjugate gradient algorithms, or a reduced set of second derivatives.

In this section we present a conjugate gradient subroutine for minimizing the energy of the system. It is a good general-purpose algorithm that is appropriate even for very large systems.

The subroutine, which is called OPTIMIZE_CONJUGATE_GRADIENT, appears in the module OPTIMIZE_COORDINATES, whose definition is

```
MODULE OPTIMIZE_COORDINATES

CONTAINS

    SUBROUTINE OPTIMIZE_CONJUGATE_GRADIENT ( PRINT_FREQUENCY, STEP_NUMBER, &
                                             FUNCTION_TOLERANCE, &
                                             GRADIENT_TOLERANCE )
        INTEGER, INTENT(IN), OPTIONAL :: PRINT_FREQUENCY, STEP_NUMBER
        REAL,    INTENT(IN), OPTIONAL :: FUNCTION_TOLERANCE, &
                                         GRADIENT_TOLERANCE
    END SUBROUTINE OPTIMIZE_CONJUGATE_GRADIENT

END MODULE OPTIMIZE_COORDINATES
```

The subroutine has four principal optional input arguments.

1. PRINT_FREQUENCY specifies the frequency at which information about the minimization procedure is printed. If its value is set to zero, a negative value or a
   value greater than the total number of steps specified for the optimization then
   no printing is done.
2. All optimization algorithms are iterative, in that they start with a given set of
   variables and then refine those variables in successive steps. The argument
   STEP_NUMBER defines the maximum number of iterations or steps that are to
   be performed during the minimization. If the convergence criteria are satisfied in
   fewer steps then the minimization will exit, otherwise it will stop after
   STEP_NUMBER steps. If STEP_NUMBER is not specified then it takes the value of
   zero.
3. GRADIENT_TOLERANCE is the main parameter that determines the convergence
   criterion for termination of the minimization process. A large variety of convergence criteria is in use but this one refers to a condition on the value of the RMS
   gradient of the energy, G_{RMS}. At a stationary point, of course, the gradient
   vector is zero. In actual calculations the gradient vector will rarely, if ever, be
   able to attain this value at the end of an optimization and so some finite value
   for the RMS gradient (equation (4.34)) is employed to indicate that a stationary
   point has been reached. If the value of the RMS gradient falls below the value
   given by GRADIENT_TOLERANCE at any step then the optimization is assumed to
   have converged and the algorithm stops.
4. FUNCTION_TOLERANCE is a second parameter that can be used to halt the optimization procedure. It specifies that, if the difference between the energies of the
   system in two successive iterations of the optimization procedure divided by the
   average energy value for the two steps is less than FUNCTION_TOLERANCE then the
   optimization has converged. By default, the value of FUNCTION_TOLERANCE is
   zero and so this test will never be satisfied.

To work, OPTIMIZE_CONJUGATE_GRADIENT requires that the energy and
gradients for a system can be calculated. It leaves all data structures in the
program unchanged, except for the values of the coordinates present in
ATMCRD. When the subroutine terminates, whether the convergence criteria
have been satisfied or not, ATMCRD will contain the coordinates for the
point with the lowest energy that was found during the minimization
procedure.

6.4 Example 6

To illustrate the use of the minimization subroutine we present a program for
the optimization of the blocked alanine molecule, bALA:

```
PROGRAM EXAMPLE6

... Declaration statements ...

! . Local scalars.
REAL :: ESTART, ESTOP

! . Local arrays.
REAL, ALLOCATABLE, DIMENSION(:,:) :: CRDOLD

! . Read in the system file.
CALL MM_SYSTEM_READ ( ''bALA.sys_bin'' )

! . Read in a first set of coordinates.
CALL COORDINATES_READ ( ''bALA1.crd'' )

! . Allocate space for the temporary coordinates and save them.
ALLOCATE ( CRDOLD(1:3,1:NATOMS) ) CRDOLD = ATMCRD

! . Calculate and print the initial energy.
CALL ENERGY ; ESTART = TOTAL_PE

! . Perform a coordinate minimization.
CALL OPTIMIZE_CONJUGATE_GRADIENT ( PRINT_FREQUENCY = 100, &
                                   STEP_NUMBER = 2000 )

! . Calculate and print the final energy.
CALL ENERGY ; ESTOP = TOTAL_PE

! . Print the energy lowering.
WRITE ( OUTPUT, ''(/A,F20.4)'' ) &
   ''Energy reduction after minimization = '', ESTOP - ESTART

! . Superimpose the two structures (transform the structure in ATMCRD).
CALL SUPERIMPOSE_KABSCH ( CRDOLD, ATMCRD, ATMMAS )

! . Recalculate the RMS coordinate deviation between the two structures.
CALL RMS_DEVIATION ( CRDOLD, ATMCRD, ATMMAS )

! . Deallocate the arrays.
DEALLOCATE ( CRDOLD )

END PROGRAM EXAMPLE6
```

The first thing the program does is to read in the binary file, bALA.sys_bin, that contains the data for the MM terms of the blocked alanine molecule. This was the file created in the program EXAMPLE4 of section 5.5. The program then reads in a suitable set of coordinates from the file bALA1.crd. It calculates and prints the energy for this set of coordinates and saves that energy in the variable ESTART. After that the minimization is performed. A maximum of 2000 steps is requested with details about the minimization to be printed at 100-step intervals. Once the minimization has terminated, the energy of the optimized structure is calculated and printed and the total reduction in the energy produced by the minimization procedure is determined and printed. In this case, the minimization performs about 700 cycles before the convergence criterion on the RMS gradient (10^{-3} kJ mol^{-1} Å^{-1}) is satisfied, resulting in a total reduction in energy between the two configurations of about 63 kJ mol^{-1}. Note that normally after a minimization the optimized coordinates would be written out to another file for further analysis or for subsequent use.

The program terminates by superimposing the coordinates of the optimized structure onto the initial structure and then calculating the RMS coordinate deviation between them. This provides a crude measure of how much the structure has changed as a result of the minimization process.

6.5 Locating saddle points

Whereas algorithms for the location of minima have been the subject of intense research by mathematicians, algorithms for the location of saddle points have been studied much less widely. As a result, many of these algorithms have been developed by chemists and other researchers directly interested in looking at potential energy surfaces. The location of saddle points is usually a much more demanding task than the location of minima. The main reason for this is intuitive. When searching for a minimum one wants to reduce always the value of the energy, so one can choose to go in any direction that gives this result. When looking for saddle points, however, one is trying to find a point that is a *maximum* in one direction but a minimum in all the others and so the algorithm has to perform a delicate balancing act between the two conflicting types of search.

A wide range of saddle-point-location algorithms has been proposed. In this section we shall discuss a *mode-following* algorithm developed by J. Baker, which is widely used and appears to be one of the most efficient methods available. It uses both first- and second-derivative information and so its application is restricted to relatively small systems, although this is a general limitation of saddle-point-location subroutines. In principle, whenever a stationary point has been obtained by an algorithm for the location of minima or saddle points, the only way to verify that the point is of the type required is to determine how many negative eigenvalues there are in the second-derivative matrix. For saddle points this is especially crucial due to the difficulties with saddle-point searches. It is less crucial, although still desirable, for minima because the algorithms for the location of minima are more robust. When treating large systems for which the second-derivative matrix cannot be diagonalized due to computational constraints it is generally assumed that the stationary point obtained is a minimum. Saddle-point searches in these cases are very difficult.

To understand the theory behind Baker's algorithm we consider a point on the potential energy surface of the system with coordinates R_0. We then approximate the energy, V, of neighbouring points on the surface using a Taylor series. If the coordinates of the displaced point on the surface are R, where $R = R_0 + D$ and D is a displacement vector, then the Taylor series to second order in the displacements is

$$V(R) = V(R_0) + G^T D + \frac{1}{2} D^T H D + \cdots \qquad (6.5)$$

where G is the vector of first derivatives and H is the matrix of second derivatives, both of which are evaluated at the point R_0. To find the displacement vector, D, which minimizes the energy of this expression, we differentiate equation (6.5) with respect to D to obtain dV/dD, set the result to zero and solve to get

$$D = -H^{-1}G \qquad (6.6)$$

This is the *Newton–Raphson (NR) step*. Using the properties of the second-derivative matrix alluded to in section 6.2, this can be rewritten as a sum over the eigenvectors of the second-derivative matrix:

$$D = -\sum_i \frac{e_i^T G}{\lambda_i} e_i \qquad (6.7)$$

The NR step is structured so that it will minimize along the eigenvectors or *modes* with positive eigenvalues and maximize along modes with negative eigenvalues. Thus, it will tend to optimize to structures that have the same number of negative eigenvalues or *curvature* as the original structure. If the starting structure is of the correct curvature, the NR step is a good one to take, but, in the general case, the NR step must be modified so that structures with the desired curvature can be obtained.

The required modification, developed by C. Cerjan and W. Miller and by J. Simons and his co-workers, is a simple one and involves altering the denominator of equation (6.7) by adding parameters, γ_i, which shift the values of the eigenvalues:

$$D = -\sum_i \frac{e_i^T G}{\lambda_i - \gamma_i} e_i \qquad (6.8)$$

Normally, only two different shift parameters are used – one for the modes along which a minimization is to be done and another along which there is to be a maximization. In the special case of searching for a minimum, only the first of these will be required. There are several prescriptions for choosing the values of the shift parameters, which need not concern us here. The important point is that, with an appropriate choice, it is possible to force a search along a particular mode opposite to that in which it would normally go. For example, when searching for a minimum the shift parameter would be negative and have a value less than that of the smallest eigenvalue. This would ensure that the denominators in equation (6.8) are always positive and so a minimization occurs along all the modes. When the optimization reaches a region of the correct curvature the value of the shift parameter can be reduced. For a saddle-point search there will be one mode along which there is a maximization and for which the shift parameter will be such that the denominator for that mode in equation (6.8) will be negative. This means that it is possible to start at a minimum on the potential energy surface and 'walk' up one of the modes until a saddle point is reached.

The above algorithm is available as another subroutine, OPTIMIZE_BAKER, in the module OPTIMIZE_COORDINATES. Its definition is

```
MODULE OPTIMIZE_COORDINATES

CONTAINS

   SUBROUTINE OPTIMIZE_BAKER ( POINT, FOLLOW_MODE, PRINT_FREQUENCY, &
                               STEP_NUMBER, GRADIENT_TOLERANCE, &
                               MAXIMUM_STEP )
      CHARACTER ( LEN = * ), INTENT(IN) :: POINT
      INTEGER, INTENT(IN), OPTIONAL :: FOLLOW_MODE, PRINT_FREQUENCY, &
                               STEP_NUMBER
      REAL,    INTENT(IN), OPTIONAL :: GRADIENT_TOLERANCE, &
                               MAXIMUM_STEP
   END SUBROUTINE OPTIMIZE_BAKER

END MODULE OPTIMIZE_COORDINATES
```

There is one essential input argument. It is a character variable that must be either of the two strings ''MINIMUM'' and ''SADDLE''. In the former case the algorithm searches for a minimum starting with the current values of the coordinates in the array ATMCRD. In the second case, a (first-order) saddle point is sought.

Of the optional arguments, PRINT_FREQUENCY, STEP_NUMBER and GRADIENT_TOLERANCE have the same meaning as the arguments of the same name in the subroutine OPTIMIZE_CONJUGATE_GRADIENT. The argument MAXIMUM_STEP specifies the size of the greatest allowed step that the algorithm can take during any iteration. For certain systems it is necessary to change this to ensure that the algorithm converges, but the default value should be sufficient in most cases.

The optional argument FOLLOW_MODE requires more explanation. It is an integer that specifies, for a saddle-point search, the mode along which the maximization is to be performed. If no value is given then the mode chosen is automatically the one with the lowest eigenvalue. The mode-following option is most useful when starting out from a minimum. In that case all the eigenvalues are positive. Often it turns out that walking up the 'softest' mode (the lowest eigenvalue mode) will lead to a saddle point. This is not always true and so it is sometimes useful to search in other directions.

There are two technical points that can be made about the implementation of the algorithm. First of all, at each step, the full second-derivative matrix is calculated analytically. It is not, in principle, necessary to do this. There are various approximate techniques to update the value of the Hessian using the values of the gradient vector calculated at the current point. If these *updating formulae* are employed then the number of exact second-derivative matrix calculations can be substantially reduced. The Hessian could, for example, be evaluated at the start of the optimization and periodically every few cycles after this. The use of these updating techniques here would probably save some time. However, for the molecular mechanics energy functions we are using, it is the time required for the diagonalization of the second-derivative matrix that limits the application of the method, not the time required for the calculation of the second derivatives.

The second technical point concerns the use of Cartesian coordinates. From the arguments at the beginning of the chapter it will be remembered that one parameter completely defines the geometry for a two-atom system, three parameters for a three-atom system and so on but, for a two-atom system, there are six Cartesian coordinates and for a three-atom system there are nine. This means that the set of Cartesian coordinates contains too many variables and is redundant. The redundancies, in fact, are related to the overall translational and rotational motions which the entire system can undergo. For first-derivative algorithms these motions are unimportant. For algorithms that use second derivatives it is necessary to modify the second-derivative matrix so that they are removed. This is done by the algorithm implemented in the subroutine OPTIMIZE_BAKER but a more detailed discussion of this point will be left to the next chapter.

6.6 Example 7

To illustrate the Baker algorithm we choose to study the molecule cyclohexane, which undergoes a well-known transition between different conformational forms. In this example we start with the chair form of the molecule and
look for a saddle point that leads to another conformer.

The program is

```
PROGRAM EXAMPLE7

... Declaration statements ...

! . Local scalars.
REAL :: ESTART, ESTOP

! . Process the MM file for cyclohexane.
CALL MM_FILE_PROCESS ( ''cyclohexane.opls_bin'', ''cyclohexane.opls'' )

! . Construct the system file for cyclohexane.
CALL MM_SYSTEM_CONSTRUCT ( ''cyclohexane.opls_bin'', ''cyclohexane.seq'' )

! . Save the system file for cyclohexane for later use.
CALL MM_SYSTEM_WRITE ( ''cyclohexane.sys_bin'' )

! . Read in the coordinates.
CALL COORDINATES_READ ( ''cyclohexane_chair.crd'' )

! . Calculate and print the initial energy.
CALL ENERGY ; ESTART = TOTAL_PE

! . Perform a coordinate minimization.
CALL OPTIMIZE_BAKER ( ''SADDLE'', FOLLOW_MODE = 1, MAXIMUM_STEP = 0.1, &
                                  PRINT_FREQUENCY = 1, STEP_NUMBER = 200, &
                                            GRADIENT_TOLERANCE = 1.0E-6 )

! . Calculate and print the final energy.
CALL ENERGY ; ESTOP = TOTAL_PE

! . Print the energy change.
WRITE ( OUTPUT, ''(/A,F20.4)'' ) &
   ''Energy change after the search = '', ESTOP - ESTART

! . Write out the coordinates.
CALL COORDINATES_WRITE ( ''cyclohexane_saddle.crd'' )

END PROGRAM EXAMPLE7
```

The program first sets up the data structures needed for the calculation of
the energy using the files, `cyclohexane.opls` and `cyclohexane.seq`, that
are provided. The coordinates of the chair structure of the molecule are then

read in and the energy of the structure is calculated and stored. Next the saddle-point search is performed following the mode with the lowest eigenvalue. A very stringent convergence criterion on the RMS gradient of 10^{-6} kJ mol^{-1} Å^{-1} is specified. Once the search has finished the energy of the final point is calculated and the difference between it and that of the chair form of cyclohexane is printed. The optimized coordinates of the saddle point are then saved in an external file.

The algorithm finds a saddle point and requires about 40 steps to reach convergence. It is about 50 kJ mol^{-1} higher in energy than the chair form. The structure of the chair conformer and the saddle point found in the search are shown in figure 6.2 together with the boat and twist-boat conformers of cyclohexane.

6.7 Following reaction paths

A reaction, which can involve either the breaking and forming of bonds or a change in the conformation of a molecule, may be defined in terms of the minima on the potential energy surface that correspond to the reactant and product configurations and any stable intermediate states and the saddle points that lie between all the minima. In principle, knowledge of all these

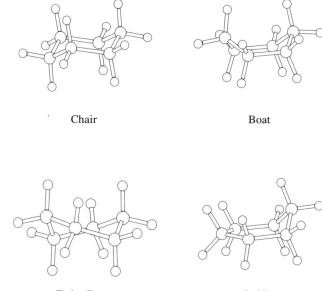

Chair   Boat

Twist–Boat   Saddle

Figure 6.2. The structures of the chair, boat and twist-boat conformers of cyclohexane and the saddle-point structure found in Example 7.

minima and saddle points is sufficient to give a detailed picture of the *mechanism* of the transition. In practice, though, it is often useful to be able to generate points intermediate between the minima and saddle points so that the geometrical and other changes occurring during the transition can be analysed in greater detail.

Although the stationary points on the potential energy surface have a unique definition, there is no single definition of a reaction path. It is not even necessary, as we shall see later, to define it so that it goes through the minima and saddle points that correspond to the transition being studied. The definition that we shall adopt in this section is that of the *steepest descent reaction path* (SDRP), which is probably the one in most common use, at least for the study of reactions in systems with small numbers of atoms. K. Fukui was one of the earliest proponents of this definition, although he called it the *intrinsic reaction coordinate* (IRC).

The SDRP is usually defined in terms of *mass-weighted* Cartesian coordinates, although unweighted coordinates can also be used. If r_i and m_i are the coordinate vector and the mass of atom i, the mass-weighted coordinate, q_i, is

$$q_i = \sqrt{m_i} r_i \tag{6.9}$$

The mass-weighted gradient of the energy for an atom, ξ_i, is defined in a similar way in terms of the unweighted gradient, g_i:

$$\xi_i = \frac{g_i}{\sqrt{m_i}} \tag{6.10}$$

It is normal to introduce a parameter, s, that denotes the distance along the reaction path. If we do this the coordinates of the atoms in the structures along the path are also functions of s and the SDRP can be formally described by the differential equation

$$\frac{dq_i(s)}{ds} = \pm \frac{\xi_i}{\sqrt{\sum_{i=1}^{N} \xi_i^2}} \tag{6.11}$$

where the plus sign refers to a path going up from a minimum and the minus sign to one descending from a saddle point. Mathematically what this equation means is that the reaction path at any point follows the direction of the mass-weighted gradient, which is equivalent to the direction of steepest descent.

Practically, the easiest way to generate structures on a reaction path and the one that will be used here, is to start at a saddle point and to take a series of finite steps in the direction of descent. If the current structure that has just been generated is denoted by the index α, the coordinates of the next structure are calculated from

$$q_i^{\alpha+1} = q_i^\alpha - \left(\frac{\xi_i^\alpha}{\sqrt{\sum_{i=1}^N (\xi_i^\alpha)^2}} \right) \delta s \qquad (6.12)$$

where δs is a small step to be taken down the path. This algorithm works reasonably well if the step size is small. Even with small step sizes, though, the path generated will have a tendency to oscillate about the true path and it will not end up exactly at the minimum, although it should terminate in its vicinity. Other more sophisticated algorithms have been developed to generate smoother reaction paths. These often start off with a step similar to that given in equation (6.12) but then try to refine the point using *line-search* or *constrained minimization* techniques. We do not present an implementation of one of these techniques here, although an alternative method for calculating smooth reaction paths is described at the end of the chapter.

The only problem with the definition of the path in equation (6.11) is that it does not hold at a stationary point. At these points the gradient is zero and the denominator on the right-hand side of the equation is undefined. Therefore, when starting at a saddle point, a different type of starting step must be chosen. This is done by calculating the eigenvector of the (mass-weighted) second-derivative matrix at the saddle point that corresponds to the mode with the negative frequency. This mode produces a reduction in energy of the system and so points downhill when the system is displaced along it away from the stationary point. Thus, the step taken from a saddle point is the one that goes along this mode (in either the plus or the minus direction because there are two downhill directions).

The algorithm described above has been implemented as the subroutine, REACTION_PATH_TRACE, in the module REACTION_PATH, which has the following definition:

```
MODULE REACTION_PATH

CONTAINS

    SUBROUTINE REACTION_PATH_TRACE ( DIRECTION, MAXSTP, PRINT_FREQUENCY, &
                                     FROM_SADDLE, USE_MASS_WEIGHTING, &
                                     ENERGY_STEP, PATH_STEP, &
                                     SAVE_FREQUENCY, FILE )
        CHARACTER ( LEN = 1 ), INTENT(IN) :: DIRECTION
        INTEGER,               INTENT(IN) :: MAXSTP
        CHARACTER ( LEN = * ), INTENT(IN), OPTIONAL :: FILE
        INTEGER, INTENT(IN), OPTIONAL :: PRINT_FREQUENCY, SAVE_FREQUENCY
```

```
        LOGICAL, INTENT(IN), OPTIONAL :: FROM_SADDLE, USE_MASS_WEIGHTING
        REAL,    INTENT(IN), OPTIONAL :: ENERGY_STEP, PATH_STEP
     END SUBROUTINE REACTION_PATH_TRACE

  END MODULE REACTION_PATH
```

The subroutine generates a sequence of structures using equation (6.12), starting with the structure defined by the coordinates present in the array ATMCRD. When the subroutine terminates, the coordinates of the final point calculated along the path are left in ATMCRD.

There are two essential arguments. The first is DIRECTION, which specifies the direction to take along the eigenvector that corresponds to the mode with the negative eigenvalue (if the starting structure is a saddle point). It is a character variable of length one and can take either of the values "+" and "−". The second argument MAXSTP is the number of steps to take when generating the path. Note that the algorithm has no other termination criteria (for example, if there is an increase in energy along the path) and so all the points specified will be calculated.

The remaining optional arguments are the following.

1. PRINT_FREQUENCY specifies the frequency at which to print out information about the path points. The default is to print information about every point.
2. FROM_SADDLE is a logical variable that indicates whether the starting structure, in ATMCRD, is a saddle point. If it is then the initial step taken is along the appropriate eigenvector at that point. If it is not then the steepest descent steps are taken immediately. If the argument is not present then the algorithm assumes that the structure in the array ATMCRD corresponds to a saddle point.
3. USE_MASS_WEIGHTING is another logical variable that specifies whether to use mass-weighted coordinates. The default is not to use mass-weighting.
4. ENERGY_STEP is the parameter which determines the size of the initial step away from the saddle point. It specifies the approximate reduction in energy required as a result of the step and is calculated using the formula $\sqrt{-2\,\Delta E/\lambda_-}$, where ΔE is the value of ENERGY_STEP and λ_- is the value of the negative eigenvalue. The default value of this parameter is 1 kJ mol^{-1}.
5. PATH_STEP is the value of δs to be taken for the steepest descent steps. Its default value is 0.1 Å if mass-weighting is not used and 0.1 $\sqrt{\text{a.m.u.}}$ Å otherwise.
6. SAVE_FREQUENCY specifies the frequency with which to write out calculated structures to an external file for subsequent analysis. If the frequency is less than unity or greater than MAXSTP then no structures are written out. The default is to write out no structures. In all cases, when structures are saved the initial structure for the steepest descent procedure is also saved.
7. If structures are to be saved then the argument FILE which contains the name of the *trajectory file* to which structures are to be written is required. The trajectory file is a special one that we will encounter several times in this book. A full

discussion will be left until a later chapter in which we consider how to analyse the data written to such files.

A final point can be made about the use of this algorithm. Having found a saddle-point structure (using the algorithm of the previous section, for example) it is not uncommon not to know to which minima it leads (either one or both)! In this case, the reaction-path-following algorithm can be used to identify them.

6.8 Example 8

In this section we illustrate the use of the reaction-path-following algorithm by tracing out the path from the saddle point for cyclohexane calculated in the previous example. The program is

```
PROGRAM EXAMPLE8

... Declaration statements ...

! . Construct the system file for cyclohexane.
CALL MM_SYSTEM_READ ( ''cyclohexane.sys_bin'' )

! . Read in the coordinates.
CALL COORDINATES_READ ( ''cyclohexane_saddle.crd'' )

! . Calculate and print the initial energy.
CALL ENERGY

! . Follow the path downwards in the plus direction.
CALL REACTION_PATH_TRACE ( ''+'', 200, ENERGY_STEP = 2.0, PATH_STEP = 0.025, &
                                   PRINT_FREQUENCY = 10, SAVE_FREQUENCY = 10, &
                                            FILE = ''cyclohexane_plus.sdp'', &
                                            USE_MASS_WEIGHTING = .TRUE)

! . Reset the coordinates.
CALL COORDINATES_READ ( ''cyclohexane_saddle.crd'' )

! . Follow the path downwards in the minus direction.
CALL REACTION_PATH_TRACE ( ''-'', 200, ENERGY_STEP = 2.0, PATH_STEP = 0.025, &
                                   PRINT_FREQUENCY = 10, SAVE_FREQUENCY = 10, &
                                            FILE = ''cyclohexane_minus.sdp'', &
                                            USE_MASS_WEIGHTING = .TRUE)
END PROGRAM EXAMPLE8
```

After the initial commands in which the system file for cyclohexane and the coordinates of the saddle point are read in from external files, the SDRP is generated in the positive direction. It is calculated with 200 structures at separations of 0.025 $\sqrt{\text{a.m.u.}}$ Å and the initial step is such that there will be a drop in energy of approximately 2 kJ mol^{-1} on displacement from the saddle point. The details about the path points are printed every ten steps and structures are written to an external file also every tenth step, giving 21 structures to be written out in all. After the path has been traced in the positive direction the saddle-point coordinates are re-read (because they have been overwritten) and the path is traced in the negative direction with the same parameters.

The paths generated by this example lead to the chair form of cyclohexane in the positive direction and to the boat form in the negative direction. In both cases 200 steps suffice to bring the path down into regions of the potential energy surface that correspond to the two different conformers, although the structures generated are relatively far from the exact ones and the path tends to meander around them. The parameters chosen here for cyclohexane give reasonable results but it is advisable to try a number of different parameter sets to see which give the best paths whenever this algorithm is being used.

6.9 Determining complete reaction paths

In section 6.7 we saw how to calculate reaction paths when a saddle-point structure was available. However, as indicated in section 6.5, the calculation of saddle-point structures, using methods that require second derivatives, is difficult or impossible for large systems and alternative approaches for determining reaction paths are needed.

Only a few workers have suggested algorithms for such calculations – it is still an area of active research – but we shall discuss an elegant algorithm of R. Elber and co-workers who have developed a range of related techniques. A different approach from that of the algorithms discussed in the previous sections is taken. The method does not attempt to find saddle points but only a series of intermediate structures along the path between the reactant and product structures. It does this by first generating a *chain* of structures between the reactants and products and then minimizing a 'target' function, $\mathcal{F}$, with respect to the coordinates of *all* the structures along the chain simultaneously. The minimization procedure cannot freely minimize the structures in the chain because there are certain *constraint conditions* that force the structures to remain roughly equidistant from one another. The target function has the form

$$\mathcal{F} = \sum_{I=1}^{M} \mathcal{V}_I + \gamma \sum_{I=0}^{M} (d_{I,I+1} - \langle d \rangle)^2 + \frac{\rho}{\lambda} \sum_{I=0}^{M-1} \exp\left[-\lambda \left(\frac{d_{I,I+2}}{\langle d \rangle} \right)^2 \right] \quad (6.13)$$

where $\mathcal{V}_I$ is the energy of the Ith structure, $d_{I,J}$ is the distance between two structures I and J, $\langle d \rangle$ is the average distance between neighbouring structures and γ, λ and ρ are constants. The number of structures in the chain is $M + 2$. The reactant's structure is labelled 0, the product's structure $M + 1$ and the intermediate structures by values ranging from 1 to M. The distances between structures are defined as

$$d_{I,J} = |\mathbf{R}_I - \mathbf{R}_J| \quad (6.14)$$

where $\mathbf{R}_I$ is the $3N$-dimensional vector containing the coordinates of the atoms of structure I. The average distance between neighbouring structures is

$$\langle d \rangle = \frac{1}{M+1} \sum_{I=0}^{M} d_{I,I+1} \quad (6.15)$$

If the system contains N atoms, the target function, $\mathcal{F}$, is a function of $3NM$ variables (note that the coordinates of the reactant and product structures are not optimized). The first constraint term in the function (the γ term) is the term that keeps adjacent structures roughly the same distance apart. The second term (the ρ term) keeps structures separated by one other structure apart. This avoids the problem of having the chain collapse or fold back on itself during the optimization procedure. Once the target function has been defined, the algorithm works by minimizing the function with respect to the atomic coordinates, using a standard technique, until the convergence criteria have been satisfied.

There is one extra complication, which arises due to the fact that we are again using Cartesian coordinates to define the atom positions in each structure. It is the same problem as that which we discussed at the end of section 6.5 and concerns the translational and rotational degrees of freedom of the structures. Once the structures in the chain to be minimized have been defined it is normal to reorient each of them with respect to a reference structure so that the distances, $d_{I,J}$, between the structures are minimized. One of the techniques discussed in section 3.6 is appropriate for the reorientation because minimization of the distance between structures is equivalent to minimization of their RMS coordinate difference. During the subsequent optimization, the relative orientations of the structures should be maintained so that the calculation of the distances between structures using equation (6.14) remains valid. This means that the structures must not be allowed to rotate or to translate with respect to each other during the minimization. A full discussion of this point and the way in which it is treated will be left until the next chapter.

The above algorithm was called the *self-avoiding walk* (SAW) method by Elber and co-workers and so it has been implemented as a series of subroutines in the module SELF_AVOIDING_WALK. There are three subroutines with the following definitions:

```
MODULE SELF_AVOIDING_WALK

CONTAINS

    SUBROUTINE CHAIN_EXPAND ( FILE_IN, FILE_OUT, NINSERT )
        CHARACTER ( LEN = * ), INTENT(IN) :: FILE_IN, FILE_OUT
        INTEGER,              INTENT(IN) :: NINSERT
    END SUBROUTINE CHAIN_EXPAND

    SUBROUTINE CHAIN_INITIALIZE ( FILE, NPOINTS, REACTANTS, PRODUCTS )
        CHARACTER ( LEN = * ), INTENT(IN) :: FILE
        INTEGER,              INTENT(IN) :: NPOINTS
        REAL, DIMENSION(1:3,1:NATOMS), INTENT(IN) :: PRODUCTS, REACTANTS
    END SUBROUTINE CHAIN_INITIALIZE

    SUBROUTINE CHAIN_OPTIMIZE ( FILE_IN, FILE_OUT, GAMMA, LAMBDA, RHO, &
                                      PRINT_FREQUENCY, STEP_NUMBER, &
                                FUNCTION_TOLERANCE, GRADIENT_TOLERANCE )
        CHARACTER ( LEN = * ), INTENT(IN) :: FILE_IN, FILE_OUT
        REAL,    INTENT(IN), OPTIONAL :: GAMMA, LAMBDA, RHO
        INTEGER, INTENT(IN), OPTIONAL :: PRINT_FREQUENCY, STEP_NUMBER
        REAL,    INTENT(IN), OPTIONAL :: FUNCTION_TOLERANCE, &
                                          GRADIENT_TOLERANCE
    END SUBROUTINE CHAIN_OPTIMIZE

END MODULE SELF_AVOIDING_WALK
```

The subroutine CHAIN_INITIALIZE creates a chain of structures for a SAW calculation given the structures of the reactants and of the products. The subroutine CHAIN_EXPAND adds structures to an existing chain and the subroutine CHAIN_OPTIMIZE carries out the minimization process.

CHAIN_INITIALIZE has four arguments, all of which are required and all of which are input arguments. The first argument, FILE, is the name of the file to which the chain of structures is to be written once it has been created. The format of the file is the trajectory file format, which is the same as that used for storing structures produced by the reaction-path-tracing procedure of section 6.7. The argument, NPOINTS, specifies the number of structures to put in the chain. This number must always be odd and, because a chain also includes the reactant and product structures, its minimum value is 3. The arguments REACTANTS and PRODUCTS are arrays that contain the coordinates of the reactant and product structures.

The determination of the chain structures is accomplished by first orienting the product and reactant structures and then using a simple linear interpolation to obtain the intermediate structures. Thus, employing the notation of the beginning of this section, the coordinates of the Ith structure would be constructed as

$$R_I = R_0 + \frac{I}{M+1}\left(R_{M+1} - R_0\right) \tag{6.16}$$

After construction, all the structures, including the reactants and products, are written to the trajectory file.

The subroutine CHAIN_EXPAND has three input arguments. FILE_IN and FILE_OUT are both names of files, FILE_IN for the file that contains the existing chain of structures and FILE_OUT for the file to which the expanded chain is to be written. The third argument, NINSERT, gives the number of structures to insert between each existing structure in the chain. If there are $M+2$ structures in the existing chain and NINSERT takes the value n, there will be $M+2+n(M+1)$ structures in the new chain. The coordinates of the inserted structures are calculated by linearly interpolating between the coordinates of adjacent pairs of structures in the old chain using a formula similar to that of equation (6.16). All the old structures, as well as the new structures, are written in the correct order to the output trajectory file.

The third subroutine, CHAIN_OPTIMIZE, is the subroutine that actually performs the optimization of the chain. There are two essential input arguments, FILE_IN and FILE_OUT, which specify the names of the files containing the starting, unoptimized, chain of structures and the name of the file to which the optimized chain is to be written. The remaining arguments are optional and are of two types. The arguments of the first type are GAMMA, LAMBDA and RHO, which specify values for the constants, γ, λ and ρ, in the target function, $\mathcal{F}$, of equation (6.13). The arguments of the second type specify parameters that are used by the minimization subroutine. They are PRINT_FREQUENCY, STEP_NUMBER, FUNCTION_TOLERANCE and GRADIENT_TOLERANCE and they have exactly the same meanings as the arguments of the same name in the subroutine OPTIMIZE_CONJUGATE_GRADIENT discussed in section 6.3. In fact, the implementation of the optimization of the chain uses exactly the same conjugate-gradient-minimization subroutine.

6.10 Example 9

We test the self-avoiding walk algorithm by applying it to the calculation of the reaction path for the transition between the chair and the twist-boat conformations of cyclohexane. Although the algorithm was designed for use on large systems it is equally applicable, as we shall see, to small systems.

The program is

```
PROGRAM EXAMPLE9

... Declaration statements ...

! . Local arrays.
REAL, ALLOCATABLE, DIMENSION(:,:) :: REACTANTS, PRODUCTS

! . Construct the system file for cyclohexane.
CALL MM_SYSTEM_READ ( ''cyclohexane.sys_bin'' )

! . Allocate space for the arrays.
ALLOCATE ( REACTANTS(1:3,1:NATOMS), PRODUCTS(1:3,1:NATOMS) )

! . Read in the coordinates of the chair form.
CALL COORDINATES_READ ( ''cyclohexane_chair.crd'' )

! . Save the reactant coordinates.
REACTANTS = ATMCRD

! . Read in the coordinates of the twist-boat form.
CALL COORDINATES_READ ( ''cyclohexane_twistboat.crd'' )

! . Save the product coordinates.
PRODUCTS = ATMCRD

! . Initialize a chain trajectory.
CALL CHAIN_INITIALIZE ( ''cyclohexane1.saw'', 11, REACTANTS, PRODUCTS )

! . Optimize the chain.
CALL CHAIN_OPTIMIZE ( ''cyclohexane1.saw'', ''cyclohexane2.saw'', &
                      GAMMA = 1000.0, PRINT_FREQUENCY = 1, &
                      STEP_NUMBER = 200 )

! . Deallocate the temporary space.
DEALLOCATE ( REACTANTS, PRODUCTS )

END PROGRAM EXAMPLE9
```

After it has read in the system file for cyclohexane the program allocates two coordinate arrays, REACTANTS and PRODUCTS, and fills them with the coordinates of the chair and twist-boat forms of cyclohexane. From these two structures, an initial chain of 11 structures is constructed by a call to CHAIN_INITIALIZE and written to a file cyclohexane1.saw. This chain is then optimized using CHAIN_OPTIMIZE and the results are written to the file cyclohexane2.saw.

This calculation produces a smooth path for the transition between the two forms of cyclohexane. The results are plotted in figure 6.3, which shows the energies of the structures along the path. The seventh structure in the chain is within about 0.1 kJ mol^{-1} of the saddle-point structure optimized in Example 7. The convergence criteria for the optimization were satisfied after about 86 cycles, i.e. after 9×86 energy and derivative calculations, which is approximately twice as many as the number required by the path-tracing algorithm of section 6.8.

Exercises

6.1 The blocked alanine molecule has a number of different minimum energy structures. Try to find some of these minima using the starting coordinate files that are provided. Try both the conjugate-gradient algorithm and the minimization option in the Baker algorithm. In both cases, use a stringent criterion on the value of the gradient tolerance (say a value of 10^{-6} kJ mol^{-1} Å^{-1}). Once some minima have been located, find the pathways that connect them, either by searching for saddle points and then using the reaction-path-tracing algorithm

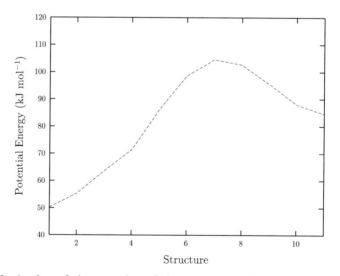

Figure 6.3. A plot of the energies of the structures along the transition pathway between the chair and twist-boat forms of the cyclohexane molecule using the SAW algorithm.

or by using the self-avoiding walk method directly between two minima. Finally, construct a schematic potential energy surface for the molecule, illustrating the various minima, the saddle points and the reaction paths connecting them.

6.2 Using the Baker algorithm, search for a saddle-point structure starting from the twist-boat form of cyclohexane. Is this saddle point the same as that obtained when starting from the chair form of cyclohexane? If not, to which reaction path does it correspond? Try to follow different modes, other than the lowest one, to see the different structures that are produced. Also try to characterize more fully the boat and twist-boat conformers of the molecule and the reaction paths that lead between them. Are the boat and twist-boat forms both minima on the potential energy surface?

7

Normal mode analysis

7.1 Introduction

A characterization of stationary points on a system's potential energy surface gives us structural and energetic information about its stable states and about possible pathways for transitions between them. However, can we get more? In particular, can we use our knowledge about these local regions to obtain dynamical and thermodynamic information about the system? The most accurate way to do this is to use methods such as *molecular dynamics* and *Monte Carlo simulations*, which will be covered in later chapters. However, there is a useful intermediate technique, *normal mode analysis*, which we shall now discuss, which can give an idea about the dynamics of a system in the neighbourhood of a stationary point. We shall also see how this information, together with other data, can be used to estimate various thermodynamic quantities.

7.2 Calculation of the normal modes

We have already met some of the concepts that underlie normal mode analysis. The method relies on being able to write an expansion of the potential energy of a system about any configuration in terms of a Taylor series. If R_0 is the coordinate vector of a reference structure and $R = R_0 + D$ is the coordinate vector of a structure displaced by a small amount, D, the Taylor series, up to terms of second order, is

$$\mathcal{V}(R) = \mathcal{V}(R_0) + G^\mathrm{T}D + \frac{1}{2}D^\mathrm{T}HD + \cdots \tag{7.1}$$

where the first-derivative vector of the energy, G, and the second-derivative matrix, $\mathbf{H}$, are determined at the reference structure, R_0.

By definition, the gradient vector is zero at a stationary point. If the reference structure, R_0, is taken to be a stationary point, equation (7.1) can be simplified. Neglecting terms after second order gives

$$\Delta V(D) = V(R) - V(R_0) = \frac{1}{2} D^{\mathrm{T}} H D \tag{7.2}$$

This equation says that the change in energy on displacement from a stationary point is a quadratic function of the displacement. This is called the *harmonic approximation* and is valid when the displacements involved are small and the terms of higher order in D can be ignored. If these terms are not small then *anharmonic* theories that include them are needed.

The important point about the expression for the change in energy in equation (7.2) is that we can solve analytically for the dynamics of a system subject to such a potential. Both classical and quantum mechanical solutions are possible, although we shall restrict ourselves to a classical description here. In this case, we can use Newton's laws to describe the motion of the atoms in the system. For each atom we have an equation of the form

$$f_i = m_i a_i \tag{7.3}$$

where f_i is the force on the atom, m_i is its mass and a_i is its acceleration. The force on an atom is defined as the negative of the first derivative of the potential energy with respect to the position vector of the atom:

$$f_i = -\frac{\partial V}{\partial r_i} = -g_i \tag{7.4}$$

while the acceleration is the second time derivative of the atom's position vector:

$$a_i = \frac{\mathrm{d}^2 r_i}{\mathrm{d}t^2} = \ddot{r}_i \tag{7.5}$$

Equation (7.3) can be rewritten for the full system as

$$F = MA \tag{7.6}$$

where F and A are the $3N$-dimensional vectors of forces and accelerations for the atoms in the system and M is a $3N \times 3N$ diagonal matrix that contains the masses of the atoms and is of the form

$$M = \begin{pmatrix} m_1 & 0 & 0 & 0 & \cdots & 0 \\ 0 & m_1 & 0 & 0 & \cdots & 0 \\ 0 & 0 & m_1 & 0 & \cdots & 0 \\ 0 & 0 & 0 & m_2 & \cdots & 0 \\ \vdots & \vdots & \vdots & \vdots & \vdots & \vdots \\ 0 & 0 & 0 & 0 & \cdots & m_N \end{pmatrix} \tag{7.7}$$

Let us apply these equations to the particular case in which the potential energy of the system is given by equation (7.2). First we note that the vector representing the configuration of the system, R, can be replaced by the displacement vector, D, because R_0 is a reference structure and so is constant. The forces on the atoms in the system are thus obtained by differentiating the energy expression of equation (7.2) with respect to the displacement for that atom, giving

$$M\frac{d^2 D}{dt^2} = -HD \tag{7.8}$$

This equation is a second-order differential equation that can be solved exactly. The solutions are of the form

$$D = \mathcal{A} \cos(\omega t + \phi) \tag{7.9}$$

where the vector $\mathcal{A}$ and the scalars ω and ϕ are to be determined. After substitution of this expression into equation (7.8) and cancelling out of the cosine factors, the equation becomes

$$H\mathcal{A} = \omega^2 M\mathcal{A} \tag{7.10}$$

It is normal to rewrite this equation by using mass-weighted Cartesian coordinates so that the dependence on the mass matrix, M, is removed from the right-hand side. This can be done by introducing the inverse square root of the mass matrix, $M^{-\frac{1}{2}}$, which is equal to the diagonal matrix of the inverse square roots of the atomic masses. With this matrix, after some rearrangement, equation (7.10) becomes

$$\left(M^{-\frac{1}{2}}HM^{-\frac{1}{2}}\right)\left(M^{\frac{1}{2}}\mathcal{A}\right) = \omega^2\left(M^{\frac{1}{2}}\mathcal{A}\right) \tag{7.11}$$

or

$$H'\mathcal{A}' = \lambda\mathcal{A}' \tag{7.12}$$

where

$$\mathbf{H}' = \mathbf{M}^{-\frac{1}{2}}\mathbf{H}\mathbf{M}^{-\frac{1}{2}} \tag{7.13}$$

$$\mathcal{A}' = \mathbf{M}^{\frac{1}{2}}\mathcal{A} \tag{7.14}$$

$$\lambda = \omega^2 \tag{7.15}$$

Equation (7.12) is a secular equation for the mass-weighted second-derivative matrix and can be solved to obtain the eigenvalues, λ, and the eigenvectors, $\mathcal{A}'$. Because the matrix has dimensions $3N \times 3N$ there will be $3N$ different solutions, each of which represents an independent displacement that the system can make. These displacements are called the *normal modes*. Associated with each mode is a *frequency* that is the square root of the mode's eigenvalue. As stated before, the eigenvalues of a symmetric matrix are all real, so the frequencies associated with each mode can be real (for a positive eigenvalue) or imaginary (for a negative eigenvalue). The motion produced by each mode with a positive eigenvalue is a simple oscillation at a characteristic frequency.

Because the normal modes are independent, the most general solution for the displacement vector, $\boldsymbol{D}$, is a linear combination of all the modes:

$$\boldsymbol{D} = \mathbf{M}^{-\frac{1}{2}}\sum_{k=1}^{3N}\alpha_k\mathcal{A}'_k\cos(\omega_k t + \phi_k) \tag{7.16}$$

where the α_k are linear expansion coefficients and the ϕ_k are arbitrary phases. Both of these sets of parameters are determined by imposing additional constraints on the solution, such as, for example, by specifying initial conditions for the motion of the atoms.

The expression for the energy can be rewritten in terms of the normal mode vectors by substituting the expression for the displacement, equation (7.16), into equation (7.2). Remembering the fact that the various eigenvectors are orthonormal gives

$$\Delta\mathcal{V} = \frac{1}{2}\sum_{k=1}^{3N}\alpha_k^2\omega_k^2\cos^2(\omega_k t + \phi_k) \tag{7.17}$$

The analysis above is important because it provides detailed information about the dynamics of a system around a stationary point. For minima, it is often true that the frequencies of motion, ω_k, can be obtained experimentally, most notably by using vibrational infrared spectroscopy, so a direct link between experiment and theory can be made.

The harmonic normal mode analysis discussed above is an important tool but there are some other related techniques that are sometimes used, especially when analysing spectroscopic data. For example, in some cases the harmonic analysis outlined above is not sufficiently precise and so extra third- and fourth-order terms are included in equation (7.2). These terms are obviously more onerous to calculate insofar as they involve the third and fourth derivatives of the energy with respect to the atomic positions. There are also other derivatives that it is possible to relate to experimental data. One of the more simple of these is the *dipole moment derivatives* whose squares are related to the *infrared intensities* of the vibrational motions of a molecule.

The dipole moment vector, $\boldsymbol{\mu}$, of a system described with the empirical energy function that we are employing can be determined from the following expression:

$$\boldsymbol{\mu} = \sum_{i=1}^{N} q_i \mathbf{r}_i \tag{7.18}$$

For a system with a total charge of zero the dipole vector is independent of the choice of origin for the coordinate system. For a non-zero total charge the dipole vector is origin-dependent and it is usual to translate the molecule to its center of mass before calculating the vector.

Because the charges on the atoms in the force field we are using are constant, the dipole moment vector's derivatives with respect to the Cartesian coordinates of an atom, i, can be trivially calculated. They are

$$\begin{pmatrix} \dfrac{\partial \mu_x}{\partial x_i} & \dfrac{\partial \mu_y}{\partial x_i} & \dfrac{\partial \mu_z}{\partial x_i} \\[2ex] \dfrac{\partial \mu_x}{\partial y_i} & \dfrac{\partial \mu_y}{\partial y_i} & \dfrac{\partial \mu_z}{\partial y_i} \\[2ex] \dfrac{\partial \mu_x}{\partial z_i} & \dfrac{\partial \mu_y}{\partial z_i} & \dfrac{\partial \mu_z}{\partial z_i} \end{pmatrix} = q_i \mathbf{I} \tag{7.19}$$

where $\mathbf{I}$ is the 3×3 identity matrix.

The intensities for each vibrational mode are proportional to the square of the dipole derivative vectors projected onto the normal mode vectors. Writing the intensity of the kth mode as $\mathcal{I}_k$, we have

$$\mathcal{I}_k \propto \sum_{\alpha = x,y,z} \left(\mathcal{A}_k^{\mathrm{T}} \frac{\partial \mu_\alpha}{\partial \mathbf{R}} \right)^2 \tag{7.20}$$

The module NORMAL_MODE implements a number of subroutines to calculate the normal modes and the infrared intensities for a system. Its definition is

```
MODULE NORMAL_MODE

CONTAINS

   SUBROUTINE NORMAL_MODE_FREQUENCIES ( HESSIAN, MODIFY, PRINT )
      REAL, DIMENSION(:),     INTENT(IN), OPTIONAL :: HESSIAN
      CHARACTER ( LEN = * ), INTENT(IN), OPTIONAL :: MODIFY
      LOGICAL,                INTENT(IN), OPTIONAL :: PRINT
   END SUBROUTINE NORMAL_MODE_FREQUENCIES

   SUBROUTINE NORMAL_MODE_INTENSITIES ( PRINT )
      LOGICAL, INTENT(IN), OPTIONAL :: PRINT
   END SUBROUTINE NORMAL_MODE_INTENSITIES

   SUBROUTINE NORMAL_MODE_PRINT ( START, STOP, ATOM_SELECTION )
      INTEGER, INTENT(IN), OPTIONAL :: START, STOP
      LOGICAL, DIMENSION(1:NATOMS), INTENT(IN), OPTIONAL :: ATOM_SELECTION
   END SUBROUTINE NORMAL_MODE_PRINT

END MODULE NORMAL_MODE
```

The subroutine NORMAL_MODE_FREQUENCIES calculates the normal modes and the vibrational frequencies for a system. It has three arguments, all of which are optional. The first, HESSIAN, is the matrix of second derivatives of the potential energy with respect to the Cartesian coordinates. If this is supplied it must be of the correct size (i.e. $[3N(3N + 1)]/2$ elements, where N is the number of atoms) and it must correspond to the information given in the data structures of the module ATOMS. If it is not present then the subroutine calls the subroutine HESSIAN from the module POTENTIAL_ENERGY to calculate the second-derivative matrix for itself. The second argument, MODIFY, is a scalar string and it specifies various modifications that can be performed on the second-derivative matrix before the normal mode analysis is carried out. These options are explained in more detail in the following section. The final argument, PRINT, specifies whether the frequencies should be printed or not. The default is to print out the values.

The frequencies and normal modes calculated by the subroutine are saved and kept by the module for subsequent use by one of the other subroutines in the module. Every time NORMAL_MODE_FREQUENCIES is recalled the old frequency and normal mode arrays are destroyed.

The second subroutine in the module is NORMAL_MODE_INTENSITIES, which calculates the infrared intensities. For this subroutine to work, a normal mode analysis must have been done previously so that the normal modes for the system exist. The subroutine calculates the infrared intensities and stores them in an array in the module so that they can be used later. This array, like the normal mode and frequency arrays, is destroyed by a call to the subroutine NORMAL_MODE_FREQUENCIES. NORMAL_MODE_INTENSITIES has only one argument, PRINT, which specifies whether the intensities are to be printed once they have been calculated. By default the intensities are printed.

The third subroutine in the module is NORMAL_MODE_PRINT, which prints the normal modes, frequencies and infrared intensities stored in the module. Obviously, for this subroutine to work, a previous normal mode analysis must have been carried out. If the infrared intensities have been calculated these will be printed too. There are three arguments, all of which are optional. The first two give the numbers of the first and last modes to print. If no numbers are given, all the modes will be printed. The third argument is a logical atom-selection array that specifies for which atoms the mode components are to be printed. If no array is given, all $3N$ elements for each vector will be printed.

The module NORMAL_MODE makes use of two other modules that have not been introduced before and will not be described in detail because users are unlikely to want to access them directly. The modules are MULTIPOLES, which contains a subroutine for the calculation of the dipole derivatives of a system, and NORMAL_MODE_UTILITIES, which contains a variety of miscellaneous procedures for manipulating the Hessian matrix and the normal modes.

7.3 Rotational and translational modes

We have already mentioned several times that the set of Cartesian coordinates is redundant. This is because, in general, only $3N - 6$ parameters are needed to determine the geometry of a molecule with N atoms completely, but there are $3N$ Cartesian coordinates. The difference between the two representations is that the Cartesian set has six extra *degrees of freedom* that define the position and the orientation of the system in space.

For a system in vacuum, which category comprises all those we have looked at so far, the position and orientation of the system are unimportant because its potential energy and other properties will be *invariant* with respect to rotations and translations applied to the system as a whole. If, however, there is a preferred direction in space, due to the presence of an external field (such as an electric field) or some other environment, the system's absolute position and orientation in space will be important and its properties will no longer be invariant with respect to rotations and translations.

In those cases in which the absolute position and orientation of a system are not of interest, it is possible to remove the redundancy inherent to the Cartesian description by defining six constraint conditions that are functions of the coordinates and reduce the number of degrees of freedom available to the system to $3N - 6$. There are several forms for the constraints but the most commonly used are the *Eckart conditions* which relate the coordinates of the atoms in a system, r_i, to their values in a reference structure, r_i^0. The Eckart condition on the translational motion is

$$\sum_{i=1}^{N} m_i(r_i - r_i^0) = 0 \qquad (7.21)$$

and that on the rotational motion is

$$\sum_{i=1}^{N} m_i r_i \wedge (r_i^0 - r_i) = \sum_{i=1}^{N} m_i r_i \wedge r_i^0 = 0 \qquad (7.22)$$

These constraints are derived by considering the dynamics of the atoms in the system when they are displaced away from their positions in the reference structure. First a Hamiltonian that describes the dynamics is defined (we shall talk about Hamiltonians further in the next chapter) and then it is manipulated so that the motions due to the rotations and translations of the entire system are separated out from those due to its internal vibrations. It turns out that it is possible to separate off completely the translational motion if the condition of equation (7.21) is satisfied. The rotational and vibrational motions, however, cannot be completely separated but their coupling can be reduced by requiring that equation (7.22) holds.

The derivation described above is a general one in that the overall rotational and translational motions of one structure with respect to another can be constrained if equations (7.21) and (7.22) are obeyed. We came across such an example when discussing the self-avoiding walk algorithm for calculating reaction paths in section 6.9. There we saw that it was important to prevent rotational and translational motions of the structures along the chain during the optimization process. The Eckart conditions provide a means of doing this. The idea is to select one of the initial structures along the path as the reference structure and then to ensure that all the other structures along the path obey a set of Eckart conditions with respect to the reference at each step in the subsequent optimization.

This procedure is straightforward to implement. The reason is that the constraints in equations (7.21) and (7.22) are linear functions of the coordinates of the atoms in the non-reference structure (note that the coordinates of the reference structure are treated as constants) and so their derivatives with respect to the coordinates will be constants. This allows us to project out of the gradient vector for each structure at each step the contributions that lie in the subspace spanned by the constraint derivative vectors. If we denote the gradient vector for a structure, I, as G_I and the orthogonalized constraint derivative vectors as Λ_α ($\alpha = 1, 6$), the modified gradient vector G_I' has the form

$$G_I' = G_I - \sum_{\alpha=1}^{6} \left(\frac{\Lambda_\alpha^T G_I}{\Lambda_\alpha^T \Lambda_\alpha} \right) \Lambda_\alpha \qquad (7.23)$$

The modified gradients can now be used in the optimization process as normal but, because of the projection, they will not produce displacements that induce rotations and translations.

Let us now return to the vibrational problem. The invariance of the potential energy of a system with respect to rotations and translations is manifested in a normal mode analysis at a stationary point by the presence of six modes with zero frequencies. Three of these modes correspond to translations and three to rotations. It can be seen from equation (7.17) that a displacement along the mode leaves the potential energy of the system unchanged if the frequency of a mode is zero.

The forms for the vectors of the rotational and translational modes can be derived from the Eckart conditions. To do this properly, though, requires that the formulae in equations (7.21) and (7.22) be re-expressed in mass-weighted coordinates because these are the natural coordinates to use when dealing with eigenvectors of the mass-weighted Hessian matrix (see equation (7.12)). If we take the Eckart condition for translation along the x axis as an example, we can write the part of the condition that depends upon the non-reference structure as

$$\sum_{i=1}^{N} m_i x_i = \sum_{i=1}^{N} \sqrt{m_i} q_i^x \qquad (7.24)$$

where q_i^x is the mass-weighted x coordinate for atom i which was defined in equation (6.9). Equation (7.24) can be re-expressed in vector form as $T_x^T Q$, where Q is the $3N$-dimensional vector of mass-weighted Cartesian coordinates for the system and T_x is the mode vector for translation in the x direction. The latter has the form

$$\mathcal{T}_x \propto \mathbf{M}^{\frac{1}{2}} \begin{pmatrix} 1 \\ 0 \\ 0 \\ 1 \\ 0 \\ 0 \\ \vdots \\ 1 \\ 0 \\ 0 \end{pmatrix} \tag{7.25}$$

That this vector is equivalent to a translation of the whole system may be confirmed by replacing $\mathcal{A}'$ by the expression for $\mathcal{T}_x$ in equation (7.16).

The remaining vectors for translation and those for rotation are constructed in the same fashion. The translation vectors, $\mathcal{T}_y$ and $\mathcal{T}_z$, have the same form as $\mathcal{T}_x$ except they have non-zero elements for their y and z components respectively. The vector for rotation about the x axis, $\mathcal{R}_x$, is

$$\mathcal{R}_x \propto \mathbf{M}^{\frac{1}{2}} \begin{pmatrix} 0 \\ -z_1 \\ y_1 \\ 0 \\ -z_2 \\ y_2 \\ \vdots \\ 0 \\ -z_N \\ y_N \end{pmatrix} \tag{7.26}$$

The $\mathcal{R}_y$ and $\mathcal{R}_z$ modes are similar but with the arrangements $(z, 0, -x)$ and $(-y, x, 0)$ for each atom, respectively.

There are circumstances under which it is important to be able to manipulate the rotational and translational modes that arise from a Hessian matrix. One such case is when normal mode analyses are done at points that are not stationary. In these instances, the arguments of section 7.2 are invalid because the gradient vector is not zero and there will no longer be six zero eigenvalues but only the three which correspond to translational motion. A second case is when performing optimizations with algorithms such as the Baker algorithm discussed in section 6.5, which employ second-derivative information. It is obviously uninteresting to search along the modes which rotate or translate the entire system and so it is preferable to have some means of removing them from the optimization process.

We shall mention two methods that can be used to modify the Hessian matrix. The first technique, which was used in the Baker algorithm, changes the eigenvalues of the rotational and translational modes. To do this for the translation in the x direction, for example, requires adding a term to the mass-weighted Hessian matrix as follows:

$$\mathbf{H}' \to \mathbf{H}' + \lambda \mathcal{T}_x \mathcal{T}_x^{\mathrm{T}} \tag{7.27}$$

where λ is the new (usually large) value for the mode's eigenvalue.

In the second technique a projection matrix, $\mathbf{P}$, is defined with the form

$$\mathbf{P} = \mathbf{I} - \mathcal{R}_x \mathcal{R}_x^{\mathrm{T}} - \mathcal{R}_y \mathcal{R}_y^{\mathrm{T}} - \mathcal{R}_z \mathcal{R}_z^{\mathrm{T}} - \mathcal{T}_x \mathcal{T}_x^{\mathrm{T}} - \mathcal{T}_y \mathcal{T}_y^{\mathrm{T}} - \mathcal{T}_z \mathcal{T}_z^{\mathrm{T}} \tag{7.28}$$

and the Hessian matrix is modified as follows:

$$\mathbf{H}' \to \mathbf{P} \mathbf{H}' \mathbf{P} \tag{7.29}$$

In contrast to the first technique, this method guarantees that the six translational and rotational modes will have zero eigenvalues even if the normal mode analysis is to be carried out away from a stationary point.

Both these techniques have been implemented for use in a normal mode analysis. They are invoked by using the optional argument MODIFY of the subroutine NORMAL_MODE_FREQUENCIES described in the last section. If modify is set to the string, ''RAISE'', the technique of equation (7.27) is used and the eigenvalues of rotational and translational modes are increased to very large values. If it is ''PROJECT'', equation (7.29) is used instead. If the modify option is not used the second-derivative matrix is left unchanged.

7.4 Generating normal mode trajectories

The motion of the atoms in the system within the harmonic approximation is given by the solution for the displacement vector of equation (7.16). To complete the determination of the solution the parameters α_k and ϕ_k remain to be specified. The values of the phases, ϕ_k, are arbitrary and do not alter the general behaviour of the dynamics but the values of the linear expansion coefficients, α_k, affect the relative importance of each mode in the displacement.

These parameters can be characterized more precisely by investigating the *total* energy of the system. Up to now, when we have discussed energies, we have been concerned exclusively with the value of the potential energy. The total energy of a system is the sum of the potential energy and its kinetic energy, which is the energy that arises due to the motions of the atoms in the system. In classical mechanics, the kinetic energy, $\mathcal{K}$, has the form

$$K = \frac{1}{2} \sum_{i=1}^{N} m_i v_i^2 \qquad (7.30)$$

where v_i is the velocity of the atom which is equal to the first time derivative of the atom position, i.e.

$$v_i = \frac{\mathrm{d}r_i}{\mathrm{d}t} = \dot{r}_i \qquad (7.31)$$

Using these definitions, we can write down the total energy of the system within the harmonic approximation as a function of the displacement vector, **D**. It is

$$E = \frac{1}{2} \dot{\boldsymbol{D}}^{\mathrm{T}} \boldsymbol{M} \dot{\boldsymbol{D}} + \frac{1}{2} \boldsymbol{D}^{\mathrm{T}} \boldsymbol{H} \boldsymbol{D} \qquad (7.32)$$

The atoms' velocities can be obtained by differentiating equation (7.16) with respect to time. Doing this, substituting into the above equation and simplifying produces the final result for the total energy:

$$E = \frac{1}{2} \sum_{k=1}^{3N} \alpha_k^2 \omega_k^2 \qquad (7.33)$$

To identify the α_k we can make use of a result from statistical thermodynamics, which says that, at equilibrium, the total energy of a system of independent harmonic oscillators will be equal to the number of oscillators multiplied by $k_{\mathrm{B}}T$ where k_{B} is Boltzmann's constant and T is the absolute temperature. To apply this result to equation (7.33), we equate the energy contributed by each oscillator to $k_{\mathrm{B}}T$, giving $\alpha_k = \sqrt{2k_{\mathrm{B}}T/\omega_k}$. Thus, the size of the amplitude of each mode will be inversely proportional to its frequency. The lower the frequency, the larger the amplitude of the motion. It is important to note that the above analysis is invalid for modes with zero or imaginary frequencies, for which other arguments have to be used.

With this result we can now investigate the motion due to a particular mode at a given temperature. To do this, an extra subroutine, NORMAL_MODE_TRAJECTORY, has been provided in the module NORMAL_MODE. It has the following definition:

```
MODULE NORMAL_MODE

CONTAINS

    SUBROUTINE NORMAL_MODE_TRAJECTORY ( FILE, MODE, NCYCLES, NFRAMES, T )
        CHARACTER ( LEN = * ), INTENT(IN) :: FILE
        INTEGER,                INTENT(IN) :: MODE, NCYCLES, NFRAMES
        REAL),                  INTENT(IN) :: T
    END SUBROUTINE NORMAL_MODE_TRAJECTORY

END MODULE NORMAL_MODE
```

For this subroutine to work, a previous normal mode calculation must have been performed so that the modes and their frequencies exist in the module. There are five arguments to the subroutine, all of which must be present. The first, FILE, is the name of the file to which the trajectory is to be written. The second, MODE, is the number of the mode for which the trajectory is to be generated. A simple check on the frequency of the mode is made. If the absolute value of the frequency is too small then the subroutine returns automatically without doing a calculation. If the frequency is imaginary, but with a magnitude that is large enough, a trajectory will be generated making the assumption that the mode has a *real* frequency. This is, of course, physically invalid but is done so that at least the displacements generated by the mode can be studied or visualized.

The third argument, NCYCLES, specifies for how many complete vibrational cycles the trajectory should be generated while the fourth argument, NFRAMES, gives the number of frames to generate for each cycle. The total number of coordinate sets stored on the trajectory will be the product of these two arguments. The last argument, T, gives the temperature in kelvins.

7.5 Example 10

The example in this section calculates the normal modes and the infrared intensities for the chair form of cyclohexane that was introduced in the last chapter. The program is

```
PROGRAM EXAMPLE10

... Declaration statements ...

! . Construct the system file for cyclohexane.
CALL MM_SYSTEM_READ ( ''cyclohexane.sys_bin'' )

! . Read in the coordinates of the chair form.
CALL COORDINATES_READ ( ''cyclohexane_chair.crd'' )

! . Calculate the frequencies.
CALL NORMAL_MODE_FREQUENCIES ( MODIFY = ''PROJECT'' )

! . Calculate the infrared intensities.
CALL NORMAL_MODE_INTENSITIES

! . Print the normal modes.
CALL NORMAL_MODE_PRINT

! . Calculate a single mode trajectory for the lowest frequency mode.
CALL NORMAL_MODE_TRAJECTORY ( ''cyclohexane_chair.mode7'',   &
                              7, 10, 21, 600.0 )
```

END PROGRAM EXAMPLE10

The program starts by reading in the system file and the coordinates for the chair structure of cyclohexane. The normal mode analysis is performed by a call to the subroutine NORMAL_MODE_FREQUENCIES. This subroutine calculates the second-derivative matrix and then carries out the analysis after it has projected out the motions corresponding to the rotational and translational degrees of freedom from the second-derivative matrix. After the normal mode calculation, the infrared intensities are determined and all the modes, of which there are 54, are printed. Finally, a trajectory for the mode with the lowest non-zero frequency is generated at a temperature of 600 K. The trajectory contains ten cycles, each of which has 21 frames, giving 210 frames in total.

The frequencies arising as a result of this calculation and for exactly equivalent ones on the twist-boat and saddle-point structures of cyclohexane are shown in figure 7.1. Some of the modes for the chair conformer are *degenerate*, which means that there are two (or more in the general case) modes with exactly the same frequency. This normally arises because the structure has a particular symmetry. The infrared intensities, which are not shown, possess equivalent symmetry properties. It is to be noted that the values of the intensities calculated with the force-field approximation used in this book are often poor and do not reproduce well those found experimentally.

The imaginary mode for the saddle-point structure is illustrated in figure 7.2. The arrows on the atoms give the direction and the magnitude of the displacement induced by the mode. It can be seen that the mode induces displacements that lead to the chair and boat structures, respectively.

7.6 Calculation of thermodynamic quantities

One of the principal aims of doing molecular simulations is to be able to compare the results of the calculations with those obtained from experiment. We have already met two ways in which this can be done in this chapter – we can calculate the vibrational frequencies and the infrared intensities for a system. In this section we consider more general ways in which macroscopic properties can be obtained from the atomic properties that we calculate when we do a simulation. To do this we use some well-known results from statistical thermodynamics, which is the branch of physics that links the microscopic and the macroscopic worlds. It is impractical in the space that we have here to provide any but the most cursory overview of this subject but a brief

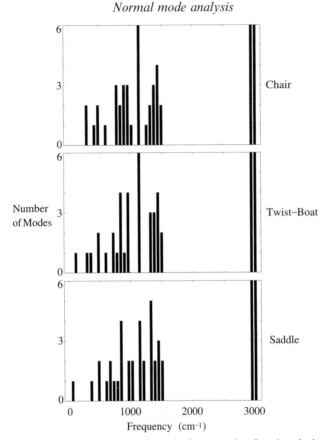

Figure 7.1. Histograms of the normal mode frequencies for the chair, twist-boat and saddle-point structures of cyclohexane. The width of each histogram bin is 60 cm^{-1}. The imaginary frequency for the saddle-point structure, which is not shown, has a value of 242i cm^{-1}.

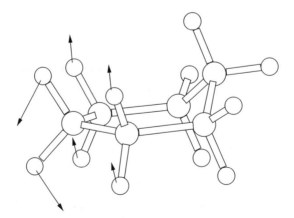

Figure 7.2. Atomic displacements generated by the normal mode with the imaginary frequency for the saddle-point structure of cyclohexane. Only the largest displacements are shown for clarity.

description is provided, whose principal purpose will be to define precisely the quantities that are calculated by the module that is to be introduced later.

There are both quantum and classical formulations of statistical mechanics, but, in the quantum case, the theory is developed in terms of the various *quantum states* that the system can occupy. A fundamental quantity in the theory is the *partition function*, z, which can be regarded as a measure of the effective number of states that are accessible to the system. Its expression is

$$z = \sum_i e^{-\epsilon_i/(k_B T)} \tag{7.34}$$

where the sum runs over all the different states of the system, each of which has an energy, ϵ_i. To make progress in calculating the partition function from molecular quantities it is necessary to make some approximations. One of the most useful is to hypothesize that the energy for a state can be written as a sum of energies, each of which comes from a different property of the system. In particular, it can be assumed that the state energy is the sum of electronic, nuclear, rotational, translational and vibrational energies, i.e.

$$\epsilon_i = \epsilon_i^{\text{elec}} + \epsilon_i^{\text{nucl}} + \epsilon_i^{\text{rot}} + \epsilon_i^{\text{trans}} + \epsilon_i^{\text{vib}} \tag{7.35}$$

where the superscripts refer to electronic, nuclear, rotational, translational and vibrational terms, respectively. This separation of the state energies into different terms and, in particular, the separation of the rotational and vibrational energies of the system is called the *rigid-rotor, harmonic oscillator approximation*.

Within the rigid-rotor, harmonic oscillator approximation the partition function for a molecule can be written as a product of the partition functions pertaining to the different types of energy. Thus, the total partition function is

$$z = z_{\text{elec}} z_{\text{nucl}} z_{\text{rot}} z_{\text{trans}} z_{\text{vib}} \tag{7.36}$$

This partition function is for a single molecule. The partition function, Z, for a collection of N_{mol} molecules will be equal to the product of the partition functions of the individual molecules, $z^{N_{\text{mol}}}$, if the molecules are distinguishable (as in a solid) or equal to $z^{N_{\text{mol}}}/N_{\text{mol}}!$ if the molecules are indistinguishable (as in a gas).

To calculate the electronic and nuclear partition functions the values of the electronic and nuclear energy levels are required. For most cases in which we will be interested we can assume that it is only the electronic and nuclear ground states that are important, the excited states being so much higher in energy that their contribution to the total partition function will be negligible. Thus, these terms can be considered to have values of 1.

The other three components of the total partition function can be readily calculated, either exactly, for the case of the vibrational function, or by approximating the sum over states by an integral for the rotational and translational functions. The results are

$$
z_{rot} = \frac{\sqrt{\pi}}{\sigma} \left(\frac{8\pi^2 \mathcal{I}_A k_B T}{h^2} \right)^{\frac{1}{2}} \left(\frac{8\pi^2 \mathcal{I}_B k_B T}{h^2} \right)^{\frac{1}{2}} \left(\frac{8\pi^2 \mathcal{I}_C k_B T}{h^2} \right)^{\frac{1}{2}} \tag{7.37}
$$

$$
z_{trans} = \left(\frac{2\pi M_T k_B T}{h^2} \right)^{\frac{3}{2}} V \tag{7.38}
$$

$$
z_{vib} = \prod_{i=1}^{3N-6} \frac{\exp[-h\omega_i/(2k_B T)]}{1 - \exp[-h\omega_i/(k_B T)]} \tag{7.39}
$$

In these equations, $\mathcal{I}_A$, $\mathcal{I}_B$ and $\mathcal{I}_C$ are the moments of inertia of the system, M_T is its total mass, V is the volume it occupies and h is Planck's constant. σ is what is known as the *symmetry number* of the molecule. For a molecule without any symmetry its value is 1. For molecules at minima that have symmetry, it is the number of different ways the molecule can be rotated into configurations that are indistinguishable from the original configuration.

The expressions for the rotational and vibrational partition functions need to be modified when the molecule is linear. The upper limit of the sum in equation (7.39) becomes $3N - 5$ and equation (7.37) must be replaced by

$$
z_{rot} = \frac{8\pi^2 \mathcal{I}_A k_B T}{h^2 \sigma} \tag{7.40}
$$

since a linear molecule has only one unique moment of inertia, $\mathcal{I}_A$. For atomic systems the rotational and vibrational partition functions have a value of 1.

The partition function can be related to thermodynamic quantities using standard relations from statistical thermodynamics. For an ideal gas, which is the case in which we are interested, the equations are

$$U = RT^2 \left(\frac{\partial \ln z}{\partial T} \right)_V \tag{7.41}$$

$$S = R \ln z + RT \left(\frac{\partial \ln z}{\partial T} \right)_V - R \ln N_{av} + R \tag{7.42}$$

$$A = U - TS \tag{7.43}$$

$$H = U + PV \tag{7.44}$$

$$G = H - TS \tag{7.45}$$

$$C_V = \left(\frac{\partial U}{\partial T} \right)_V \tag{7.46}$$

$$C_P = \left(\frac{\partial H}{\partial T} \right)_P \tag{7.47}$$

$$PV = RT \tag{7.48}$$

where A is the Helmholtz free energy, G is the Gibbs free energy, H is the enthalpy, S is the entropy, U is the internal energy and C_V and C_P are the heat capacities at constant volume and pressure, respectively. P, T and V denote the pressure, temperature and volume of the system. R is the molar gas constant and is equal to the product of Avogadro's constant, N_{av}, and Boltzmann's constant, k_B. The subscripts on the brackets surrounding the partial derivatives indicate that these quantities are assumed to be constant for the differentiation.

The values of these quantities can be determined experimentally in many cases and it is found that the rigid-rotor, harmonic oscillator approximation often gives values that are in good agreement with experiment. It is also possible to combine the values for several structures to calculate equilibrium and rate constants. For example, if we consider the equilibrium between two states of a system, A and B:

$$A \underset{}{\overset{K_{eq}}{\rightleftharpoons}} B \tag{7.49}$$

the equilibrium constant (at constant pressure), K_{eq}, can be written as

$$K_{eq} = \exp \left[-\left(\frac{G_B - G_A}{RT} \right) \right] \tag{7.50}$$

where G_A and G_B are the free energies of states A and B, respectively.

To calculate rate constants for a reaction it is possible to use *transition state theory*. In its simplest version, this theory assumes that there is an *activated complex* or *transition state structure* that is in equilibrium with the reactant molecules and is transformed into products. If $A^{\ddagger}$ is this complex, the reaction can be written as

$$A \overset{K^\ddagger}{\rightleftharpoons} A^\ddagger \overset{k'}{\rightarrow} \text{products} \tag{7.51}$$

The rate for this process, k_f, is then written as

$$k_f = k' K^\ddagger = \frac{k_B T}{h} \exp\left[-\left(\frac{G_{A^\ddagger} - G_A}{RT}\right)\right] \tag{7.52}$$

where the factor $k_B T/h$ gives the rate at which the activated complex goes to products and the exponential factor is the equilibrium constant for the equilibrium between the activated complex and reactants. It is normal to equate the activated complex structure to the saddle-point structure on the reaction path between reactants and products. In these cases, the imaginary frequency of the saddle point is omitted from the calculation of the vibrational partition function of equation (7.39).

Note that, for the determination both of the equilibrium and of the rate constants in equations (7.50) and (7.52), the free energy values are calculated with respect to the same reference value on the potential energy surface. In other words, the differences between the free energy values include the difference in potential energy between the two structures in addition to the terms depending directly upon the partition function.

To calculate the thermodynamic quantities defined in equations (7.41)–(7.48), a module has been provided, THERMODYNAMICS_RRHO, with one subroutine, THERMODYNAMICS. Its definition is

```
MODULE THERMODYNAMICS_RRHO

CONTAINS

    SUBROUTINE THERMODYNAMICS ( TEMPERATURES, RESULTS, SYMMETRY_NUMBER, &
                                                 PRESSURE, PRINT )
        REAL, DIMENSION(:),    INTENT(IN)            :: TEMPERATURES
        INTEGER,               INTENT(IN),  OPTIONAL :: SYMMETRY_NUMBER
        LOGICAL,               INTENT(IN),  OPTIONAL :: PRINT
        REAL,                  INTENT(IN),  OPTIONAL :: PRESSURE
        REAL, DIMENSION(:,:),  INTENT(OUT), OPTIONAL :: RESULTS
    END SUBROUTINE THERMODYNAMICS

END MODULE THERMODYNAMICS_RRHO
```

The subroutine has one essential argument, TEMPERATURES, which is a real array that contains the values of the temperatures at which the thermodynamic quantities are to be calculated. The length of the array is arbitrary. The four remaining arguments are optional. RESULTS is a real array that, if it is present, will return the values of the thermodynamic quantities that have been calculated. It *must* have the dimension (1:7,1:SIZE(TEMPERATURES)). For each value of the temperature the values for A, C_P, C_V, G, H, S and U will be returned in that order. The units for A, G, H and U are kJ mol^{-1} and those for C_P, C_V and S are kJ mol^{-1} K^{-1}. The argument SYMMETRY_NUMBER gives the symmetry number for the system. If no value is input the subroutine automatically uses a value of 1. The argument PRESSURE specifies the pressure of the system in atmospheres with a default value of 1. The pressure is used to determine the value of the volume of the system in the calculation of the translational partition function in equation (7.38) using equation (7.48). Finally PRINT specifies whether the thermodynamic quantities are to be printed. The default is to print them.

The subroutine THERMODYNAMICS uses the information from the modules ATOMS and NORMAL_MODE to calculate the partition functions. Thus, the coordinates and masses for the atoms in ATOMS should correspond to the values of the frequencies stored in the module NORMAL_MODE. It should also be noted that, for this subroutine to return sensible values, the system should be at a stationary point because the calculation of the vibrational partition function leaves out the six (five for a linear system) frequencies with the lowest absolute magnitudes as well as any imaginary frequencies.

7.7 Example 11

We can use the module presented in the last section to calculate the thermo-dynamic quantities for the chair and twist-boat forms of cyclohexane at a series of temperatures and then to determine the equilibrium constant for the process

$$\text{Chair} \rightleftharpoons \text{Twist-boat} \tag{7.53}$$

The program is

```
PROGRAM EXAMPLE11

... Declaration statements ...

! . Local parameters.
INTEGER, PARAMETER :: NT

! . Local scalars.
INTEGER :: I
REAL :: ETBOAT, ECHAIR

! . The temperature array.
REAL, DIMENSION(1:NT)      :: K, T = (/ ( 100.0 * I, I = 1,NT ) /)
REAL, DIMENSION(1:7,1:NT) :: RESULTS_TBOAT, RESULTS_CHAIR
```

```
! . Construct the system file for cyclohexane.
CALL MM_SYSTEM_READ ( ''cyclohexane.sys_bin'' )

! . Read in the coordinates of the chair form.
CALL COORDINATES_READ ( ''cyclohexane_chair.crd'' )

! . Calculate the energy and its derivatives.
CALL HESSIAN ; ECHAIR = TOTAL_PE

! . Calculate the frequencies.
CALL NORMAL_MODE_FREQUENCIES ( HESSIAN = ATMHES, MODIFY = ''PROJECT'' )

! . Calculate the thermodynamic quantities for the chair form.
CALL THERMODYNAMICS ( T, RESULTS_CHAIR, 6, 1.0, .TRUE. )

! . Read in the coordinates of the twist-boat form.
CALL COORDINATES_READ ( ''cyclohexane_tboat.crd'' )

! . Calculate the energy and its derivatives.
CALL HESSIAN ; ETBOAT = TOTAL_PE

! . Calculate the frequencies.
CALL NORMAL_MODE_FREQUENCIES ( HESSIAN = ATMHES, MODIFY = ''PROJECT'' )

! . Calculate the thermodynamic quantities for the twist-boat form.
CALL THERMODYNAMICS ( T, RESULTS_TBOAT, 4, 1.0, .TRUE. )

! . Calculate the equilibrium constants.
K = EXP ( - 1.0E+3 * ( RESULTS_TBOAT(4,1:NT) - RESULTS_CHAIR(4,1:NT) + &
                       ETBOAT - ECHAIR ) / ( NAVOGADRO * KBOLTZ * T ) )

! . Write out the results.
WRITE ( OUTPUT, ''(/A)'' ) ''Equilibrium constants (chair -> twist-boat):''
WRITE ( OUTPUT, ''(/9X,A,19X,A)'' ) ''Temperature'', ''K''
WRITE ( OUTPUT, ''(F20.4,G20.6)'' ) ( T(I), K(I), I = 1,NT )

END PROGRAM EXAMPLE11
```

The program starts by defining, as a parameter, the number of different temperatures for which the thermodynamic quantities are to be calculated. In the current example this parameter, NT, is assigned a value of 9. There are two array declaration statements. The first defines an array K that is to hold the values of the equilibrium constant for the process (7.53) at each temperature and an array T that contains the temperature values. The latter array is initialized immediately in the declaration statement so that its elements have values from 100 to 900 at intervals of 100. In the second array declaration statement, two arrays are defined that will hold the calculated thermodynamic quantities for the chair and twist-boat structures.

After the declarations, the system file for cyclohexane is read in and the energy and normal mode calculations are carried out for the chair form of cyclohexane, followed by a calculation of the thermodynamic quantities at the temperatures specified in the array, T. The results are returned in the array RESULTS_CHAIR, the calculations being carried out at a pressure of 1 atm. The chair form of cyclohexane has point group symmetry, D_{3d}, so its symmetry number is 6. The calculations are then repeated in exactly the same way for the twist-boat form except that the symmetry number of the structure is 4 because the twist-boat form has D_2 symmetry.

To finish, the equilibrium constants are calculated using the formula

$$K = \exp\left[-\left(\frac{G_{\text{twist-boat}} - G_{\text{chair}}}{RT}\right)\right] \qquad (7.54)$$

For this calculation the Gibbs free energy values which are stored in element 4 of the first dimension of the results arrays are used. It should also be noted that the Gibbs free energies that are returned by the thermodynamics sub-routine need to be corrected by the difference in the potential energy between the two structures, whence the term ETBOAT - ECHAIR.

The results of the calculation are displayed in figure 7.3. As expected the equilibrium constant for the two species increases with the temperature. It should be remarked that the results of these calculations are somewhat ficti-tious and have been presented for illustrative purposes only. In particular, note that this molecule liquifies at temperatures below about 350 K and so

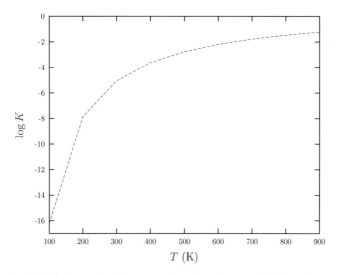

Figure 7.3. A plot of the equilibrium constant for the equilibrium between the chair and twist-boat forms of cyclohexane as a function of temperature.

the gas-phase equilibrium constants below this will be unattainable experimentally.

Exercises

7.1 Calculate the rate constants for the interconversion of the chair and twist-boat forms of cyclohexane using the transition state theory expression given in equation (7.52). The program should be similar to that of Example 11 except that a third normal mode calculation needs to be done for the saddle-point structure. Calculate both the forwards and the reverse rate constants.

7.2 The results of a normal mode analysis depend upon the masses of the atoms. Thus, if different isotopes are used for atoms, the frequencies of vibration will shift. These *isotope effects* can be an important analytic tool for the investigation of the mechanisms of reactions. Choose a molecule, such as cyclohexane, and investigate the effect of using different isotopes (exchanging H for D, for example) on the vibrational frequencies of each structure and the values of the equilibrium and rate constants.

8

Molecular dynamics simulations I

8.1 Introduction

We saw in chapter 6 how it was possible to explore relatively small parts of a potential energy surface and in chapter 7 how to use some of this information to obtain approximate dynamical and thermodynamic information about a system. These methods, though, are local – they consider only a limited portion of the potential energy surface and the dynamics of the system within it. It is possible to go beyond these 'static' approximations to study the dynamics of a system directly. Some of these techniques will be introduced in the present chapter.

8.2 Molecular dynamics

As we discussed in chapter 4, complete knowledge of the behaviour of a system can be obtained, in principle, by solving its time-dependent Schrödinger equation (equation (4.1)), which governs the dynamics of all the particles in the system, both electrons and nuclei. To progress in the solution of this equation we introduced the Born–Oppenheimer approximation, which allows the electronic and the nuclear problems to be treated separately. This separation leads to the concept of a potential energy surface, which is the effective potential that the nuclei experience once the electronic problem has been solved. In principle, once this separation has been effected, it is possible to study the dynamics of the nuclei under the influence of the effective electronic potential using an equivalent equation to equation (4.1) but for the nuclei only. This can be done for systems consisting of a very small number of particles but proves impractical otherwise.

Fortunately, whereas it is difficult or impossible to study the dynamics of the system quantum mechanically, a classical dynamical study is relatively straightforward and provides much useful information. Although they constitute an approximation to the real dynamics, classical dynamical simulation techniques are believed to provide accurate descriptions in many cases. They do, however, omit a number of effects that could be important in certain circumstances. For example, they may fail in the treatment of certain aspects of the dynamics of light particles (especially hydrogen), for which quantum mechanical *tunneling effects* can be important, and they do not include *zero-point motion*, which is the vibrational motion that all quantum mechanical systems undergo even at the absolute zero of temperature (0 K).

There are various formulations for the classical dynamical analysis of a system. We shall, however, adopt the description which starts off by defining the *classical Hamiltonian*, $\mathcal{H}$, for the system. This is essentially equivalent to the system's total energy, kinetic plus potential, and can be written as

$$\mathcal{H}(\boldsymbol{p}_i, \boldsymbol{r}_i) = \sum_{i=1}^{N} \frac{1}{2m_i} p_i^2 + \mathcal{V}(\boldsymbol{r}_i) \tag{8.1}$$

where $\boldsymbol{p}_i$ is the momentum of particle i and $\mathcal{V}$ is the effective potential (in our case the empirical energy function). Note that the Hamiltonian is a function of $6N$ independent variables, the $3N$ particle momenta and the $3N$ particle positions.

It is possible to derive equations of motion for the variables from equation (8.1) using Hamilton's equations, which are

$$\dot{\boldsymbol{p}}_i = -\frac{\partial \mathcal{H}}{\partial \boldsymbol{r}_i}$$

$$= -\frac{\partial \mathcal{V}}{\partial \boldsymbol{r}_i}$$

$$= \boldsymbol{f}_i \tag{8.2}$$

$$\dot{\boldsymbol{r}}_i = \frac{\partial \mathcal{H}}{\partial \boldsymbol{p}_i}$$

$$= \frac{\boldsymbol{p}_i}{m_i} \tag{8.3}$$

These equations are first-order differential equations. A second-order differential equation can be obtained by noting from equation (8.3) that the momentum of a particle is equal to the product of the mass of the particle and its velocity (the time derivative of its position). Substitution of this expression, equation (8.3), into equation (8.2) gives Newton's equation of motion for the particle:

$$m_i \ddot{\boldsymbol{r}}_i = \boldsymbol{f}_i \tag{8.4}$$

To study the dynamics of a system, or to perform a molecular dynamics simulation, the equations of motion (either equations (8.2) and (8.3) or equation (8.4)) must be solved for each particle. This is another example of a well-studied mathematical problem, that of the integration of a set of ordinary differential equations, for which a large variety of algorithms exists. The choice of algorithm depends upon the exact nature of the equations of motion and the accuracy of the solution required, but the great majority of algorithms are *initial value algorithms*, which means that they start off with initial values for the particles' positions and momenta (or velocities) and integrate the equations for a specific length of time.

For very-high-accuracy solutions of the equations of motion it is usually advantageous to solve the system of first-order differential equations for each particle (equations (8.2) and (8.3)). Suitable algorithms are the *predictor–corrector integrators* and the method due to R. Burlisch and J. Stoer. It turns out that, owing to the special form of Newton's equation, it is more efficient for normal use to solve the set of second-order differential equations directly (equation (8.4)). Methods for solving these equations, such as Stoermer's rule, have been known for a long time, but they are generally called *Verlet methods* after L. Verlet, who was one of the first people to apply them to molecular simulations.

The standard Verlet method is easy to derive. If, at a time t, the positions of the atoms in the system are $\mathbf{R}(t)$, then the positions of the atoms at a time $t + \Delta$ can be obtained from a Taylor expansion in terms of the *timestep*, Δ, and the positions and their derivatives at time t. The expansion is

$$\mathbf{R}(t + \Delta) = \mathbf{R}(t) + \Delta \dot{\mathbf{R}}(t) + \frac{\Delta^2}{2} \ddot{\mathbf{R}}(t) + O(\Delta^3) \tag{8.5}$$

Similarly, the positions at a time $t - \Delta$ are obtained from the expansion

$$\mathbf{R}(t - \Delta) = \mathbf{R}(t) - \Delta \dot{\mathbf{R}}(t) + \frac{\Delta^2}{2} \ddot{\mathbf{R}}(t) - O(\Delta^3) \tag{8.6}$$

Adding these equations and rearranging gives an expression for the positions of the particles at $t + \Delta$ in terms of the positions and forces on the particles at earlier times:

$$\mathbf{R}(t + \Delta) = 2\mathbf{R}(t) - \mathbf{R}(t - \Delta) + \Delta^2 \ddot{\mathbf{R}}(t) + O(\Delta^4)$$
$$\simeq 2\mathbf{R}(t) - \mathbf{R}(t - \Delta) + \Delta^2 \mathbf{M}^{-1} \mathbf{F}(t) \tag{8.7}$$

where on going from the first to the second equation we have made use of Newton's equations for the particles (equation (8.4)).

Subtracting equation (8.6) from equation (8.5) gives an equation for the velocities of the particles, V, at the current time, t:

$$V(t) = \dot{R}(t)$$

$$\simeq \frac{1}{2\Delta}(R(t + \Delta) - R(t - \Delta)) \tag{8.8}$$

Equations (8.7) and (8.8) are sufficient to integrate the equations of motion but they are slightly inconvenient. This is because the velocities at time t are available only once the positions at time $t + \Delta$ have been calculated, which means that, at the start of the simulation, i.e. when $t = 0$, it is necessary to use another formula. A slight modification of these equations produces an algorithm called the *velocity Verlet method* that avoids these problems and can be shown to produce results that are equivalent to those of the standard Verlet method. The equations are

$$R(t + \Delta) = R(t) + \Delta V(t) + \frac{\Delta^2}{2}M^{-1}F(t) \tag{8.9}$$

$$V(t + \Delta) = V(t) + \frac{\Delta}{2}M^{-1}(F(t) + F(t + \Delta)) \tag{8.10}$$

It should be noted that there are a wide variety of Verlet-type algorithms in use, including the so-called *leapfrog* methods which calculate the positions on the full step (i.e. at $t + \Delta$) and the velocities on the half step (i.e. at $t + \Delta/2$). We will use the velocity Verlet algorithm in this chapter because it will be sufficient for our needs, but it may be that in other situations one of the other algorithms may be more appropriate.

It is important to be able to check the accuracy of any integration algorithm. Some of the more useful measures of the precision of a simulation are the *conservation conditions* on certain properties of the system, notably the momentum, angular momentum and the energy. The total momentum, $\mathcal{M}$, and angular momentum, $\mathcal{L}$, of a system are defined as

$$\mathcal{M} = \sum_{i=1}^{N} p_i \tag{8.11}$$

$$\mathcal{L} = \sum_{i=1}^{N} r_i \wedge p_i \tag{8.12}$$

It is straightforward to show that the total energy of a system described by a classical Hamiltonian that is independent of time, such as the one given in equation (8.1), is conserved or constant. This can be done by differentiating equation (8.1) with respect to time and then substituting the expressions for

Hamilton's equations of motion (equations (8.2) and (8.3)) to show that the total derivative is zero. This means that throughout the simulation the total energy should be the same as the energy at the beginning. Of course, this will not exactly be the case, owing to there being errors in the integration algorithm, but the size of the deviations in the total energy or the drift away from its initial value will give a measure of the precision of the integration. The conservation conditions for the momentum and angular momentum can be derived in a similar way by differentiating equations (8.11) and (8.12). On doing this it can be seen that there will be conservation of momentum and angular momentum if there is no nett force, $\sum_{i=1}^{N} f_i$, or torque, $\sum_{i=1}^{N} r_i \wedge f_i$, on the system, respectively. This will be the case for a system in vacuum but it will not necessarily hold if the system experiences some exterior influence due to an external field, for example.

Having defined the algorithm for the integration of the equations of motion it is straightforward to devise a scheme to perform a molecular dynamics simulation. The one that we shall use is as follows.

1. Define the composition of the system, including the number and type of atoms, their masses and their interaction potential.
2. Assign initial values ($t = 0$) to the particles' positions, R, and velocities, V.
3. Define the timestep, Δ, for the integration and the number of integration steps (i.e. the duration of the simulation).
4. Perform the simulation. Initially the positions and the velocities of the particles are known, but the forces at $t = 0$ must be calculated. At subsequent integration steps do the following.
   (a) Calculate the positions at the current step $t + \Delta$ using equation (8.9).
   (b) Calculate the forces at $t + \Delta$.
   (c) Calculate the velocities at $t + \Delta$ using equation (8.10).
   (d) Do any analysis that is required with the positions and velocities at the current step $t + \Delta$. This can include the calculation, the printing of intermediate results or the storage of the position and velocity data on an external file to create a molecular dynamics trajectory.
   (e) Increment the time by the timestep.
5. Analyse the results.

There are a number of issues raised by this scheme that need elaboration. The first is that of how to choose the value of the timestep. In general, we would like a timestep that is as large as possible so that the simulation is as long as possible, but not so large that the accuracy of the integration procedure is jeopardized. The factor which normally limits the upper size of the timestep is the nature of the highest frequency motions in the system. In organic molecules, these are typically motions involving hydrogen (because it is light) and include, for example, the stretching associated with the vibrations of carbon—hydrogen bonds. To integrate accurately over these motions the timestep needs to be small with respect to the period of the vibration. So, for example, if the highest frequency vibrations have values of around 3000 cm^{-1}, their characteristic timescales will be of the order of a few femtoseconds (1 fs $= 10^{-15}$ s), which means that the timestep will need to be less than this. In practice, values of about 1 fs are found to be the largest reasonable for systems possessing these types of motion when Verlet algorithms are employed.

The second point that needs discussion is that of how to choose the initial values of the velocities for the atoms (we shall assume that a starting set of coordinates is available). One of the most convenient ways is to choose the velocities so that the system will have a particular temperature at the start of the simulation. From statistical thermodynamics it is known that the velocities of the atoms in a classical system are distributed according to the Maxwell–Boltzmann distribution. This says that, if the temperature of the system is T, the probability of each component of the velocity of the ith atom having a value between v and $v + dv$ is

$$f(v)\, dv = \sqrt{\frac{m_i}{2\pi k_B T}} \exp\left(-\frac{m_i}{2k_B T} v^2\right) dv \tag{8.13}$$

The values of the velocities on the atoms can be assigned by treating them as independent *Gaussian random variables* drawn from the distribution defined in equation (8.13) which has a mean value of zero and a standard deviation of $\sqrt{k_B T/m_i}$. If this is done the temperature of the system will not be exactly T because the values are assigned randomly but it is easy to scale the velocities obtained in this way uniformly so that the *instantaneous temperature* of the system does correspond to the value desired. There is a well-known result from statistical thermodynamics that relates the average of the kinetic energy of the system to the temperature. It is

$$T = \frac{2}{N_{df} k_B} \langle \mathcal{K} \rangle \tag{8.14}$$

The average in this equation is a thermodynamic, *ensemble* average that must be done over all the configurations that are accessible to the system. An instantaneous temperature, $\mathcal{T}$, can be defined using the same equation but by removing the average. Thus

$$\mathcal{T} = \frac{2}{N_{df} k_B} \mathcal{K} \tag{8.15}$$

This expression allows the instantaneous temperature to be defined exactly once the initial velocity values have been chosen.

The quantity N_{df} in equations (8.14) and (8.15) is the number of degrees of freedom accessible to the system. As we saw in section 7.3, the number of internal degrees of freedom that a molecule has is $3N - 6$ ($3N - 5$ if the molecule is linear). Classically, as can be deduced from equation (8.14), each degree of freedom contributes on average $k_B T/2$ to the kinetic energy. The remaining degrees of freedom, six for non-linear and five for linear molecules, correspond to the overall rotational and translational degrees of freedom for the system. These will contribute to the kinetic energy – the translational motion giving an overall momentum to the system and the rotational motion an angular momentum. When assigning velocities, it is possible not only to scale the velocities such that the correct instantaneous temperature is obtained, but also to ensure that the overall translational and rotational motions are removed. If this is done, the number of degrees of freedom used in the equation for the temperature should be the one with these degrees of freedom removed. If the overall translational and rotational motions are left in, the number of degrees of freedom should be $3N$. In any case, for large systems the difference between the two will be small.

During a simulation it is often useful to be able to control the temperature of the system. This is particularly so during the initial stages of a simulation study and can be done straightforwardly by scaling the velocities to obtain the required instantaneous temperature after they have been calculated in step 4c of the molecular dynamics scheme outlined above. This scaling procedure is simple, although not entirely rigorous, and more sophisticated temperature control schemes that perturb the dynamics of the system less will be discussed in a later chapter.

There are two modules that are provided to perform molecular dynamics simulations using the velocity Verlet algorithm. The first does the integration and it has the following definition:

```
MODULE VELOCITY_VERLET_DYNAMICS

CONTAINS

   SUBROUTINE DYNAMICS ( DELTA, NSTEP, NPRINT, ASSIGN_VELOCITIES, &
                         INITIAL_TEMPERATURE, TARGET_TEMPERATURE, &
                                SCALE_FREQUENCY, SCALE_OPTION, &
                                          SAVE_FREQUENCY, &
                               COORDINATE_FILE, VELOCITY_FILE )
      INTEGER, INTENT(IN) :: NPRINT, NSTEP
      REAL,    INTENT(IN) :: DELTA
      CHARACTER ( LEN = * ), INTENT(IN), OPTIONAL :: COORDINATE_FILE, &
```

```
                                        SCALE_OPTION, &
                                        VELOCITY_FILE
         INTEGER, INTENT(IN), OPTIONAL :: SCALE_FREQUENCY, SAVE_FREQUENCY
         LOGICAL, INTENT(IN), OPTIONAL :: ASSIGN_VELOCITIES
         REAL,    INTENT(IN), OPTIONAL :: INITIAL_TEMPERATURE, &
                                        TARGET_TEMPERATURE
      END SUBROUTINE DYNAMICS

   END MODULE VELOCITY_VERLET_DYNAMICS
```

There is a single subroutine, DYNAMICS, with three essential and a number of optional arguments. The three essential arguments are DELTA, which gives the value of the timestep in picoseconds (1 ps is equivalent to 10^{-12} s or 1000 fs), NSTEP, which gives the number of steps for which to run the simulation, and NPRINT, which gives the step frequency at which information about the dynamics simulation should be printed.

The optional arguments concern two aspects of the functioning of the dynamics subroutine – the handling of the velocities and the saving of data. There are five arguments concerning the handling of velocities. The first is a logical argument, ASSIGN_VELOCITIES, which tells the subroutine whether to assign new velocities to the atoms. If ASSIGN_VELOCITIES is .TRUE. then the subroutine assigns values to the velocities of the atoms using the value of the temperature given by the argument INITIAL_TEMPERATURE. If this argument is not present in this case there is an error. If ASSIGN_VELOCITIES is .FALSE. then the velocities that already exist in the module VELOCITY are used (see below). These could have been created, for example, by a previous call to the subroutine DYNAMICS or by reading them in from an external file. If new velocities are not to be assigned and if the argument INITIAL_TEMPERATURE is present, the existing velocities will be scaled so that they have the temperature given by the value of that argument. If no initial temperature is specified, no scaling will be done. Note that, if ASSIGN_VELOCITIES is .FALSE. and there are no existing velocities, an error will occur.

The remaining three velocity-handling arguments detail how the velocities are to be scaled during the molecular dynamics simulation. The argument SCALE_FREQUENCY specifies the frequency at which the velocities are to be scaled. If it is zero (the default value if no argument is present) then no scaling is done and the other two arguments, TARGET_TEMPERATURE and SCALE_OPTION, are ignored. If scaling is to be done then a scaling option must be chosen. The various options for the argument SCALE_OPTION are a blank string, which is the same as no scaling, ''CONSTANT'', ''EXPONENTIAL'' and ''LINEAR''. If ''CONSTANT'' is requested then the temperature of the system is kept at its initial value throughout the simulation (and the argument TARGET_TEMPERATURE is ignored). If ''EXPONENTIAL'' or ''LINEAR'' is specified then the argument TARGET_TEMPERATURE is needed and the scaling of the velocities is done such that the temperature exponentially or linearly tends to the target temperature as the simulation proceeds.

There are three arguments that concern the saving of data. The first is the integer variable, SAVE_FREQUENCY, which gives the frequency at which sets of coordinates or velocities are to be saved. The remaining two arguments, COORDINATE_FILE and VELOCITY_FILE, identify the names of the trajectory files to which the coordinates or the velocities are to be written. Note that no data will be written if SAVE_FREQUENCY is not specified even if the file name arguments are given. If SAVE_FREQUENCY is given, but the file names are omitted, then an error will occur. Note too that the only means of analysing coordinate and velocity data from a simulation using the subroutine DYNAMICS is by storing intermediate information in a trajectory and then performing the analysis separately afterwards.

The module VELOCITY_VERLET_DYNAMICS has no internally defined variables that it stores between calls to the subroutine DYNAMICS. It does, however, change the values of the coordinates in the array ATMCRD in the module ATOMS and the values of the velocities which are stored in an array in the module VELOCITY which is discussed next. On exit from the dynamics subroutine, these arrays will contain the latest values of the coordinate and velocity data from the dynamics integration.

The second module we discuss here is the one that handles the atom velocities and calculates the temperature and kinetic energy of the system. Most of these functions are taken care of transparently by the subroutine DYNAMICS but there are two subroutines that the user may need to invoke if a dynamics simulation is to be continued in a subsequent program. The two subroutines are

```
MODULE VELOCITY

CONTAINS

   SUBROUTINE VELOCITY_READ ( FILE )
      CHARACTER ( LEN = * ), INTENT(IN) :: FILE
   END SUBROUTINE VELOCITY_READ

   SUBROUTINE VELOCITY_WRITE ( FILE )
      CHARACTER ( LEN = * ), INTENT(IN) :: FILE
   END SUBROUTINE VELOCITY_WRITE

END MODULE VELOCITY
```

The subroutines are self-explanatory. The subroutine VELOCITY_READ reads the velocities from a formatted coordinate file and the subroutine VELOCITY_WRITE writes the velocities to a file. Both subroutines have a single, essential argument, which is the name of the file to be read or written. VELOCITY_WRITE will write the values of the velocities that are currently stored in the VELOCITY module arrays. If no velocities are defined there will be an error. The subroutine VELOCITY_READ will overwrite the velocities currently stored in the module array if the array exists or will create it and fill it otherwise.

To restart a dynamics run, it is necessary first to save the atom coordinates and velocities from the previous run using the COORDINATES_WRITE and VELOCITY_WRITE subroutines, respectively, and then to read in the values using the COORDINATES_READ and VELOCITY_READ subroutines. The next call to the subroutine DYNAMICS should have the ASSIGN_VELOCITIES flag set to .FALSE., so that the velocities that have just been read in are not overwritten, and the argument INITIAL_TEMPERATURE should be missing.

There is a third subroutine from the module that may occasionally be useful. It is the subroutine VELOCITY_ASSIGN that assigns the velocities to the atoms from a Maxwell–Boltzmann distribution. As discussed above, it is possible to ensure that the center of mass translation and rotation of a system after assignment of the velocities are zero. By default, VELOCITY_ASSIGN removes the center of mass motion and defines the number of degrees of freedom for the temperature calculations to be $3N - 3$. It is possible to change these defaults by calling the subroutine directly. Its definition is

```
MODULE VELOCITY

CONTAINS

    SUBROUTINE VELOCITY_ASSIGN ( TNEEDED, DEGREES_OF_FREEDOM, REMOVE )
        REAL,    INTENT(IN)              :: TNEEDED
        INTEGER, INTENT(IN), OPTIONAL :: DEGREES_OF_FREEDOM
        LOGICAL, INTENT(IN), OPTIONAL :: REMOVE
    END SUBROUTINE VELOCITY_ASSIGN

END MODULE VELOCITY
```

There are three arguments. The first, TNEEDED, is an essential real argument that gives the temperature at which the velocities are to be assigned. The second, DEGREES_OF_FREEDOM, is an optional integer argument that sets the number of degrees of freedom for the temperature calculations while the third, REMOVE, is an optional logical argument that indicates whether the center of mass motion of the system is to be removed after the velocity assignment. To manipulate manually the number of degrees of freedom in this fashion,

VELOCITY_ASSIGN should be called before a dynamics simulation with the appropriate values for DEGREES_OF_FREEDOM and REMOVE. Then, when the subroutine DYNAMICS is called, the ASSIGN_VELOCITIES flag should be set to .FALSE.. It should be noted that, although the subroutine VELOCITY_ASSIGN can remove translational motion, it does not, at least in the current implementation, perform the similar operation for the rotational motion.

8.3 Example 12

In this section we illustrate the use of the DYNAMICS subroutine by performing a short simulation on the blocked alanine molecule. The program is

```
PROGRAM EXAMPLE12

... Declaration statements ...

! . Read in the system file.
CALL MM_SYSTEM_READ ( ''bALA.sys_bin'' )

! . Read in a first set of coordinates.
CALL COORDINATES_READ ( ''bALA1.crd'' )

! . Perform a coordinate minimization.
CALL OPTIMIZE_CONJUGATE_GRADIENT ( GRADIENT_TOLERANCE = 0.1, &
                                   PRINT_FREQUENCY = 50,  &
                                   STEP_NUMBER = 1000 )

! . Initialize the random number seed.
CALL RANDOM_INITIALIZE ( 314159 )

! . Heat the system to 300K from 10K.
CALL DYNAMICS ( 0.001, 1000, 100, INITIAL_TEMPERATURE = 10.0, &
                                   TARGET_TEMPERATURE  = 300.0, &
                                   SCALE_FREQUENCY     = 100,   &
                                   SCALE_OPTION        = ''LINEAR'' )

! . Equilibrate the system at 300K (reassign the velocities).
CALL DYNAMICS ( 0.001, 5000, 500, INITIAL_TEMPERATURE = 300.0, &
                                   SCALE_FREQUENCY     = 100,   &
                                   SCALE_OPTION        = ''CONSTANT'' )

! . Run dynamics at about 300K (keep the old velocities).
CALL DYNAMICS ( 0.001, 10000, 500, ASSIGN_VELOCITIES = .FALSE., &
                                   SAVE_FREQUENCY    = 100,    &
                                   COORDINATE_FILE   = ''bALA1.traj'' )

END PROGRAM EXAMPLE12
```

The program starts off in the usual way by reading in the system definition file for the molecule and a starting set of coordinates. Both of these must have been defined if the subsequent commands are to work. The next step involves optimizing the coordinates so that the RMS gradient for the system is not too high. This is standard procedure before starting a molecular dynamics simulation study because the integration algorithm can become unstable if the forces on some of the atoms are large due to strain in the molecule or unfavourable non-bonding contacts.

The next command is a call to initialize the random number module with a particular random number seed, in this case 314159. The random number module is used by the subroutine VELOCITY_ASSIGN when determining values for the velocities of each atom. It is a good idea to call RANDOM_INITIALIZE explicitly with a particular value of the random number seed because the sequence of random numbers generated is then reproducible. In other words, the use of a particular random number seed will generate a particular sequence of random numbers and, hence, a particular set of velocities.

The next three commands are calls to the subroutine DYNAMICS. The time-step for each call is 1 fs (10^{-3} ps). Each call does a different phase of a dynamics calculation. In the first call, the *heating phase*, the temperature of the system is increased from an initial temperature of 10 K to a final temperature of 300 K. In the second call, the *equilibration phase*, the temperature of the system is maintained constant at 300 K. In the third call, the *data-collection phase*, no temperature modification is performed. The heating and equilibration phases of the dynamics are done to prepare the system for the data-collection phase and are necessary to ensure that the kinetic energy in the system is partitioned roughly equally between all the available degrees of freedom. For a small system, such as blocked alanine, a heating period of 1 ps and then an equilibration period of 5 ps (1000 and 5000 steps, respectively) are probably adequate. For larger systems longer periods will be necessary. Note that the scaling in the heating phase is done at 100-step intervals and is performed so that the temperature of the system increases linearly between 10 and 300 K. In the equilibration phase the velocities for the atoms are reassigned initially at a temperature of 300 K – the velocities from the heating phase are not kept – and again at 100-step intervals after that.

For the data-collection stage, 10 ps of dynamics is performed and the coordinates for the molecule are saved at 100-step intervals (i.e. every 0.1 ps). The velocities from the last step of the equilibration phase of the dynamics are used to start the dynamics in the data-collection run. These will automatically have a temperature of 300 K. No velocity modification is done during this phase.

In all three calls information about the dynamics is printed out at reasonable intervals just to check that there are no anomalies in the integration. The information consists of the total energy, the kinetic and potential energies and the temperature of the system. In addition, at the end of the simulation the averages and the RMS deviations of these quantities from their averages for the complete run are printed. These values are especially useful when no velocity modification has been done because the average total energy and its RMS deviation will give an indication of how well the energy was conserved during the simulation.

The values of the energies and temperature from the data-collection phase of the simulation are plotted in figures 8.1 and 8.2. The total energy is reasonably well conserved with an RMS deviation for the entire simulation of about $0.26\,\text{kJ mol}^{-1}$. Owing to the fact that the total energy is conserved, there is a constant transfer of energy between the kinetic and the potential degrees of freedom. These can be seen to be large because the temperature varies between a value as low as $240\,\text{K}$ and a value as high as $430\,\text{K}$ about the average value which, here, is about $320\,\text{K}$.

8.4 Trajectory analysis

We have seen how to generate trajectories of coordinate or velocity data for a system from a molecular dynamics simulation. The point of performing a simulation is, of course, that we want to use the data to calculate properties of the system that can be related to those observable experimentally. In this

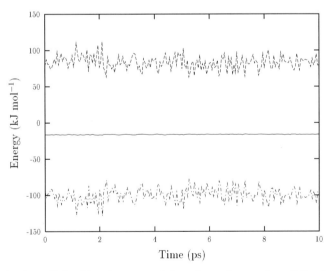

Figure 8.1. A plot of the kinetic, potential and total energies of the bALA system during the data-collection phase of the dynamics of Example 12. Kinetic energy, dashed line; potential energy, dash–dot line; total energy, solid line.

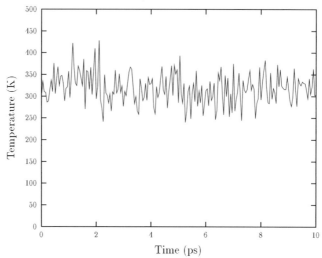

Figure 8.2. A plot of the temperature of the bALA system during the data-collection phase of the dynamics of Example 12.

section only some very simple analyses of trajectory data are described. More advanced techniques will be left to section 10.2.

At the simplest level, the analysis of a dynamics trajectory consists of calculating a property for each *frame* of the trajectory and seeing how it changes as a function of time. The sequence of data created in this way is called a *time series* for the property. Direct inspection of time-series data could be useful, for example, to chemists wanting a qualitative view of the change in the structural or other properties of a system. Usually, however, a more rigorous analysis needs to be undertaken if meaningful conclusions are to be extracted.

Here we consider two of the most useful statistical quantities that can be calculated from simulation data. These are *averages* and *fluctuations*. If $\mathcal{X}$ is the property under consideration, $\mathcal{X}_n$ is the nth value of the property in the time series and n_t is the total number of elements in the series, the average of the property is

$$\langle \mathcal{X} \rangle = \frac{1}{n_t} \sum_{n=1}^{n_t} \mathcal{X}_n \tag{8.16}$$

and the fluctuation is

$$\langle (\delta \mathcal{X})^2 \rangle = \langle (\mathcal{X} - \langle \mathcal{X} \rangle)^2 \rangle \tag{8.17}$$

$$= \langle \mathcal{X}^2 \rangle - \langle \mathcal{X} \rangle^2 \tag{8.18}$$

The importance of these two types of function is that, when they are calculated for specific properties, they can be related to experimentally observable quantities for the system. The formulae which link the two – the microscopic calculated data and the macroscopic observable data – are all derivable from statistical thermodynamics. We shall discuss this aspect in more detail in a later chapter, but we have already met one such relation in equation (8.14), which equates the temperature of the system to the average of its kinetic energy.

The sole module that we introduce in this section is one that allows the data to be extracted from a trajectory file. This is a module that we have encountered indirectly before because it is used whenever a trajectory file is read or written by another module in the program library. No module is introduced at this stage to perform statistical analyses of time-series data because the calculation of averages and fluctuations is easily expressed with the powerful array syntax of FORTRAN 90.

The module for reading and writing trajectory files is TRAJECTORY_IO. It has the following definition:

```
MODULE TRAJECTORY_IO

! . Trajectory type definitions.
TYPE TRAJECTORY_TYPE
   CHARACTER ( LEN = 32 ) :: HEADER
   INTEGER                :: NATOMS, NFRAMES, POSITION, UNIT
   LOGICAL                :: QWRITE
   REAL                   :: STEP
END TYPE TRAJECTORY_TYPE

CONTAINS

   SUBROUTINE TRAJECTORY_ACTIVATE_READ ( FILE, TRAJECTORY )
      CHARACTER ( LEN = * ), INTENT(IN)    :: FILE
      TYPE(TRAJECTORY_TYPE), INTENT(INOUT) :: TRAJECTORY
   END SUBROUTINE TRAJECTORY_ACTIVATE_READ

   SUBROUTINE TRAJECTORY_ACTIVATE_WRITE ( FILE, TRAJECTORY, HEADER, &
                                          NATOMS, NFRAMES, STEP )
      CHARACTER ( LEN = * ), INTENT(IN)    :: FILE, HEADER
      INTEGER,               INTENT(IN)    :: NATOMS, NFRAMES
      REAL,                  INTENT(IN)    :: STEP
      TYPE(TRAJECTORY_TYPE), INTENT(INOUT) :: TRAJECTORY
   END SUBROUTINE TRAJECTORY_ACTIVATE_WRITE

   SUBROUTINE TRAJECTORY_DEACTIVATE ( TRAJECTORY )
      TYPE(TRAJECTORY_TYPE), INTENT(INOUT) :: TRAJECTORY
   END SUBROUTINE TRAJECTORY_DEACTIVATE
```

```
     SUBROUTINE TRAJECTORY_INITIALIZE ( TRAJECTORY )
        TYPE(TRAJECTORY_TYPE), INTENT(OUT) :: TRAJECTORY
     END SUBROUTINE TRAJECTORY_INITIALIZE

     SUBROUTINE TRAJECTORY_READ ( TRAJECTORY, ATMDAT )
        TYPE(TRAJECTORY_TYPE), INTENT(INOUT) :: TRAJECTORY
        REAL, DIMENSION(:,:),  INTENT(OUT)   :: ATMDAT
     END SUBROUTINE TRAJECTORY_READ

     SUBROUTINE TRAJECTORY_WRITE ( TRAJECTORY, ATMDAT )
        TYPE(TRAJECTORY_TYPE), INTENT(INOUT) :: TRAJECTORY
        REAL, DIMENSION(:,:),  INTENT(IN)    :: ATMDAT
     END SUBROUTINE TRAJECTORY_WRITE


  END MODULE TRAJECTORY_IO
```

This module is relatively complicated in that it contains a definition of a new data type in addition to six subroutines. The type, TRAJECTORY_TYPE, defines a group of scalar variables that gives information about the trajectory. A separate TRAJECTORY_TYPE variable *must* be defined for each trajectory that is being used. The fields in the type are as follows.

HEADER gives a character string that identifies the type of data in the trajectory. Possible examples are ''COORDINATE'', ''VELOCITY'' and ''NORMAL MODE''.

NATOMS gives the number of atoms in each data set.

NFRAMES gives the number of data sets or frames in the trajectory.

POSITION contains the number of the frame at which the trajectory file is currently positioned. The trajectory file is defined as an unformatted sequential file and so it is not possible to access the data within the file randomly – they can only be accessed sequentially from the beginning.

UNIT is the FORTRAN unit number associated with the trajectory.

QWRITE is .TRUE. for a file that is being written to and .FALSE. for one that is being read from.

STEP is a parameter whose meaning will differ depending upon the type of trajectory file. In a dynamics trajectory, STEP refers to the difference in time between each data set, whereas in a reaction path trajectory file it will be the difference between the reaction coordinates of consecutive structures.

It is to be noted that these fields should never be altered explicitly but only through the use of the subroutines in the module TRAJECTORY_IO. The only time they will usually need to be accessed directly is after a call to TRAJECTORY_ACTIVATE_READ in order to ascertain, for example, the number of frames in the file or the number of atoms in each frame.

Once a trajectory type has been defined for a trajectory it is first initialized using the subroutine TRAJECTORY_INITIALIZE which takes the trajectory type variable for the trajectory as its sole argument. After initialization the trajectory needs to be activated using the subroutines TRAJECTORY_ACTIVATE_READ for a trajectory that is being read or TRAJECTORY_ACTIVATE_WRITE for a trajectory that is being written. TRAJECTORY_ACTIVATE_WRITE takes six arguments, all of which are essential. The first, FILE, gives the name of the trajectory file. The second, TRAJECTORY, is the trajectory type variable for the trajectory. The remaining arguments are input arguments that are used to fill the various fields in the trajectory type. TRAJECTORY_ACTIVATE_READ takes only two arguments, the name of the trajectory file to be read in FILE and its associated trajectory type variable in TRAJECTORY. Both of the activation subroutines will give errors if the trajectory type variable has not been initialized beforehand.

Data are read from and written to a trajectory file using the subroutines TRAJECTORY_READ and TRAJECTORY_WRITE once the trajectory has been activated. Both subroutines take two essential arguments, the first of which is the trajectory type variable, TRAJECTORY, and the second an array, ATMDAT. For TRAJECTORY_READ, ATMDAT is an array to which the atom data for the frame are transferred once they have been read, whereas for TRAJECTORY_WRITE it is an array that contains the data which are to be written to the trajectory file. ATMDAT in both cases is a two-dimensional array with a first dimension of 3 and a second dimension equal to the number of atoms, i.e. ATMDAT(1:3,1:NATOMS). If the dimensions are wrong there will be an error. An error will also occur if an attempt to read from a trajectory activated for writing or to write to one activated for reading is made. Note too that an error will occur if an attempt to read or write past the last data set in the file is made.

The last subroutine in the module is TRAJECTORY_DEACTIVATE, which deactivates a trajectory once it is no longer needed. There is a single argument, which is the trajectory type variable. A trajectory activated for reading can be deactivated at any time, but a trajectory activated for writing can be deactivated only when it contains the correct number of data sets, otherwise an error occurs.

8.5 Example 13

In this section we illustrate the analysis of a molecular dynamics trajectory using the trajectory file that was generated in the previous example. The program is

```
PROGRAM EXAMPLE13

... Declaration statements ...

! . Local scalars.
INTEGER :: I, NFRAMES
REAL :: AVERAGE, RMSDEV

! . Local arrays.
INTEGER,          DIMENSION(1:4,1:2) :: LIST
REAL, ALLOCATABLE, DIMENSION(:,:)      :: VALUES

! . Local types.
TYPE(TRAJECTORY_TYPE) :: TRAJECTORY

! . Read in the system file.
CALL MM_SYSTEM_READ ( ''bALA.sys_bin'' )

! . Initialize the dihedral list array.
LIST(1,1) = 5 ; LIST(2,1) = 7 ; LIST(3,1) =  9 ; LIST(4,1) = 15
LIST(1,2) = 7 ; LIST(2,2) = 9 ; LIST(3,2) = 15 ; LIST(4,2) = 17

! . Activate the trajectory.
CALL TRAJECTORY_INITIALIZE ( TRAJECTORY )
CALL TRAJECTORY_ACTIVATE_READ ( ''bALA1.traj'', TRAJECTORY )

! . Get the number of frames in the trajectory.
NFRAMES = TRAJECTORY%NFRAMES

! . Allocate some temporary space.
ALLOCATE ( VALUES(1:2,1:NFRAMES) )

! . Loop over the frames in the trajectory.
DO I = 1,NFRAMES

   ! . Read in the frame.
   CALL TRAJECTORY_READ ( TRAJECTORY, ATMCRD )

   ! . Calculate the phi and psi dihedral angles for the structure.
   VALUES(1:2,I) = GEOMETRY_DIHEDRALS ( ATMCRD, LIST )

END DO

! . Deactivate the trajectory.
CALL TRAJECTORY_DEACTIVATE ( TRAJECTORY )

! . Write out the calculated dihedrals.
WRITE ( OUTPUT, ''(/A)'' ) ''Calculated Dihedral Angles:''
WRITE ( OUTPUT, ''(/5X,A,2(7X,A))'' ) ''Frame'', ''Phi'', ''Psi''
WRITE ( OUTPUT, ''(I10,2F10.2)'' ) ( I, VALUES(1:2,I), I = 1,NFRAMES )
```

```
! . Write out the averages and RMS deviations for each dihedral.
AVERAGE =       SUM ( VALUES(1,1:NFRAMES) )     / REAL ( NFRAMES )
RMSDEV = SQRT ( SUM ( VALUES(1,1:NFRAMES)**2 ) / REAL ( NFRAMES ) - &
                                                AVERAGE * AVERAGE )
WRITE ( OUTPUT, ''(/A,F10.2)'' ) ''Phi: Average   = '', AVERAGE
WRITE ( OUTPUT, ''(A,F10.2)'' ) '' RMS Deviation   = '', RMSDEV

AVERAGE =       SUM ( VALUES(2,1:NFRAMES) )     / REAL ( NFRAMES )
RMSDEV  = SQRT ( SUM ( VALUES(2,1:NFRAMES)**2 ) / REAL ( NFRAMES ) - &
                                                AVERAGE * AVERAGE )
WRITE ( OUTPUT, ''(/A,F10.2)'' ) ''Psi: Average      = '', AVERAGE
WRITE ( OUTPUT, ''(A,F10.2)'' )  ''     RMS Deviation = '', RMSDEV

! . Deallocate the temporary space.
DEALLOCATE ( VALUES )


END PROGRAM EXAMPLE13
```

This program is more complicated than the others that we have discussed up to now but it illustrates a relatively simple analysis. The program calculates the values of two of the dihedral angles, ϕ and ψ, in the blocked alanine molecule for each frame of the molecular dynamics trajectory.

The first action in the program is to read in the system file for the molecule. Note that this is the only data file that is needed by the program apart from the trajectory that is read later. Once the system's composition has been specified, the next lines fill the array LIST with the sequence numbers of the atoms in the dihedrals that are to be calculated. This array has the same format as that required by the subroutine GEOMETRY_DIHEDRALS which is called later. It is defined with the correct dimensions at the start of the program because the number of dihedrals to be calculated is known initially. The first dihedral in the list defines the angle ϕ and the second dihedral defines the angle ψ.

The next command initializes the trajectory type variable, TRAJECTORY, that was declared at the start of the program. The trajectory that was produced in Example 12 is then activated for reading. In the next statement, the number of frames in the trajectory is obtained by interrogating the field NFRAMES in the variable TRAJECTORY. Once the number of frames in the trajectory is known, the array that will contain the calculated values of the dihedrals is allocated to have the correct dimensions and format required by the subroutine GEOMETRY_DIHEDRALS.

After the array allocation there is a loop within which the individual frames in the trajectory are read and the corresponding dihedrals calculated. Each iteration of the loop contains two commands, the first that reads the frame

coordinates into the array ATMCRD and the second that calculates the dihedrals. On exit from the loop the trajectory is immediately deactivated.

The program finishes by printing out the values of each dihedral for all the
frames and then printing the average and the *RMS deviation*, which is the
square root of the fluctuation, for each dihedral. Both quantities require only
single-line commands for them to be calculated. Finally the array VALUES is
explicitly deallocated.

The results of this program for the dihedrals are plotted in figure 8.3. Each
point in the plot represents a single point along the trajectory and gives the
values of both dihedrals. In this example the range of angles sampled is small
because the trajectory is limited to one small region of ϕ–ψ space.

In figure 8.4 the RMS coordinate deviations between the starting structure
for bALA and the subsequent structures in the trajectory are shown. Each
structure has been oriented using the methods discussed in section 3.6 so as to
minimize the value of the RMS deviation. The average value of the coordinate deviation is about 0.4 Å but there are deviations from this average of up
to 0.25 Å along the trajectory.

As a final point in this section we note that the same framework for
analysis as that we have used here can be used to analyse velocity trajectories
and also coordinate trajectories generated in other applications. We have
already met two of these, for normal modes (section 7.4) and reaction
paths (sections 6.7 and 6.9).

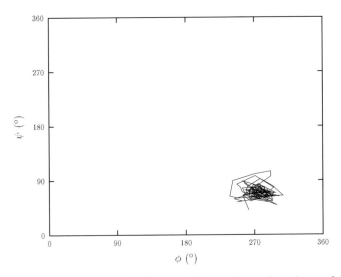

Figure 8.3. A plot of the values of the ϕ and ψ angles as functions of time for the
simulation of bALA of Example 12.

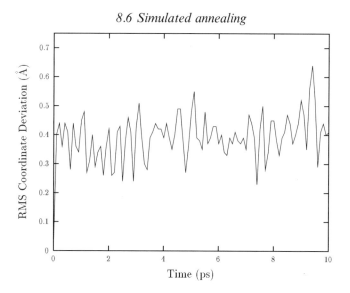

Figure 8.4. A plot of the RMS coordinate deviations between the starting and the subsequent bALA structures along the trajectory generated in the dynamics simulation of Example 12.

8.6 Simulated annealing

We leave for a moment the use of molecular dynamics simulation as a tool for the investigation of the equilibrium and dynamical properties of a system and discuss its application in another context. In chapter 6, algorithms for exploring the potential energy surface of a system were discussed. These algorithms had in common the fact that they were local and searched the region of the surface in the neighbourhood of the starting configuration. Such procedures are useful in many cases but in other applications knowledge of the global minimum or near-global minima is required, for which local search algorithms are inappropriate.

Because of their practical importance, global optimization algorithms have been the subject of intense research and several global search strategies have been developed. One of the earliest and still one of the most useful is the method of *simulated annealing*, which was introduced by S. Kirkpatrick, C. D. Gelatt and M. P. Vecchi during the early 1980s. This method is based upon the correspondence between a statistical mechanical system and an optimization problem in which a minimum of a function that depends on many parameters is to be found.

The essential idea behind simulated annealing and the fact that distinguishes it from local optimization algorithms is the consideration of the temperature of the system. When a system has a non-zero temperature its total energy is no longer just the potential energy because there is a kinetic energy component too. An image of this is shown in figure 8.5, in which the addition of the temperature 'lifts' the system off the potential energy surface and makes a much larger number of configurations accessible. The higher the temperature the larger the number of configurations that are accessible because the system has energy to surmount larger barriers.

A simulated annealing calculation proceeds by giving the system a high temperature, allowing it to equilibrate and then cooling until the system has been *annealed* to the potential energy surface, i.e. until the temperature is zero. The way in which the cooling is done, the *cooling schedule*, determines the effectiveness of the simulated annealing method. In general, the cooling needs to be done slowly so that the system can thoroughly explore the potential energy surface and avoid becoming trapped in regions of high potential energy. Of course, there is no guarantee that this will not happen, but it is known from statistical mechanics that the probability that a particular configuration will be favoured is proportional to its Boltzmann factor, $\exp[-\mathcal{V}/(k_\mathrm{B}T)]$, where $\mathcal{V}$ is the potential energy of the configuration. Thus,

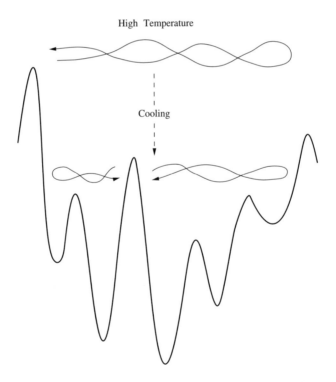

Figure 8.5. A schematic diagram of a simulated annealing calculation on a potential energy surface.

the lower the potential energy of the configuration the more probable it is. The optimal cooling schedule cannot be found in most problems of interest and so the choice of schedule is to some extent a matter of experimentation. Fortunately, however, even relatively crude schedules can give good results.

Molecular dynamics methods, because they employ a temperature, can be used for simulated annealing optimization calculations and it is this approach that we shall illustrate in the next example. It should be noted, though, that Monte Carlo algorithms, which are discussed later, are equally viable for performing simulated annealing calculations. In fact, it was with these that the first simulated annealing applications were performed.

8.7 Example 14

As an example of a simulated annealing calculation we consider calculations on a Lennard-Jones cluster of size 13. There are two reasons for this. First, it is relatively easy to obtain the global minimum for the bALA molecule using local minimization techniques (try it and see!) so this does not represent a particularly interesting test case. Second, the Lennard-Jones clusters, as mentioned in section 6.2, represent something of a benchmark for the evaluation of global optimization methods for molecular systems.

The program is

```
PROGRAM EXAMPLE14

... Declaration statements ...

! . Local parameters.
INTEGER, PARAMETER :: NTRIES = 100, SEED0 = 314159

! . Local scalars.
INTEGER :: ITRY, TMPBND
REAL :: TSTART

! . Local arrays.
REAL, DIMENSION(1:NTRIES) :: PE0, PE1, PE2, PE3

! . Local arrays.
REAL, ALLOCATABLE, DIMENSION(:,:) :: TMPCRD

! . Process the MM file for the cluster.
CALL MM_FILE_PROCESS ( ''LJ.opls_bin'', ''LJ.opls'' )

! . Construct the system file for the cluster.
CALL MM_SYSTEM_CONSTRUCT ( ''LJ.opls_bin'', ''LJ13.seq'' )

! . Add constraints.
```

```
CALL ADD_CONSTRAINTS

! . Allocate the temporary coordinate array.
ALLOCATE ( TMPCRD(1:3,1:NATOMS) )

! . Loop over the number of attempts.
DO ITRY = 1,NTRIES

   ! . Set the number of constraints.
   NBONDS = TMPBND

   ! . Generate the coordinates for the cluster atoms.
   CALL STRUCTURE

   ! . Initialize the coordinates.
   ATMCRD = TMPCRD

   ! . Do a short dynamics at TSTART to relieve bad contacts
   ! . between atoms.
   CALL DYNAMICS ( 0.001, 100, 0, INITIAL_TEMPERATURE = TSTART, &
                   SCALE_FREQUENCY = 1, SCALE_OPTION = ''CONSTANT'' )

   ! . Calculate the starting energy.
   CALL ENERGY ; PE0(ITRY) = TOTAL_PE

   ! . Save the coordinates.
   TMPCRD = ATMCRD

   ! . Directly minimize the coordinates (with constraints).
   CALL OPTIMIZE_CONJUGATE_GRADIENT ( GRADIENT_TOLERANCE = 1.0E-4, &
                                      PRINT_FREQUENCY    = 0,      &
                                      STEP_NUMBER        = 10000 )

   ! . Remove the constraints and reminimize.
   NBONDS = 0
   CALL OPTIMIZE_CONJUGATE_GRADIENT ( GRADIENT_TOLERANCE = 1.0E-4, &
                                      PRINT_FREQUENCY    = 0,      &
                                      STEP_NUMBER        = 10000 )
   CALL ENERGY ; PE1(ITRY) = TOTAL_PE

   ! . Re-introduce the constraints.
   NBONDS = TMPBND

   ! . Reset the coordinates.
   ATMCRD = TMPCRD

   ! . Cool the system from TSTART to TSTOP.
   CALL DYNAMICS ( 0.001, 40000, 0, INITIAL_TEMPERATURE = TSTART,              &
                                    TARGET_TEMPERATURE  = TSTART * EXP ( -10.0 ), &
                                    SCALE_FREQUENCY     = 10,                  &
                                    SCALE_OPTION        = ''EXPONENTIAL'' )
```

```
   ! . Remove the constraints and reminimize.
   NBONDS = 0
   CALL ENERGY ; PE2(ITRY) = TOTAL_PE
   CALL OPTIMIZE_CONJUGATE_GRADIENT ( GRADIENT_TOLERANCE = 1.0E-4, &
                                      PRINT_FREQUENCY    = 0,      &
                                      STEP_NUMBER        = 10000 )
   CALL ENERGY ; PE3(ITRY) = TOTAL_PE

END DO

! . Deallocate the temporary space.
DEALLOCATE ( TMPCRD )

! . Write out the results.
CALL WRITE_RESULTS

!===============================================================================
CONTAINS
!===============================================================================

   !------------------------
   SUBROUTINE ADD_CONSTRAINTS
   !------------------------

   ! . Local scalars.
   INTEGER :: I, II, J

   ! . Calculate the number of bond constraints.
   NBONDS = ( NATOMS * ( NATOMS - 1 ) ) / 2

   ! . Deallocate the bond array.
   IF ( ALLOCATED ( BONDS ) ) DEALLOCATE ( BONDS )

   ! . Allocate the bond array.
   ALLOCATE ( BONDS(1:NBONDS) )

   ! . Fill the atom bond indices.
   II = 0
   DO I = 1,(NATOMS-1)
        DO J = (I+1),NATOMS
        II = II + 1
        BONDS(II)%I = I
        BONDS(II)%J = J
     END DO
   END DO

   ! . Fill the remaining bond arrays.
   BONDS(1:NBONDS)%EQ = 2.0
   BONDS(1:NBONDS)%FC = 0.2

   ! . Store the number of bonds.
```

```
TMPBND = NBONDS

END SUBROUTINE ADD_CONSTRAINTS

!--------------------
SUBROUTINE STRUCTURE
!--------------------

! . Local scalars.
INTEGER :: I, J
REAL :: SIDE

! . Initialize the random number seed.
CALL RANDOM_INITIALIZE ( SEED0 + ITRY )

! . Get the side of the box.
SIDE = 5.0 * ( RANDOM ( ) + 1.0 )

! . Loop over the atoms.
DO I = 1,NATOMS

   10 CONTINUE

   ! . Choose coordinates for the atoms.
   DO J = 1,3
      TMPCRD(J,I) = SIDE * RANDOM ( )
   END DO

   ! . Verify that the coordinates are not too close to any others.
   DO J = 1,(I-1)
      IF ( SUM ( ( TMPCRD(1:3,I) - TMPCRD(1:3,J) )**2 ) < 6.25 ) GO TO 10
   END DO

END DO

! . Generate the starting and stopping temperatures.
TSTART = 300.0 * ( RANDOM ( ) + 1.0 )

END SUBROUTINE STRUCTURE

!-----------------------
SUBROUTINE WRITE_RESULTS
!-----------------------

! . Local scalars.
INTEGER :: NMIN1, NMIN3
REAL    :: EMIN1, EMIN3

! . Find the lowest energies and their frequencies.
EMIN1 = MINVAL ( PE1 ) ; NMIN1 = COUNT ( ABS ( PE1 - EMIN1 ) < 1.0E-2 )
EMIN3 = MINVAL ( PE3 ) ; NMIN3 = COUNT ( ABS ( PE3 - EMIN3 ) < 1.0E-2 )
```

```
!  . Write out the header.
WRITE ( OUTPUT, ''(/A)'' ) ''Optimization Results:''
WRITE ( OUTPUT, ''(/3X,A,6X,A,4X,A,5X,A,8X,A)'' ) &
                ''Attempt'', ''Initial Energy'', ''Minimized Energy'', &
                ''Annealed Energy'', ''Final Energy''

!  . Write out the results.
DO ITRY = 1,NTRIES
   WRITE ( OUTPUT, ''(I10,4F20.3)'' ) ITRY, PE0(ITRY), PE1(ITRY), &
                                      PE2(ITRY), PE3(ITRY)
END DO

!  . Write out the lowest values.
WRITE ( OUTPUT, ''(/A,F20.3,I6)'' ) &
   ''Lowest minimized energy and frequency = '', EMIN1, NMIN1
WRITE ( OUTPUT, ''(/A,F20.3,I6)'' ) &
   ''Lowest annealed energy and frequency = '', EMIN3, NMIN3

END SUBROUTINE WRITE_RESULTS

END PROGRAM EXAMPLE14
```

This program is the most complex that we have encountered up to now but
it conveys well the nature of a simulated annealing study. First, the system
file is set up for the 13-atom cluster using the standard commands and the
OPLS and sequence files for the cluster. It is to be noted that, when studying
Lennard-Jones clusters, in general, special *reduced units* are used, in which
the Lennard-Jones radius and well depth and the particle mass all have values
of 1. In the programs that we use it is possible to set the Lennard-Jones
parameters to 1 so that the energies calculated will be not in kJ mol^{-1} but
in reduced units. This is not possible for the temperature and time parameters
because there is an automatic conversion internally to the appropriate units
and so we treat the atoms as having the same mass as argon atoms and use
the standard units for the temperature (K) and the time (ps).

After the system file has been set up, a subroutine ADD_CONSTRAINTS is
called, which is defined after the CONTAINS statement. This subroutine adds
weak harmonic bond terms between each atom in the molecule. The equilibrium bond distance and force constant are set to 2 and 0.2 reduced units,
respectively. The bonds are added by explicitly allocating the array BONDS
that is found in the module MM_TERMS. Normally, of course, it is the subroutine MM_SYSTEM_CONSTRUCT that will do this. The reason that weak bonds
are added between the atoms is that they act as constraints to prevent the
atoms from moving too far apart. If there were no constraints it is possible
that some atoms would leave the cluster during the minimization and simulated annealing procedures.

After the setting up of the constraints, a temporary coordinate array is allocated. Then there is a loop, each iteration of which corresponds to a distinct simulated annealing calculation. The first command in the loop is a call to an internal subroutine STRUCTURE that generates values of the coordinates for the atoms in the cluster and a value of the initial temperature. The coordinates are chosen randomly so that all the atoms fit into a box of side 5–10 Å. The box size is itself also chosen randomly. If the coordinates of an atom are such that it is less than $\sqrt{6.25}$ reduced units from an atom whose position has previously been determined, its coordinates are reselected until a hole of the proper size is found. The initial temperature is selected randomly in the range 300–600 K. The other thing to note about the subroutine is that right at the start there is an explicit call to RANDOM_INITIALIZE, which resets the seed for the generation of random numbers. This is done so that the coordinates of the configuration selected can be regenerated in subsequent programs if that is required.

After the generation of the coordinates a short dynamics calculation of 100 steps is done at the starting temperature. The velocities of the atoms are scaled every step. This is done to ensure that any high-energy contacts in the structure are relaxed.

Subsequently, two separate types of calculations are done. The first consists of a straightforward minimization from the starting configuration. The minimization is done with a very low gradient tolerance to ensure that a very-well-defined minimum is reached. A minimization is done first with constraints and then with no constraints (which is achieved by setting the variable NBONDS from the module MM_TERMS to zero). The second calculation consists of a dynamic simulated annealing calculation starting at the initial temperature, TSTART, and finishing at a temperature, TSTOP, calculated as TSTART $\times$ e^{-10}. The simulation is done for 40 ps and the temperature scaling is done at 10 fs intervals so that it decreases exponentially towards the target value. The annealing is done with constraints but it is followed by a minimization without constraints to refine the structure to a true minimum.

The program terminates by deallocating any temporary space and then printing the energies of the clusters at each stage of the calculation. The lowest energy and the number of times it was found using both the local minimization and the global simulated annealing procedures are also written out.

Note that the annealing procedure needs to be repeated many times, for there is no guarantee of finding the global minimum in one attempt. Only after many simulations, in this case 100, can a reasonable guess be made regarding whether the global minimum has been located. It should also be noted that there are many ways of generating the initial structures of the configurations. In the example this was done randomly because of the simplicity of the structure of the Lennard-Jones clusters. For other molecules the structures could be taken, for example, at sufficiently large intervals from a high-temperature molecular dynamics trajectory.

For the 13-atom cluster studied in this example, the global minimum has an energy of -44.327 reduced units (see table 6.1). The minimization procedure found this value 18 times out of 100 simulations whereas the simulated annealing protocol found it 58 times. This is typical for these sizes of clusters for a wide range of different cooling schedules. In addition, the spectrum of energies of the remaining, non-global, minima is significantly lower in value for the annealed structures than it is for the minimized ones.

Exercises

8.1 In Example 12 a timestep of 1 fs was used to integrate the dynamics equations. Repeat the simulations using different timesteps, but for the same total simulation time (i.e. 15 ps) to see how the results change. Is energy conservation substantially better with a shorter timestep? How do the values of the dihedral angles change? Can you think of any better ways of comparing the different trajectories?

8.2 Do several long simulations of bALA starting from different structures and repeat the analysis of the ϕ and ψ angles of section 8.5. How do the ϕ–ψ maps compare with the schematic potential energy surface for bALA of Exercise 6.1?

8.3 The simulated annealing calculations in Example 14 were done with a Lennard-Jones cluster of 13 atoms. Repeat the calculations with other cluster sizes and using different annealing schemes. In particular try the 'magic number' clusters with sizes of 19, 55 and 147. Does the simulated annealing procedure stay as efficient as the size of the cluster increases? One property of interest in cluster studies is the value of the energies of the states which are less stable than the ground state. How does the energy of the second minimum change as a function of the cluster's size?

9

More on non-bonding interactions

9.1 Introduction

Up until now we have considered a variety of standard techniques for the simulation of molecular systems. All the systems we have looked at, however, have been in vacuum and we have not, as yet, considered any extended condensed phase systems such as liquids, solvated molecules or crystals. This is because special techniques are needed to evaluate the non-bonding interactions in such systems. In the present chapter we introduce a number of the simpler methods for determining these interactions which will allow us to treat some condensed phase problems.

9.2 Cutoff methods for the calculation of non-bonding interactions

As we discussed in detail in section 4.4.2, the non-bonding energy for a molecular system for the type of force field that we are using can be written as a sum of electrostatic and Lennard-Jones contributions. The expression for the energy, $\mathcal{V}_{\text{nb}}$, is

$$\mathcal{V}_{\text{nb}} = \sum_{ij \text{ pairs}} \left(\frac{q_i q_j}{4\pi\epsilon_0 \epsilon r_{ij}} + \frac{A_{ij}}{r_{ij}^{12}} - \frac{B_{ij}}{r_{ij}^6} \right) \tag{9.1}$$

The crucial aspect of this equation is that the sum runs over all pairs of interacting atoms in the system, which comprise all possible pairs of atoms except those 1–2, 1–3 and (possibly) 1–4 interactions that are specifically excluded. In all the simulations we have done to date we have used a version of the module ENERGY_NON_BONDING that evaluates the non-bonding energy in the simplest way possible, by calculating the interaction for all pairs of atoms explicitly. Because the number of pairs increases as $O(N^2)$, where N is the number of atoms in the system, the calculation of the non-bonding energy using this technique rapidly becomes unmanageable. In particular, it becomes difficult or impossible to use for condensed phase systems.

To overcome this problem and to increase the efficiency of the non-bonding energy evaluation, a number of different techniques of varying sophistication can be used. They can be broadly divided into two categories – those that attempt to evaluate equation (9.1) exactly (or, at least, to within a certain estimated precision) and those that modify the form of the expression for the non-bonding interaction in some way so that it is more readily evaluated. Here, we shall mostly discuss the latter methods because, although they are less rigorous, they are easily implemented and they have been widely employed for condensed phase simulations. We shall return to the more exact methods at the end of the chapter.

The principal problem with the non-bonding energy is the long range electrostatic interaction which decays as the reciprocal of the distance between the atoms. The long-range nature of the interaction means that many pairs have to be included in the sum of equation (9.1) to obtain a non-bonding energy of a given precision. The most widely used approximate methods for the evaluation of the non-bonding energy overcome the long-range nature of the electrostatic interaction by modifying its form so that the interactions between atoms are zero after some finite distance. These are the *cutoff* or *truncation* methods. The fact that the interactions are truncated means that the complexity of the calculation is formally reduced from $O(N^2)$ to $O(N)$. That this is so can be seen by the following argument. Suppose that a spherical truncation scheme is used and the cutoff distance for the interaction is r_c. Then each atom within the system will interact with all the atoms within a volume of $4\pi r_c^3/3$. If the mean number density of atoms within the system is ρ, the total number of interactions for the system will be $4\pi r_c^3 \rho N/3$ or $\propto N$. Obviously the cost of the calculation will depend upon the size of the cutoff. The trick is to use as small a cutoff as possible while still providing an adequate treatment of the non-bonding interactions.

There are several subtleties that have to be addressed when using cutoff schemes. The first is that of how the truncation is to be effected. The easiest way is to use an abrupt truncation and simply ignore all interactions that are beyond the cutoff distance. This is equivalent to multiplying each term in the non-bonding energy expression (equation (9.1)) by a truncation function, $S(r)$, of the form

$$S(r) = \begin{cases} 1 & r \leq r_c \\ 0 & r > r_c \end{cases} \tag{9.2}$$

The problem with this type of truncation is that the energy and its derivatives are no longer continuous functions of the atomic coordinates and that there will be jumps in the energy during a minimization or a dynamics simulation as atoms move in and out of each other's cutoff distance. These can disrupt a minimization process or lead to unwanted effects (such as heating) in a dynamics simulation.

An alternative to abrupt truncation is to use a smoothing function, $S(r)$, that tapers the interaction continuously to zero at a given distance. Many smoothing functions have been proposed. One example is a *switch function* that is cubic in the square of the interaction distance, r^2:

$$S(r) = \begin{cases} 1 & r \leq r_{on} \\ \dfrac{\left(r_{off}^2 - r^2\right)^2 \left(r_{off}^2 + 2r^2 - 3r_{on}^2\right)}{\left(r_{off}^2 - r_{on}^2\right)^3} & r_{on} < r \leq r_{off} \\ 0 & r > r_{off} \end{cases} \qquad (9.3)$$

It has the property that the interaction is not modified for distances less than an inner cutoff distance, r_{on}, and is smoothed to zero at the outer cutoff, r_{off}. The function is constructed so that its first derivative is continuous in the full range $r \leq r_{off}$, which is necessary if problems in minimizations and in dynamics simulations are to be avoided. The second derivatives, though, are discontinuous.

A second example of a smoothing function is a *shift function*:

$$S(r) = \begin{cases} \left[1 - \left(\dfrac{r}{r_c}\right)^2\right]^2 & r \leq r_c \\ 0 & r > r_c \end{cases} \qquad (9.4)$$

Like the switch function, the shift function has continuous first derivatives but, unlike the switch function, it relies on just one cutoff distance and it modifies the form of the interaction throughout its entire range.

Graphs of these functions and their first and second derivatives are illustrated in figures 9.1, 9.2 and 9.3, respectively. In figure 9.4, the products of the smoothing functions and a Coulomb interaction between two unit positive charges are shown. The first derivatives of these interactions are plotted in figure 9.5. It can be seen that the use of truncation techniques can introduce radical differences in the form of the interaction potential.

The second problem that needs to be addressed is that of how the truncation or smoothing function is to be employed. The simplest way is to apply the smoothing procedure to each interaction separately. This means that, for each pair of atoms, ij, the smoothing function is calculated and the interaction for that pair is the product of the function and the pair's non-bonding

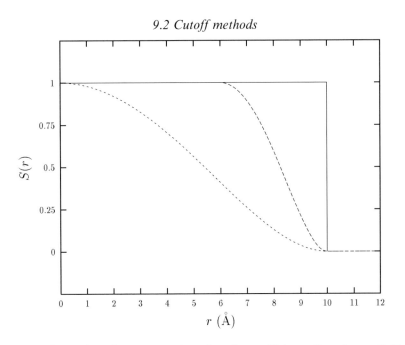

Figure 9.1. Plots of various truncation functions, $S(r)$, as functions of distance. Direct truncation, solid line; switch function, long-dash line; shift function, short-dash line. The values of the cutoffs are $r_c = r_{off} = 10$ Å and $r_{on} = 6$ Å.

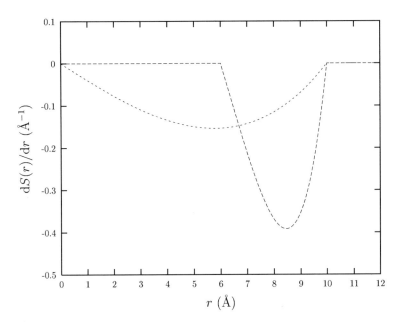

Figure 9.2. Plots of the first derivatives of the switch and shift truncation functions displayed in figure 9.1.

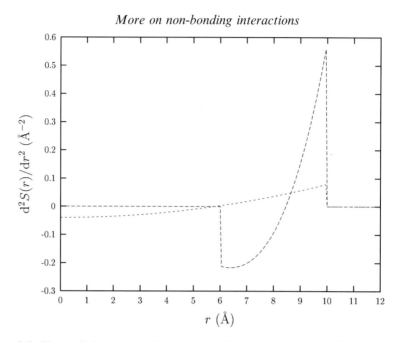

Figure 9.3. Plots of the second derivatives of the switch and shift truncation functions displayed in figure 9.1.

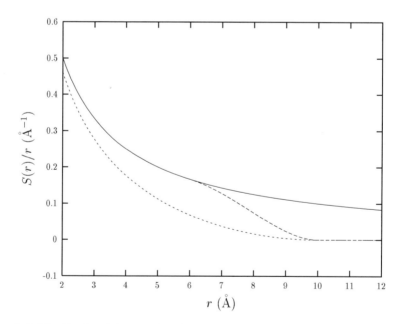

Figure 9.4. The Coulomb interaction between two unit positive charges as a function of distance and as modified by the application of the switch and shift truncation functions. Full interaction, solid line; switched interaction, long-dash line; shifted interaction, short-dash line.

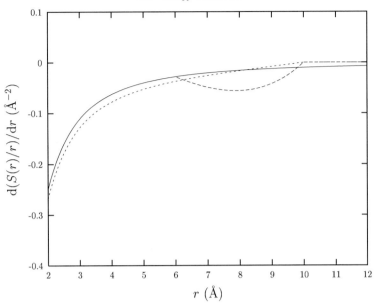

Figure 9.5. The derivatives of the Coulomb interactions displayed in figure 9.4.

energy, $S(r_{ij})V_{nb}^{ij}$. This is adequate for uncharged systems, in which there are only Lennard-Jones interactions, or for systems with small charges but, for charged systems, an *atom-based* truncation scheme can lead to a problem that is colloquially known as *splitting of the dipoles*. For such systems *residue-based* truncation schemes can provide better behaviour. In these methods the non-bonding interactions between all the atoms of two residues are calculated in full. The full residue–residue interaction energy is then multiplied by a single truncation or smoothing function that is calculated using a characteristic distance between the two sets of atoms such as, for example, the distance between their centers of geometry.

To illustrate the dipole-splitting problem, consider the interaction between an ion with a unit charge and a simple point-charge model for a water molecule. The water molecule is neutral overall but has a dipole moment with a value of, say, μ. The energy of the interaction between the charge and the dipole is proportional to the charge, the dipole and the inverse of the distance between the two species squared. Thus, the interaction decays more rapidly with distance than does a charge–charge interaction. Now, within a simple point-charge model, the water dipole can be represented by charges of $-2q$ on the oxygen and $+q$ on each of the hydrogens and so the charge–dipole interaction will be represented by three charge–charge interactions. To reproduce this interaction accurately it is necessary to include *all* these interactions fully within the calculation. It is also important to note that each charge–charge interaction is of a larger magnitude and is of longer range than the total charge–dipole interaction. Thus, if a cutoff model that splits these interactions is used (if, for example, two of the interactions are within the cutoff and one is not) large distortions in the energy and forces are likely to take place. Of course, these effects will be reduced if a smoothing function is employed rather than straight truncation but they will persist nevertheless. Although the dipole-splitting effect was illustrated with a charge–dipole interaction, it occurs generally. For example, the interaction of two water molecules is a dipole–dipole interaction that scales as the inverse cube of the distance between the dipoles. Thus, splitting of the dipoles in this case could lead to larger errors than those for the charge–dipole interaction.

The final point to be addressed in connection with cutoff schemes is that of how to evaluate which interactions are within the cutoff and which are not. After all, determining the distances between all atom pairs in order to find which to calculate and which not is exactly the operation that we are trying to avoid! The way around this problem is to keep a list of the non-bonding interactions. This is generated using a cutoff distance, the list cutoff, r_{list}, that is greater than the interaction cutoff, r_c or r_{off}. This means that the list will contain more interactions than are necessary for a single energy calculation, but it also means that the list does not need to be regenerated for each energy evaluation. This is done only when the atoms have moved by an amount of the order of $r_{list} - r_c$ or $r_{list} - r_{off}$ so that the current list is invalidated.

Of the many possible truncation schemes that exist, the method that will be used in this work for the evaluation of the non-bonding interactions is one that has been described by P. Steinbach and B. R. Brooks and is called the *atom-based force-switching truncation scheme*. In this method it is not the interaction energy that is truncated directly, but its first derivative (and, hence, its force). Thus, if $f_{true}(r)$ is the force between two particles, the modified force, $f(r)$, has the form

$$f(r) = S(r)f_{true}(r) \tag{9.5}$$

For the electrostatic interactions, Steinbach and Brooks used the same switching function as that in equation (9.3). Because this function has continous first derivatives, it implies that the second derivatives of the energy will be continuous throughout the range of the modified interaction as well. The potential energy of each interaction is obtained by integrating equation (9.5). If $V(r)$ is the modified interaction and V_{true} the true interaction ($\equiv c/r$), one has

$$
V(r) = \begin{cases}
V_{\text{true}} + \dfrac{8c}{\gamma}\left[r_{\text{on}}^2 r_{\text{off}}^2 (r_{\text{off}} - r_{\text{on}}) - \tfrac{1}{5}\left(r_{\text{off}}^5 - r_{\text{on}}^5\right)\right] & r \le r_{\text{on}} \\[2ex]
c\left[A\left(\dfrac{1}{r} - \dfrac{1}{r_{\text{off}}}\right) + B(r_{\text{off}} - r) + C(r_{\text{off}}^3 - r^3) \right. & \\[1ex]
\left. \qquad + D(r_{\text{off}}^5 - r^5)\right] & r_{\text{on}} < r \le r_{\text{off}} \\[2ex]
0 & r > r_{\text{off}}
\end{cases}
\tag{9.6}
$$

where the following constants have been defined:

$$
\gamma = (r_{\text{off}}^2 - r_{\text{on}}^2)^3
$$

$$
A = \frac{r_{\text{off}}^4(r_{\text{off}}^2 - 3r_{\text{on}}^2)}{\gamma}
$$

$$
B = \frac{6r_{\text{on}}^2 r_{\text{off}}^2}{\gamma}
$$

$$
C = -\frac{r_{\text{off}}^2 + r_{\text{on}}^2}{\gamma}
$$

$$
D = \frac{2}{5\gamma}
$$

For the Lennard-Jones interactions, a simpler truncation function is used. This does not ensure continuity of the second derivatives of the energy but this should be acceptable because the magnitudes of the Lennard-Jones terms are so much smaller. If the form of each part of the Lennard-Jones interaction is c/r^n, where $n = 6$ or 12, the modified interaction energy is

$$
V(r) = \begin{cases}
V_{\text{true}} - \dfrac{c}{(r_{\text{on}}r_{\text{off}})^{n/2}} & r \le r_{\text{on}} \\[2ex]
\dfrac{cr_{\text{off}}^{n/2}}{(r_{\text{off}} - r_{\text{on}})^{n/2}}\left[\left(\dfrac{1}{r}\right)^{n/2} - \left(\dfrac{1}{r_{\text{off}}}\right)^{n/2}\right] & r_{\text{on}} < r \le r_{\text{off}} \\[2ex]
0 & r > r_{\text{off}}
\end{cases}
\tag{9.7}
$$

It can be seen, in both cases, that the interaction energies at less than the inner cutoff distance, r_{on}, are unaltered except that a constant that ensures that the potential is continuous at $r = r_{\text{on}}$ has been added. The interaction energy of two unit charges given by the scheme of equation (9.6) and its derivative are plotted in figures 9.6 and 9.7. Evidently the forces are only slightly distorted.

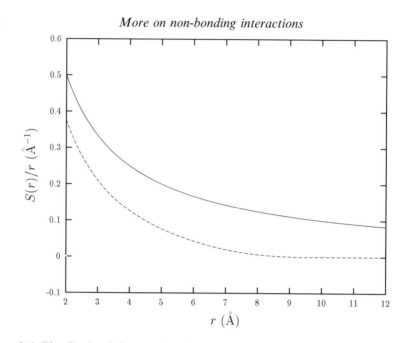

Figure 9.6. The Coulomb interaction between two unit positive charges as a function of distance and as modified by the application of the force-switching function. Full interaction, solid line; force-switched interaction, dashed line. The values of the cutoffs are $r_{off} = 10$ Å and $r_{on} = 6$ Å and the constant $c = 1$.

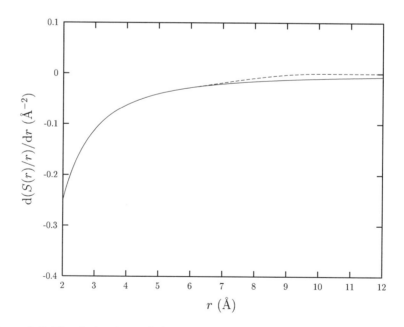

Figure 9.7. The derivatives of the Coulomb interactions displayed in figure 9.6.

Steinbach and Brooks found that this scheme was one of the most effective for calculating non-bonding interactions – i.e. it reproduced well the results of calculations in which the full non-bonding interaction was calculated – as long as a reasonably long inner cutoff distance and a broad switching region were used (the latter to minimize the dipole-splitting effect). They suggested that minimum values of 8 and 12 Å were suitable for the inner and outer cutoffs, respectively. The method, although atom-based, was found to give results that were as good as or better than those from residue-based methods even for highly charged systems.

The method of calculating the non-bonding interaction energy described above has been implemented in a second version of the module, ENERGY_NON_BONDING. The subroutines that are concerned with the calculation of the energy and its derivatives have the same names and calling sequences as those in the first version of the module so the two modules can be used interchangeably. The subroutine that defines the options for the non-bonding energy calculation has some additional arguments and there is an extra subroutine that can be accessed to provide information about the evaluation of the non-bonding energy. Their interfaces are

```
MODULE ENERGY_NON_BONDING

CONTAINS

    SUBROUTINE ENERGY_NON_BONDING_OPTIONS ( LIST_CUTOFF, OUTER_CUTOFF, &
                                            INNER_CUTOFF, DIELECTRIC, &
                                            MINIMUM_IMAGE, PRINT )
      LOGICAL, INTENT(IN), OPTIONAL :: MINIMUM_IMAGE, PRINT
      REAL,    INTENT(IN),   OPTIONAL :: LIST_CUTOFF, OUTER_CUTOFF, &
                                         INNER_CUTOFF, DIELECTRIC
    END SUBROUTINE ENERGY_NON_BONDING_OPTIONS

    SUBROUTINE ENERGY_NON_BONDING_STATISTICS
    END SUBROUTINE ENERGY_NON_BONDING_STATISTICS

END MODULE ENERGY_NON_BONDING
```

The subroutine specifying the non-bonding energy options has six optional arguments. One gives the value of the dielectric constant for the electrostatic energy calculation as in the corresponding subroutine of the first version of the module. Three more specify the cutoffs for the generation of the atom-interaction list and the calculation of the energy. These are LIST_CUTOFF, INNER_CUTOFF and OUTER_CUTOFF. The defaults are to use very large values for all these quantities so that, in effect, all the interactions in the system are calculated without truncation. The fifth argument, PRINT, is a logical one that specifies whether data are to be written out at each update of the non-bonding list. The default is not to do any printing. The final argument, MINIMUM_IMAGE, is also a logical one but its explanation will be left until later.

The subroutine ENERGY_NON_BONDING_STATISTICS takes no arguments but prints out information about the non-bonding interaction list and its creation, such as the average number of interactions in the list and the number of times the non-bonding interaction list has been updated. The period for which the statistics are printed out is determined by a previous call to ENERGY_NON_BONDING_STATISTICS or to ENERGY_NON_BONDING_OPTIONS because these subroutines re-initialize the counters that are stored internally in the module to calculate the statistics.

The module generates the atom non-bonding interaction list automatically and stores it internally using a procedure similar to that described in section 3.2 for the determination of the connectivity of a system. The list is generated when the energy is first called and subsequently when any atom has moved by more than half the difference between the cutoff distance used to generate the non-bonding energy interaction list (LIST_CUTOFF) and the outer cutoff distance (OUTER_CUTOFF). This ensures that the list is always up to date. The calculation of the energy using the atom-based list will scale as $O(N)$ for a given cutoff distance. The creation of the non-bonding list, however, is an $O(N_r^2)$ process, where N_r is the number of residues in the system, because all residue pairs have to be searched. If the cutoff distances are set correctly then the update procedure should have to be done only once every 10–20 energy calculations so the proportionate cost is not too great. A value for the variable LIST_CUTOFF of 1–2 Å greater than the outer cutoff distance should give a reasonable update frequency.

To finish this section we emphasize a couple of general points. First, it is the electrostatic interactions that cause the biggest problems because of their long range and large size. The effects of truncation on the Lennard-Jones energies and forces are less crucial, although they can still be sizeable. The second point is that no truncation method is ideal. The one that has been chosen here should give reasonable results in most cases, if it is properly used, but there will be others that will give results that are as valid in particular circumstances. According to Steinbach and Brooks, the most important lesson is that the cutoff should be as large as possible. The differences among the best alternative truncation methods are then less significant.

9.3 Example 15

To illustrate how the use of a truncation method can alter the calculation of the energy of a system it is interesting to calculate the energy for a large system using different cutoff schemes. The example program in this section does this for a small protein, crambin.

The program is

```
PROGRAM EXAMPLE15

... Declaration statements ...

! . Local parameters.
INTEGER,  PARAMETER :: NCALC  = 40
REAL,     PARAMETER :: BUFFER = 4.0

! . Local scalars.
INTEGER :: I
REAL    :: CUT = 0.0

! . Local arrays.
REAL, DIMENSION(1:NCALC) :: CUTOFF, ENERGY_EL, ENERGY_LJ

! . Process the MM file for bALA.
CALL MM_FILE_PROCESS ( ''protein.opls_bin'', ''protein.opls'' )

! . Construct the system file for bALA.
CALL MM_SYSTEM_CONSTRUCT ( ''protein.opls_bin'', ''crambin.seq'' )

! . Read in some coordinates for crambin.
CALL COORDINATES_READ ( ''crambin.crd'' )

! . Calculate and print an initial energy.
CALL ENERGY

! . Loop over the number of energy calculations.
DO I = 1,NCALC

   ! . Increment CUT.
   CUT = CUT + 1.0

   ! . Set the cutoff value.
   CALL ENERGY_NON_BONDING_OPTIONS ( LIST_CUTOFF  = CUT + BUFFER, &
                                     OUTER_CUTOFF = CUT + BUFFER, &
                                     INNER_CUTOFF = CUT, PRINT = .TRUE. )

   ! . Calculate and print the energy.
   CALL ENERGY
```

```
    ! . Save the inner cutoff value.
    CUTOFF(I) = CUT

    ! . Save the energy values.
    ENERGY_EL(I) = EELECT
    ENERGY_LJ(I) = ELJ

END DO

    ! . Write out the results.
    WRITE ( OUTPUT, ''(/8X,A,6X,A,9X,A)'' ) &
          ''Inner Cutoff'', ''Energy - Elec.'', ''Energy - LJ''
    WRITE ( OUTPUT, ''(F20.1,2F20.4)'' ) &
          ( CUTOFF(I), ENERGY_EL(I), ENERGY_LJ(I), I = 1,NCALC )


END PROGRAM EXAMPLE15
```

The program starts by setting up the MM file and system data structures for the protein and reading in an initial set of coordinates. After this, an energy is calculated using the default values for the cutoffs that are stored in the non-bonding energy module. These are very large ($\simeq 9999$ Å) and so the calculated energy should correspond to that obtained with a full non-bonding energy calculation.

After the reference-energy calculation a series of energies is evaluated with different cutoff options. In each iteration of the loop the inner cutoff value is incremented by 1 Å and the energy is calculated and stored. Note that the value of the list cutoff specified is the same as that for the outer cutoff. In this example program, in which only a single energy calculation is being done for each cutoff value, this is sufficient but in a minimization or dynamics study a larger value of the list cutoff would be needed. Finally, the results of the calculation are printed.

The electrostatic energies as a function of the cutoff distance, r_{off}, are shown in figure 9.8. Also shown are the electrostatic energies produced when the inner cutoff distance in Example 15 is given the same value as the outer cutoff, i.e. $r_{on} = r_{off}$. The importance of a smoothing region is clear insofar as the energy tends smoothly to its limiting value if one is employed and oscillates wildly otherwise. This is a manifestation of the dipole-splitting problem. The effect of straight truncation is much less marked for the Lennard-Jones energies, which are shown in figure 9.9. The number of non-bonding interaction pairs is plotted versus the cutoff distance in figure 9.10. Above 32 Å all the possible interactions between atoms within the protein are included in the energy calculation.

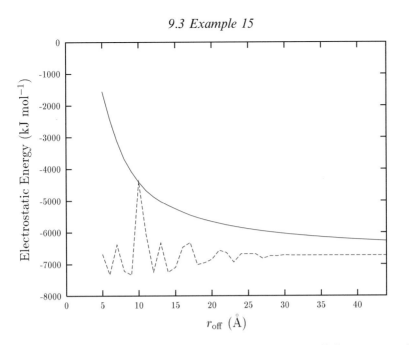

Figure 9.8. The electrostatic energies as a function of the cutoff distance, r_{off}. With a smoothing region, solid line; straight truncation, dotted line.

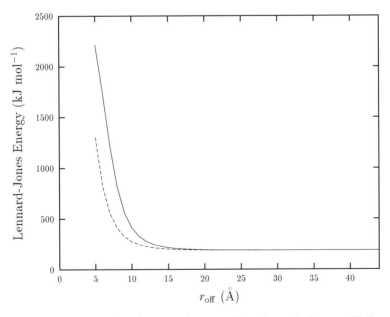

Figure 9.9. The Lennard-Jones energies as a function of the cutoff distance, r_{off}. With a smoothing region, solid line; straight truncation, dotted line.

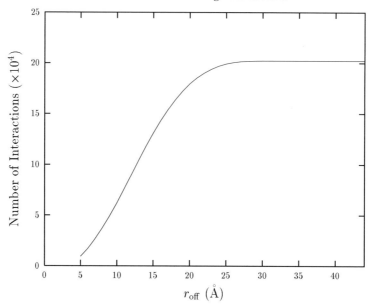

Figure 9.10.The number of non-bonding interaction pairs as a function of the cutoff distance, r_{off}.

9.4 Including an environment

One of the most interesting and important applications of molecular simulations is the study of systems in the condensed phase. After all, it is in the condensed phase that the great majority of chemical and biochemical processes occur. Unlike the systems that we have been studying up to now, condensed phase systems are effectively infinite in extent. It is obviously impractical to try to simulate such systems directly because, no matter how large and fast the computer, the limits of its computational ability would be attained very rapidly! Currently, the only really feasible way of simulating condensed phase systems at an atomic level is to select a small part of the system to study in detail – for example, a small volume of a liquid or a crystal or a single solvated protein molecule – and then use methods that imitate the effect of the remainder of the system or the *environment*. The fact that an infinite system is being studied by using a finite one means, of necessity, that there are limitations to the types of process that can be studied. It is evident, for example, that one needs to be careful about drawing conclusions for properties of a system that have length scales larger than the size of the finite simulation system. However, if these concerns are borne in mind, simulation approaches can be powerful tools for investigating processes in the condensed phase.

There is a wide range of approximations in use to model the environment of a system and we shall discuss only a few. Probably the most widely used model and arguably the most reliable, if not the cheapest, is the method of *periodic boundary conditions* (PBCs). In this model, one chooses a small part of the full system to simulate and then makes the assumption that the complete condensed phase system can be modeled as an infinite series of copies of the central box. The assumption of periodicity immediately makes the simulation of such a system tractable even though all the molecules in it are modeled explicitly. Because of its importance the PBC method is the one that we shall use and it will be described in more detail in the remaining sections of this chapter.

The PBC method works well in many cases but it has some drawbacks. First, the assumption of periodicity means that an order is being imposed upon the system that would not normally be present. This can lead to artefactual results for the structural and dynamical properties of a system obtained from a simulation. Second, the method is often expensive, especially for large molecules. To see this, consider a large molecule, such as a protein, in solution (see, for example, figure 2.1). It takes only a little reflection to realize that, to immerse the molecule fully in the solvent, a volume of solvent much larger than that of the molecule itself will be required. This adds considerably to the size of the system and means that most of the time during a simulation will be spent dealing with the solvent rather than with the solvated molecule which is the object of principal interest.

As a result of the limitations of the PBC method, alternative techniques have been sought. One series of methods has been developed primarily to mimic effects of solvent on molecules. These *implicit solvent methods* replace the *explicit* description of the solvent molecules of the PBC approach by simpler models. It is usual to use different strategies to model the electrostatic and non-polar (Lennard-Jones) interactions between the solute and the solvent because these interactions are different in nature.

For the electrostatic interactions a common approach is to use a *reaction field model* in which the solvent is replaced by a medium that has a dielectric constant that is appropriate to the solvent being modeled. The solute is assumed to be located in a cavity (of a different dielectric – often unity) within the continuum. The solute's charge distribution polarizes the solvent which in turn acts back upon the solute's charges with a reaction field. In this model the energy of the interaction between the solute and solvent is determined by first solving the *Poisson–Boltzmann equation* for the electrostatic potential, ϕ, in the system. This equation has the form

$$\nabla^{T}(\epsilon\nabla\phi) - \epsilon\kappa^{2}\sinh\phi + 4\pi\rho = 0 \tag{9.8}$$

where ϵ is the dielectric, κ is the *Debye–Hückel parameter*, which is related to the type and concentration of ions in the solution, and ρ is the charge distribution in the system. It is to be noted that all these parameters, as well as the potential, are functions of position. If the κ function is everywhere zero the Poisson–Boltzmann equation reduces to the Poisson equation. Once the potential has been obtained, the total electrostatic energy for the system, $\mathcal{V}_{el}$, can be calculated. For a simple point-charge model of the solute's charge distribution and for the case in which the potential is small (so that the sinh term in equation (9.8) can be linearized) this is

$$\mathcal{V}_{el} = \frac{1}{2}\sum_{i=1}^{N} q_{i}\phi(\mathbf{r}_{i}) \tag{9.9}$$

Although the solution to the Poisson–Boltzmann equation appears to give good results for solvation energies, it is expensive to solve. As a result it has been employed primarily to calculate the potentials and the energies of single structures and not in minimization or molecular dynamics calculations.

For these longer types of calculations other more approximate, *ad hoc* methods have been developed. The simplest include models that reduce the charges on charged groups to account for the screening of charges by the solvent and the use of a dielectric constant other than unity in the calculation of the electrostatic interactions (equation (9.1)). This can be a constant with a larger value or it can be a function of the distance between the particles. Thus, for example a *distance-dependent dielectric function*, $\epsilon(r) \propto r$, has often been used in the simulation of biomacromolecules. These methods are, however, of dubious accuracy and are better avoided if viable alternatives are available. More precise methods, albeit still approximate, are based upon generalizations of the *Born expression* for the electrostatic solvation energy of a charged sphere in a medium of a different dielectric constant. This energy, ΔG_{Born}, is

$$\Delta G_{Born} \propto \frac{q^{2}}{2a}\left(\frac{1}{\epsilon_{o}} - \frac{1}{\epsilon_{i}}\right) \tag{9.10}$$

where q is the charge on the sphere, a is its radius and ϵ_{i} and ϵ_{o} are the dielectric constants inside and outside the sphere, respectively. To account for interactions between charged spheres, equation (9.10) can be generalized to

$$\Delta G_{\text{solv}} \propto \left(\frac{1}{\epsilon_o} - \frac{1}{\epsilon_i}\right) \sum_{i=1}^{N} \sum_{j=1}^{N} \frac{q_i q_j}{f(r_{ij}, a_i, a_j)} \tag{9.11}$$

where $f(r_{ij}, a_i, a_j)$ is a function such that $f \to 1/r_{ij}$ as $r_{ij} \to \infty$ (i.e. when the spheres are very far apart and do not overlap) and $f \to a_i$ or a_j as $r_{ij} \to 0$ (i.e. when the spheres coalesce). The accuracy of the representation depends upon the form of the function chosen for f and upon the way in which the *effective Born radii*, a_i, are calculated. These will obviously be different depending upon whether the atom is completely buried in the interior of a molecule (and so has no exposure to solvent) or is at the surface.

The continuum dielectric models and their equivalents account for the electrostatic interactions between solute and solvent. To treat the non-polar interactions (dispersion and repulsion) other models are necessary. The commonest is the *surface free energy description*, which relates the interaction energy to the accessible surface area of the molecule. Like the Born model, it has been used primarily for solvation studies. The non-polar energy, $\mathcal{V}_{\text{np}}$, is

$$\mathcal{V}_{\text{np}} = \sum_{i=1}^{N} \gamma_i A_i \tag{9.12}$$

where A_i is the surface area of atom i that is accessible to solvent and γ_i is a constant that depends upon the chemical type of the atom and must be parametrized to reproduce the non-polar surface free energy. At the present time the combination of *generalized Born* and *solvent-accessible surface area* models probably provides the most accurate and cost-effective implicit solvation models for molecular simulations. One of the more widely used models of this type is the one developed by W. C. Still and co-workers.

An alternative class of methods mixes elements of explicit and implicit models. Although they are not as effective as PBCs, they are probably more reliable than purely continuum approaches. Examples of these methods include the various *boundary approximations* which focus on a small, often spherical, region of the system of interest and treat that with an atomic level model. The remainder of the system is removed and replaced by a *boundary potential*. In the simplest cases the potential can be neglected or it can be a hard wall but in more sophisticated algorithms it will have a form that accounts for non-polar interactions as well as reaction-field-type terms. Taking the solvation of a solute as an example again, the solute molecule would be placed at the center of the sphere, surrounded with a few shells of solvent molecules and then the boundary potential would be placed outside this. These methods can reduce quite substantially the time required for a simulation with explicit solvent molecules but the choice of boundary potential is not always evident and can significantly affect the results obtained.

9.5 Minimum image periodic boundary conditions

In the periodic boundary condition approximation an infinite system is constructed as a periodically repeated array of the finite system that is being studied. A schematic diagram of a two-dimensional system is shown in figure 9.11. Obviously, for the infinite system to be continuous, the finite system must be of a sufficiently regular shape that, when it is replicated, it can fill space. The most common option, by far, in three dimensions is to use finite systems that are cubic. Also common are *orthorhombic* boxes whose angles are all right angles but that have sides of different lengths. Other shapes that are possible are *triclinic, hexagonal, truncated octahedral* and various sorts of *dodecahedral* boxes. These shapes are geometrically more complicated but they can be required when studying certain types of system, such as crystals.

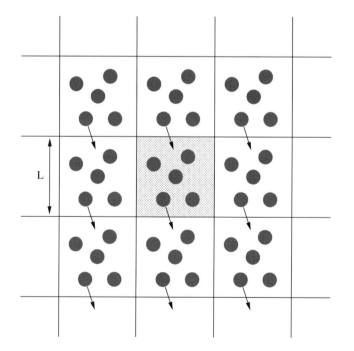

Figure 9.11. The periodic boundary condition approximation in two dimensions. The central, square, shaded box of side L is replicated in both dimensions. A particle moving out of a box will be replaced by one moving in from the opposite side.

The fact that the system is infinitely replicated means that some new techniques must be used to calculate the non-bonding interaction energy. There are algorithms for evaluating the non-bonding energy exactly for periodic systems. A well-known example is *Ewald's lattice summation method*, which we shall discuss in detail later. Truncation methods can also be applied and we shall consider one that employs the *minimum image convention* here. The major assumption of this convention is that a particle will only interact with, at most, the nearest copy of another particle in the system (see figure 9.12). This automatically imposes an upper limit to the distance at which the non-bonding interactions are truncated at one half of the length of the box. The advantages of the minimum image method are that it is relatively inexpensive and that it is easy to implement – a subroutine to calculate interactions with the minimum image convention is very similar to a standard one. Note that, if it is wished to simulate systems in which the atoms of the central box interact with those of many other boxes (for example, because the desired cutoff is large relative to the size of the central box), more sophisticated techniques are needed.

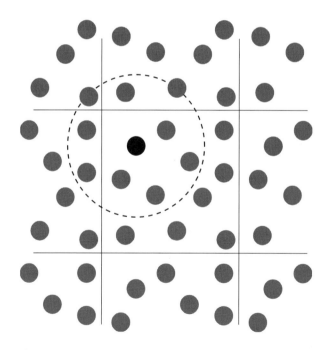

Figure 9.12. The minimum image approximation in two dimensions. The black particle will interact with all the particles within the circle, assuming a circular truncation of the interparticle interactions. The circle has a radius of half the box length.

The implementation of the PBC approximation requires a means of defining the type of periodic box and its shape as well a method of calculating the non-bonding interactions within the minimum image convention. To do the former a new module, SYMMETRY, is introduced while to do the latter an extension of the module ENERGY_NON_BONDING is required.

The use of a periodic box to replicate a system is an aspect of the use of symmetry in a system (in this case translational symmetry) and so these operations are handled by the module SYMMETRY. This module is a fundamental one and ranks in importance with the modules ATOMS and SEQUENCE in a system's definition. The public parts of the module that we shall need are

```
MODULE SYMMETRY

! . Module arrays.
REAL, DIMENSION(1:3) :: BOXL

CONTAINS

   SUBROUTINE SYMMETRY_CUBIC_BOX ( A )
      REAL, INTENT(IN) :: A
   END SUBROUTINE SYMMETRY_CUBIC_BOX

   SUBROUTINE SYMMETRY_INITIALIZE
   END SUBROUTINE SYMMETRY_INITIALIZE

END MODULE SYMMETRY
```

The array, BOXL, contains the lengths of the sides of the periodic box. This allows, in principle, the treatment of orthorhombic as well as cubic boxes, although all the applications we shall consider are cubic. The two subroutines are SYMMETRY_CUBIC_BOX, which defines a cubic box for a system, and SYMMETRY_INITIALIZE, which deactivates the use of periodic boundary conditions. The first subroutine takes a single argument that specifies the length of the side of the cubic box. The second subroutine takes no arguments. To use periodic boundary conditions a periodic box must normally be explicitly defined with this command. The edges of the cubic box are taken to be along the Cartesian axes.

To calculate the energy of a system within the minimum image convention the version of the module ENERGY_NON_BONDING described in section 9.2 is used. To activate this facility there is an extra argument to the subroutine ENERGY_NON_BONDING_OPTIONS. It is a logical input argument with the name MINIMUM_IMAGE, which must be set to .TRUE. to activate the minimum image convention. The default (as we saw in section 9.3) is not to use it. Note that, if a periodic box is present and the minimum image option is not set, there will be an error. Likewise there is an error in the case that the minimum image option has been requested but no periodic box has been defined.

The calculation of the non-bonding interactions (and the associated processing, such as the generation of the non-bonding list) with the minimum image convention is very similar to that for a system in vacuum. The only significant difference is that every time the distance vector between two particles is calculated it is adjusted to ensure that the interaction is the shortest possible that is consistent with the periodicity of the system – i.e. the particles lie in the same box or in adjacent boxes. Note that, because the interaction distance between particles is adjusted in this fashion during the calculation of the energy, there is no need to alter the coordinates of the atoms themselves to ensure that they lie within the central periodic box. Hence, because the coordinates of the atoms are unmodified, it is possible to have atoms that are separated by several box lengths at the end of a long simulation. This, of course, must be taken into account when analysing the results of a minimum image simulation. We shall discuss this point in more detail in the next chapter.

As a final point we emphasize that the condition under which the minimum image convention is valid is that the cutoff for the non-bonding interactions is less than or equal to half the length of the side of the box. This can impose a serious constraint on the size of the system if a large non-bonding interaction cutoff distance is desired. Thus, if a 12 Å cutoff is deemed necessary, the size of the system's cube must be of the order of at least 26 Å by the time a 1 Å buffer region has been included for the generation of the non-bonding interaction list. The non-bonding energy module will automatically check for this and give an error if this condition is not satisfied.

9.6 Example 16

As an example of a simulation using periodic boundary conditions we study a liquid system consisting of a cubic box of water molecules. The program is

```
PROGRAM EXAMPLE16

... Declaration statements ...

! . Process the MM file for water.
CALL MM_FILE_PROCESS ( ''solvent.opls_bin'', ''solvent.opls'' )

! . Construct the system file for water.
CALL MM_SYSTEM_CONSTRUCT ( ''solvent.opls_bin'', ''h2o_box729.seq'' )

! . Read in the coordinates.
```

```
CALL COORDINATES_READ ( ''h2o_box729.crd'' )

! . Initialize the non-bonding interaction list cutoff options.
CALL ENERGY_NON_BONDING_OPTIONS ( LIST_CUTOFF = 13.5, OUTER_CUTOFF = 12.0, &
                                  INNER_CUTOFF = 8.0, MINIMUM_IMAGE = .TRUE. )

! . Calculate an energy.
CALL ENERGY

! . Initialize the random number seed.
CALL RANDOM_INITIALIZE ( 314159 )

! . Equilibrate the system at 300K for 1ps.
CALL DYNAMICS ( 0.001, 1000, 500, INITIAL_TEMPERATURE = 300.0, &
                SCALE_FREQUENCY = 50, SCALE_OPTION = ''CONSTANT'' )

! . Run dynamics for 10ps using the old velocities.
CALL DYNAMICS ( 0.001, 10000, 500, ASSIGN_VELOCITIES = .FALSE., &
          SAVE_FREQUENCY = 50, COORDINATE_FILE = ''h2o_box729.traj'' )

! . Print some statistics about the non-bonding energy calculation.
CALL ENERGY_NON_BONDING_STATISTICS

END PROGRAM EXAMPLE16
```

An important point to note about this program and any others that use periodic boundaries is that there is little difference from programs that use no boundary conditions. The method of the calculation of the energy is changed but the energy and the forces can be used in exactly the same way for minimization and dynamics calculations.

The program starts off normally with the processing and the construction of the molecular mechanics data structures. The box of water that we are considering has 729 molecules ($\equiv 9^3$). In a second step the coordinate file for the box of water molecules is read. There are a couple of points to note about the file. The first is that it has the usual format except that there is an extra section at the top that defines the symmetry information in the system. In this case, the section reads

```
Symmetry   1
CUBIC           27.9620383437
```

where Symmetry is the keyword that reveals that symmetry is being used for the molecule and the integer 1 denotes the number of lines that follow which contain the symmetry definition. On the following line is the definition which states that the system has cubic symmetry with a repeat distance of about 28 Å. All the information concerning symmetry will be read from or written to coordinate (and trajectory) files whenever symmetry has been defined for a system. The second point to note about the coordinate file is that it already contains the coordinates for a system that has been equilibrated (in a previous molecular dynamics simulation) and so it can be used immediately in a simulation in which data are to be collected for analysis.

After the coordinates have been read the options for the calculation of the non-bonding energies are defined. The values of the outer and list cutoffs are 12 and 13.5 Å, respectively, and the minimum image flag is set to .TRUE.. As mentioned in the previous section, the use of a cutoff of 13.5 Å means that the box size must be at least 27 Å. That is the reason why 729 molecules of water were used, because this is near the minimum number that is necessary to fill a box of about this size. The length of the box was calculated assuming that the density of water was 997.5 kg m^{-3}.

After the definition of the energy options, an energy is calculated (to verify that everything works properly) and then a short dynamics simulation of 1 ps is performed to equilibrate the system further at 300 K before a longer simulation of 10 ps is carried out. Coordinate sets are saved every 50 steps, giving 201 sets of coordinates for analysis in all by the end of the simulation. The program finishes by printing some statistics about the non-bonding energy calculation.

Perhaps the most notable thing about this program is the length of time that it takes to run. Up to now the examples we have considered have needed relatively little resources in terms of memory or CPU time. In constrast, this program performs a simulation for about 2000 atoms and at each timestep about 10^6 non-bonding interactions are processed. It demands a lot more time than the previous examples, although it should run in less than a few days on a reasonably fast personal computer or workstation. A discussion of the analysis of the results of this simulation will be left until the next chapter.

9.7 Ewald summation techniques

In the previous sections we discussed methods for the calculation of non-bonding interactions that truncate the interactions beyond a certain cutoff distance. This is an approximation that could (and does!) have important consequences for the behaviour of a system during a molecular simulation. It would obviously be better to have methods that allow the non-bonding interactions to be calculated fully for a periodic system. A class of such approximations exists and they are called Ewald summation techniques in honour of one of their originators. These methods have been used widely for studying such systems as ionic crystals but less so for molecular simulations. However, recent algorithmic advances (see the next section) are likely to

make these methods the best choice for the calculation of non-bonding interactions in the simulation of periodic molecular and macromolecular systems.

The derivation of the Ewald summation formulae involves a number of
subtleties and so only a brief description will be presented here to give a
flavour of what is required. The literature on Ewald and related techniques
is large, but the exposition below mostly follows that due to D. Williams.
Consider a periodic atomic system within which the particles interact with a
potential of the form $\lambda_i \lambda_j / r_{ij}^p$. This is appropriate for the Coulomb interaction
and both for the repulsive and for the dispersive parts of the Lennard-Jones
interaction if geometrical mean combination rules are used. The non-bonding
interaction energy of a single box of the periodic system will be the sum of the
interactions between the atoms within the box and between its atoms and
those of all the remaining boxes. The expression for this sum, S_p, is

$$S_p = \frac{1}{2} \sum_n{}' \sum_i \sum_j \frac{\lambda_i \lambda_j}{|r_i - r_j + t_n|^p} \tag{9.13}$$

where the sum over n indicates a summation over all periodic boxes (including $n = 0$, the central box) and the sums over i and j are over the atoms in
each box. The prime on the summation over boxes means that the self-inter-
action (i.e. with $i = j$ and $n = 0$) is omitted because this is divergent. The
vector t_n is the vector that indicates the displacement between the interacting
boxes. If the sides of the boxes are along the Cartesian axes it will have the
form $(n_x a, n_y b, n_z c)$, where n_x, n_y and n_z are the integer components of the
vector n and a, b and c are the lengths of the box in the directions of the x, y
and z axes, respectively. For a cubic box $a = b = c$.

The trick common to the Ewald summation techniques is to employ a
convergence function, $\phi(r)$, to split the sum given in equation (9.13) into
two parts. This function has the property that it decays rapidly to zero as r
increases and takes the value 1 for $r = 0$. Using it, the sum S_p can be rewritten
as

$$S_p = S_p^{(1)} + S_p^{(2)} \tag{9.14}$$

$$S_p^{(1)} = \frac{1}{2} \sum_{nij}{}' \frac{\lambda_i \lambda_j \phi(r_{nij})}{r_{nij}^p} \tag{9.15}$$

$$S_p^{(2)} = \frac{1}{2} \sum_{nij}{}' \frac{\lambda_i \lambda_j (1 - \phi(r_{nij}))}{r_{nij}^p} \tag{9.16}$$

where the shorthand r_{nij} has been used to denote $|r_i - r_j + t_n|$. Because of the
properties of the convergence function the sum $S_p^{(1)}$ is a short-range one and
can be calculated directly once the form for the convergence function has

been defined. The second sum is a long-range one but it can be evaluated if its *Fourier transform* is taken and the sum performed in *reciprocal space*. Details of this transformation will not be given here but the final form of the expression depends crucially upon the form chosen for ϕ. It is usual to follow a suggestion of B. R. A. Nijboer and F. W. De Wette and use

$$\phi(r) = \frac{\Gamma(p/2, \kappa^2 r^2)}{\Gamma(p/2)} \tag{9.17}$$

where $\Gamma(x)$ and $\Gamma(x, y)$ are the complete and incomplete gamma functions, respectively:

$$\Gamma(x) = \int_0^\infty t^{x-1} \exp(-t)\, dt \tag{9.18}$$

$$\Gamma(x, y) = \int_y^\infty t^{x-1} \exp(-t)\, dt \tag{9.19}$$

The sums S_p for $p > 3$ are absolutely convergent, which means that they converge no matter what the values of λ and no matter in which order the summation is done. Using the form for ϕ given in equation (9.17) for $p = 6$ results in an expression that is suitable for the evaluation of the dispersive part of the Lennard-Jones interaction. It is

$$
\begin{aligned}
S_6 = {} & \frac{1}{2} \sum_{nij}{}' \frac{\lambda_i \lambda_j}{r_{nij}^6} \left[1 + (\kappa r_{nij})^2 + \frac{(\kappa r_{nij})^4}{2} \right] \exp\left[-(\kappa r_{nij})^2 \right] \\
& + \frac{(\sqrt{\pi}\kappa)^3}{6V} \sum_k \left\{ \left[1 - 2\left(\frac{k}{2\kappa}\right)^2 \right] \exp\left[-\left(\frac{k}{2\kappa}\right)^2 \right] + 2\left(\frac{k}{2\kappa}\right)^3 \sqrt{\pi}\, \mathrm{erfc}\left(\frac{k}{2\kappa}\right) \right\} \\
& \times \left\{ \left[\sum_i \lambda_i \cos(k^\mathsf{T} r_i) \right]^2 + \left[\sum_i \lambda_i \sin(k^\mathsf{T} r_i) \right]^2 \right\} \\
& - \frac{\kappa^3}{12} \sum_i \lambda_i^2 \tag{9.20}
\end{aligned}
$$

In this equation, V is the volume of a periodic box ($V = abc$) and erfc is the complementary error function ($\mathrm{erfc}(x) = \Gamma(\frac{1}{2}, x^2)/\sqrt{\pi}$). The second summation is over a set of vectors, k, which are called the *reciprocal space vectors* or the k *vectors*. For an orthorhombic box with the box sides along the Cartesian axes, they take the form $k = 2\pi(n_x/a, n_y/b, n_z/c)$. This sum will be rapidly convergent as k ($=|k|$) increases in size due to the presence of the exp and erfc terms.

The sums S_p for $p \leq 3$ are only conditionally convergent, which means that the value of the sum can depend upon the way in which the summation is done. In addition, the sum will not converge at all unless the condition $\sum_i \lambda_i = 0$ is satisfied. For the Coulomb interaction this means that the simulation box must have no nett charge. The derivation of the summation formula is more complicated in this case, but it is

$$
S_1 = \frac{1}{2} \sum_{nij}{}' \frac{\lambda_i \lambda_j \operatorname{erfc}(r_{nij})}{r_{nij}}
$$
$$
+ \frac{2\pi}{V} \sum_{k \neq 0} \frac{\exp\left[-\left(\frac{k}{2\kappa}\right)^2 \right]}{k^2} \left\{ \left[\sum_i \lambda_i \cos(k^{\mathrm{T}} r_i) \right]^2 + \left[\sum_i \lambda_i \sin(k^{\mathrm{T}} r_i) \right]^2 \right\}
$$
$$
- \frac{\kappa}{\sqrt{\pi}} \sum_i \lambda_i^2
$$
$$
+ J(\boldsymbol{\mu}) \tag{9.21}
$$

It is to be noted that the sum over the k vectors in this expression explicitly excludes the $k = 0$ term because this term is divergent. There is also an extra term in the sum, $J(\boldsymbol{\mu})$, often called the *surface correction* term, which is a function of the dipole moment of the periodic box, $\boldsymbol{\mu}$. It turns out, when deriving this term, that it is necessary to specify various macroscopic boundary conditions for the ensemble of periodic boxes for which the electrostatic energy is being computed. In particular, the shape of the macroscopic crystal and the dielectric constant of the medium surrounding the crystal are important. If the medium has an infinite dielectric constant the term disappears, which corresponds to *tin-foil boundary conditions*. For a spherical crystal in vacuum (with a dielectric constant of 1) the expression is

$$
J(\boldsymbol{\mu}) = \frac{2\pi}{3V} |\boldsymbol{\mu}|^2 \tag{9.22}
$$

where the box's dipole-moment vector is

$$
\boldsymbol{\mu} = \sum_{i=1}^{N} \lambda_i r_i \tag{9.23}
$$

The first three terms on the right-hand sides of the expressions for the energies in equations (9.20) and (9.21) are often called the *real space*, the *reciprocal space* and the *self-energy* terms, respectively. The derivatives of all these terms and the surface correction term in the electrostatic term with respect to the atomic positions are straightforward to determine by direct differentiation.

There is an extra complication that is not apparent in the formulae for the non-bonding energies given in equations (9.20) and (9.21). Both these formulae apply to the case in which *all* the interactions between particles are calculated. As we have seen, this is not the case for empirical force fields, which often exclude certain interactions, notably those between bonded atoms. For these excluded interactions (the 1–2, 1–3 and sometimes the 1–4 terms) it is necessary to calculate their energies using the normal expressions (i.e. without the convergence functions) and subtract them from the sums in equations (9.20) and (9.21). Care should also be taken to ensure that the 1–4 interactions are treated properly, for these are not always fully excluded but sometimes only partially so (such as with the OPLS force field used in this book).

To calculate the non-bonding interaction energy for a periodic system using the Ewald algorithm, we introduce the third (and last!) version of the module ENERGY_NON_BONDING. It has the same structure as the other modules that calculate the non-bonding energy and, hence, can be used interchangeably with them. The major difference from the previous modules is in the arguments that can be passed to the subroutine ENERGY_NON_BONDING_OPTIONS. The definition for this subroutine is

```
MODULE ENERGY_NON_BONDING

CONTAINS

   SUBROUTINE ENERGY_NON_BONDING_OPTIONS ( LIST_CUTOFF, OUTER_CUTOFF,   &
                                           KAPPA, DIELECTRIC, NMAXIMUM, &
                                                           TINFOIL, PRINT )
      LOGICAL, INTENT(IN), OPTIONAL :: PRINT, TINFOIL
      REAL,    INTENT(IN), OPTIONAL :: LIST_CUTOFF, OUTER_CUTOFF, &
                                       KAPPA, DIELECTRIC
      INTEGER, DIMENSION(1:3), INTENT(IN), OPTIONAL :: NMAXIMUM
   END SUBROUTINE ENERGY_NON_BONDING_OPTIONS

   SUBROUTINE ENERGY_NON_BONDING_STATISTICS
   END SUBROUTINE ENERGY_NON_BONDING_STATISTICS

END MODULE ENERGY_NON_BONDING
```

The subroutine ENERGY_NON_BONDING_OPTIONS has seven optional arguments. Four of them, LIST_CUTOFF, OUTER_CUTOFF, DIELECTRIC and PRINT, have exactly the same meanings as in the equivalent subroutine of the version of the module described in section 9.2. The meanings of the other three arguments are as follows.

KAPPA is an argument that sets the value of the parameter κ to be used in the
   Ewald summation procedure.

NMAXIMUM is an integer array with three components that gives the maximum
   and minimum values of the components of the vector, $\boldsymbol{n}$, to use in the
   calculation of the $\boldsymbol{k}$ vectors. Thus, if the first component of the vector
   was n_{max}, the sum in the x direction would run from $-n_{\text{max}}$ to $+n_{\text{max}}$
   and include $2n_{\text{max}} + 1$ terms.

TINFOIL is a logical flag that determines the type of boundary conditions to
   use in the calculation of the electrostatic interactions. If it takes the
   value .TRUE. then 'tin-foil' boundary conditions are used and the
   surface term of equation (9.22) is not calculated. If it is .FALSE.
   then this term is included when the electrostatic energy is calculated.

The subroutine ENERGY_NON_BONDING_STATISTICS has the same purpose
as the subroutine described in section 9.2 and, when it is called, prints out
information about the number of interactions in the real space non-bonding
list and the number of $\boldsymbol{k}$ vectors and their frequency of regeneration since the
last call to ENERGY_NON_BONDING_OPTIONS or to ENERGY_NON_BONDING_
STATISTICS.

A number of points should be made about the implementation of the
Ewald method in this module.

- In the module both the electrostatic interactions and the dispersive part of the
  Lennard-Jones interactions are calculated using an Ewald technique (equations
  (9.21) and (9.20), respectively) but the interactions due to the repulsive, r^{-12}, part
  of the Lennard-Jones potential are calculated using a truncation approximation.
  It assumes, which should be reasonable in most cases, that the value of the argument OUTER_CUTOFF in the subroutine ENERGY_NON_BONDING_OPTIONS is large
  enough (> 10 Å) that all the interactions that are left out in this way will be
  negligible in size.
- The implementation in the module assumes that the minimum image convention
  will be obeyed. In other words, all real space interactions must occur between
  atoms that are within the central simulation box. This imposes the constraint that
  the value of the argument OUTER_CUTOFF must be less than half the simulation
  box length and means, in practice, that there is a lower limit to the size of the
  system that can be treated.
- In principle it would be possible to dispense with the argument OUTER_CUTOFF
  because an appropriate value could be estimated by using the value of κ which
  determines how quickly the real space interactions decay towards zero. In practice, it is more flexible to have two parameters, although care should be taken to
  ensure that their values are consistent. Thus, it is important to verify that the

value of κ is not so small compared with the value of the cutoff distance that important interactions are being ignored.

- The Ewald summation technique is slower than the truncation methods discussed earlier. The real space part of the calculation will take about the same length of time for a given value of the cutoff (actually it will take slightly longer due to the necessity of calculating the exp and erfc terms in the energy expression). The self-energy and surface terms are easy to calculate and, like the real space summation, are of $O(N)$. The most time-consuming part of the calculation will be the reciprocal space term, which is $O(Nn_k)$, where n_k is the number of k vectors included in the reciprocal space sums.
- In the current implementation the second derivatives of the non-bonding energies cannot be calculated.

When using the Ewald summation method the values of the parameters, κ, the non-bonding list cutoff and the values of n_{max} need to be checked for each system to obtain a compromise between the speed of calculation of the non-bonding energies and derivatives and the precision required in the calculation. The aim is to use as big a value of κ as possible that reduces the number of real space interactions that need to be calculated without increasing too much the number of reciprocal space terms.

9.8 Fast methods for the evaluation of non-bonding interactions

Owing to the importance of the non-bonding interactions and particularly the long-range electrostatic interactions, much research has gone into algorithms that can be used to evaluate these terms in as efficient a way as possible. The aim is to obtain methods that scale linearly with the size of the system, i.e. as $O(N)$, rather than as the $O(N^2)$ of the direct summation techniques.

In this book, no modules that implement such methods are given, but some algorithms are finding increasing use in molecular simulations. They can be divided roughly into two categories – those that were developed for calculating the non-bonding interactions in aperiodic and periodic systems, respectively.

The fast algorithms for aperiodic systems (large numbers of particles in vacuum, for example) rely on the fact that the electrostatic potential due to a collection of charges with large separations can be well approximated as a limited *multipole expansion*, i.e. as a charge, dipole, quadrupole, octupole, etc. Thus, instead of calculating the electrostatic interaction as a sum over point charges of distant sets of charges, it can be approximated to within a certain precision as an interaction between multipoles (remembering from our previous discussion of the dipole-splitting problem that interactions between multipoles decay more rapidly than do those between charges). Algorithms using this principle typically divide space into small regions or cells, evaluate the multipole expansions due to the charges within each and then calculate the interactions between cells using a direct summation for cells that are near to each other and the multipole approximation for cells that are at longer range. L. Greengard and V. Rokhlin were the first to do systematic work upon these fast multipole approaches and it was they who developed an algorithm that scales as $O(N)$. They also showed that these algorithms can be adapted for use with periodic systems. These methods are complex to implement and it appears that the methods to be discussed below are currently to be preferred for condensed phase molecular systems with 10^4–10^5 particles in the central simulation box.

In more widespread use in molecular simulations are methods developed for accelerating the non-bonding energy calculation for periodic systems. One of the earlier algorithms of this type was due to R. Hockney and J. Eastwood. Just like in the Ewald method, the sum for the energy of a particle is split into short-range and long-range parts with the short-range interactions handled explicitly. To calculate the long-range interactions, a grid or mesh is introduced into the simulation box and the electrostatic potential at each mesh point is calculated by solving the Poisson equation using a fast Fourier transform (FFT) technique. Once the potential on the mesh points is known, the potential at each atom can be determined by interpolating the values at the mesh points and the electrostatic energy for the full system can be calculated. The efficiency of the algorithm arises from the use of the FFT algorithm which scales as $O(n \ln n)$, where n is the number of points in the mesh. This number is, in turn, proportional to the number of atoms in the system, N, for a given mesh size. The algorithm is efficient but it has been criticized for being difficult to use to obtain energies and forces of high accuracy.

A more recent algorithm that overcomes these problems and appears to be the method of choice for the simulation of periodic molecular systems at present is the *particle mesh Ewald* method. It was introduced by T. Darden, D. York and L. Pedersen and, like the previous algorithm, it uses FFT techniques to achieve its computational efficiency. In contrast to the previous approach, though, it takes as its starting point the Ewald expression for the total energy (equations (9.20) and (9.21)) and uses the FFT method together with a spline interpolation technique to enhance the efficiency of the calculation of the reciprocal space terms.

Exercises

9.1 In section 9.3 the non-bonding energies were calculated as a function of the cutoff distance for the non-bonding interactions. Perform a similar analysis but look at properties other than energies. For example, how do the structures of a molecule change as the cutoff distance is increased and how are the forms of the normal modes altered? Is the cutoff approximation a reasonable one?

9.2 Repeat the simulation of Example 16 but this time using the Ewald summation technique. What are the optimal parameters for the summation procedure? How different are the trajectories? Can a smaller dimension be used for the periodic box and equivalent results obtained? *Hint: try values of* 0.2 $Å^{-1}$ *for* κ, 12 $Å$ *for the outer cutoff and 5 for each component of* n_{max}.

10

Molecular dynamics simulations II

10.1 Introduction

In chapter 8 we saw how to perform molecular dynamics simulations, although they were not very sophisticated because there was no means of including the effect of the environment. Ways in which to overcome this limitation were introduced in the last chapter, in which we discussed methods for calculating the energy and its derivatives for a system with the periodic boundary condition approximation. As an example, a molecular dynamics simulation for a periodically replicated cubic box of water molecules was performed. Here, we shall start by describing in more detail the type of information that can be computed from molecular dynamics trajectories and also by indicating how the quality of that information can be assessed. Later we shall talk about more advanced molecular dynamics techniques that allow simulations to be carried out for various thermodynamic ensembles and can be used to calculate free energies.

10.2 Analysis of molecular dynamics trajectories

We have seen how to generate trajectories of coordinate or velocity data for a system, either in vacuum or with an environment, with the molecular dynamics technique. The point of performing a simulation is, of course, that we want to use this data to calculate some interesting quantities, preferably those which can be compared with those measured experimentally. The aim of this section is to give a brief overview of some of the techniques which can be used to analyse molecular dynamics trajectories and some of the types of quantities that can be calculated.

In section 8.4 we defined a time series for a property as a sequence of values for the property obtained from successive frames of a molecular dynamics trajectory. Of the many possible statistical quantities that can be calculated from a time series, we shall focus on three types. These are averages and fluctuations, which we met in section 8.4, and *time correlation functions*. Let $\mathcal{X}$ be the property, $\mathcal{X}_n$ the nth value of the property in the time series and n_t the total number of elements in the series. The average of the property, denoted by $\langle \mathcal{X} \rangle$, is

$$\langle \mathcal{X} \rangle = \frac{1}{n_t} \sum_{n=1}^{n_t} \mathcal{X}_n \tag{10.1}$$

The fluctuation is the average of the square of the deviation of the property from its average. If the deviation is denoted by $\delta \mathcal{X} = \mathcal{X} - \langle \mathcal{X} \rangle$, the fluctuation is

$$\langle (\delta \mathcal{X})^2 \rangle = \langle \mathcal{X}^2 \rangle - \langle \mathcal{X} \rangle^2$$
$$= \frac{1}{n_t} \sum_{n=1}^{n_t} \mathcal{X}_n^2 - \langle \mathcal{X} \rangle^2 \tag{10.2}$$

The RMS deviation of the property, denoted $\sigma(\mathcal{X})$, is the square root of the fluctuation, i.e. $\sigma^2(\mathcal{X}) = \langle (\delta \mathcal{X})^2 \rangle$.

The method of calculation of the averages and fluctuations of a time series is immediately obvious by inspection of equations (10.1) and (10.2). The calculation of correlation functions is more complicated because they are functions of time. The *autocorrelation* function for the property $\mathcal{X}$ is denoted by $\mathcal{C}_{\mathcal{X}\mathcal{X}}(t)$ and has the expression

$$\mathcal{C}_{\mathcal{X}\mathcal{X}}(t) = \langle \delta \mathcal{X}(t)\, \delta \mathcal{X}(0) \rangle$$
$$= \langle \mathcal{X}(t) \mathcal{X}(0) \rangle - \langle \mathcal{X} \rangle^2 \tag{10.3}$$

It is common to normalize the function by dividing it by the fluctuation of the property. The normalized function, $\hat{\mathcal{C}}_{\mathcal{X}\mathcal{X}}(t)$, is

$$\hat{\mathcal{C}}_{\mathcal{X}\mathcal{X}}(t) = \mathcal{C}_{\mathcal{X}\mathcal{X}}(t)/\sigma^2(\mathcal{X}) \tag{10.4}$$

A *cross-correlation function* for two different properties, $\mathcal{X}$ and $\mathcal{Y}$, can be defined as

$$\mathcal{C}_{\mathcal{X}\mathcal{Y}}(t) = \langle \delta \mathcal{X}(t)\, \delta \mathcal{Y}(0) \rangle$$
$$= \langle \mathcal{X}(t) \mathcal{Y}(0) \rangle - \langle \mathcal{X} \rangle \langle \mathcal{Y} \rangle \tag{10.5}$$

It can be normalized in the same way as the autocorrelation function by dividing it by the product of the RMS deviations of properties $\mathcal{X}$ and $\mathcal{Y}$, i.e. by $\sigma(\mathcal{X})\sigma(\mathcal{Y})$.

To calculate a correlation function it is necessary to discretize the formulae given above. The important point to note is that the correlation functions are *stationary*, which means that the value of the function is independent of the origin chosen for the calculation. In other words, for all times, τ:

$$\langle \mathcal{X}(t)\mathcal{X}(0) \rangle = \langle \mathcal{X}(\tau + t)\mathcal{X}(\tau) \rangle \tag{10.6}$$

This implies that, for a given time, t, products $\mathcal{X}(\tau + t)\mathcal{X}(\tau)$, for all possible values of τ, will contribute to the average used to calculate $\mathcal{C}_{\mathcal{X}\mathcal{X}}(t)$. For the autocorrelation function, this translates into an equation of the form

$$\langle \mathcal{X}(t_n)\mathcal{X}(0) \rangle = \frac{1}{n_{\max}} \sum_{i=1}^{n_{\max}} \mathcal{X}_i \mathcal{X}_{i+n} \tag{10.7}$$

where t_n is the time corresponding to the nth element in the time series and $n_{\max} = n_t - n$ is the number of intervals that is used to calculate the average. Note that, if $n = 0$, the expression reduces to that for the fluctuation given in equation (10.2), as it should. Note too that, as n gets large, $n_{\max}$ becomes small, so that for $n \simeq n_t$ there are very few intervals that can be used for the calculation. In practice, this means that, for large n, the values of the calculated correlation function are not reliable because there are not enough values contributing to the average to get a statistically significant result.

The importance of these three statistical quantities is that they are necessary for many of the formulae, derivable from statistical mechanics, that allow the microscopic properties calculated from a simulation and the macroscopic quantities that can be obtained experimentally to be linked. We have already come across some of these formulae, notably the one equating the temperature of the system to the average of its kinetic energy (equation (8.14)) and those that permit the calculation of some thermodynamic quantities for a molecule within the rigid-rotor, harmonic oscillator approximation (section 7.6). Another important relation involving an average, which we shall discuss in detail in section 10.4, allows the pressure for a condensed phase system to be obtained from a simulation. Examples of expressions involving fluctuations are those that relate the specific heats (either at constant volume or at constant pressure) to fluctuations in the potential and kinetic energies of the system arising in a simulation. Time correlation functions are important because they are fundamental to the derivation of formulae that permit *transport coefficients* for a system to be calculated. Examples include the *diffusion coefficient*, the *bulk* and *shear viscosities* and the *thermal conductivity*.

We shall consider the formulae for two properties in detail because these will be calculated in the example program of the next section. The first property is the diffusion coefficient for a species i, D_i, which is proportional to the time integral of its velocity autocorrelation function:

$$D_i = \frac{1}{3} \int_0^\infty dt \, \langle v_i^T(t) v_i(0) \rangle \tag{10.8}$$

This equation can be integrated by parts to give the following expression which is valid at long times, t:

$$6t D_i = \langle (r_i(t) - r_i(0))^2 \rangle \tag{10.9}$$

Equation (10.9) is an example of an *Einstein relation* for a transport coefficient. The average on the right-hand side of equation (10.9) is closely related to that of a time correlation function and it can be calculated in a very similar fashion. The only difference is that, instead of taking the average of the product of the property at two different times as in equation (10.7), the averaging is performed over the square of the difference of the property at the two times.

The second property is the *pair distribution function* or the *radial distribution function*, which is important in the theory of simple fluids (gases and liquids) because many thermodynamic quantities can be determined from it. The function, often denoted $g(r)$, can be thought of as a measure of the structure in a system because it gives the probability of finding a pair of particles a distance r apart, relative to the probability that would be expected for a random distribution with the same density. In practice $g(r)$ is not calculated for single values of the distance r, but for discrete intervals with a width of, say, δr. Denoting the value of the radial distribution function in the range $[r, r + \delta r]$ as $g(r + \frac{1}{2} \delta r)$ (which is the mid-point of the interval), one has

$$g\left(r + \frac{1}{2} \delta r\right) = \frac{n_{\text{sim}}([r, r + \delta r])}{n_{\text{random}}([r, r + \delta r])} \tag{10.10}$$

where the functions n give the average number of particles whose distances from a given particle lie within the range $[r, r + \delta r]$ and the subscripts 'sim' and 'random' denote the simulation and the random values, respectively. The expression for n_{random} is

$$n_{\text{random}}([r, r + \delta r]) = \frac{4\pi N}{3V} \left[(r + \delta r)^3 - r^3\right] \tag{10.11}$$

where N is the number of particles for which $g(r)$ is to be calculated and V is the system's volume. To determine n_{sim} it is necessary that all the distances between particles be calculated for each frame in the trajectory and a histogram kept that records the number of distances that fall within any particular range $[r, r + \delta r]$. n_{sim} is then equal to the number of distances found for the interval from the histogram divided by the number of frames in the trajectory, n_f, and by N. The division by n_f gives the average number of distances per frame and that by N the average number of distances per particle.

A crucial part of any simulation study is how to estimate the accuracy of the results that have been obtained. Errors can be introduced at several stages. At the most basic level, they arise when a model is chosen to describe the system. We know, for example, that the empirical force field used to calculate energies and the assumption that atoms are classical particles for the dynamics are approximations that will affect the generality of the simulation methodology. There is not much we can do about these errors except improve the physical basis of the model. At the next level, errors arise due to the way in which the model we have chosen is applied. Examples include whether the truncation scheme for the determination of the non-bonding interactions is adequate, whether the size of the central box for a periodic boundary condition simulation is large enough that the effects of imposing periodicity are unimportant and whether the starting configuration of the system (the atomic positions and velocities) has been sufficiently well prepared. These types of errors are characterized by the fact that, for a given model, it is possible to investigate them by changing some of the parameters of the simulation and then repeating the study to see how the results change. In practice, how well this can be done will depend upon the type of system being studied. It will often be feasible to determine systematically the importance of these effects for a system comprised of relatively few atoms or simple molecules, but it can be difficult if the system is so large that each simulation is expensive or if the system is a complicated one containing, for example, flexible molecules with multiple conformations.

A third type of errors is statistical errors, which occur in any quantity that is calculated from a simulation of finite length. The problem is to estimate how close the quantities that have been calculated from the simulation are to the values that would have been obtained from a simulation of infinite length. A related question, although one that is more difficult to answer, is that we would like to determine, if possible, the minimum length of a simulation that is necessary to obtain results of a given precision. A great variety of sophisticated statistical techniques is available for such estimates, although we shall not deal with any of them in this book. Instead, we shall limit ourselves to a number of *very basic* points. First, it is always a good idea to monitor the values of averages and fluctuations as a simulation proceeds to see whether they converge to a limiting value. It is also usually necessary to repeat the simulation several times, with different starting configurations, to see whether the same limiting values are obtained. Second, for quantities that depend on some variable, such as time correlation functions and radial distribution functions, the curves obtained should be smooth. Any roughness indicates that not enough data have been used in their determination. Finally, a useful approximate rule of thumb is that the length of simulation, t, required to obtain sufficient data for the calculation of a particular property, $\mathcal{X}$, must be much longer than the *correlation time*, $\tau_{\mathcal{X}}$, associated with that property, i.e. $t \gg \tau_{\mathcal{X}}$. This rule comes from the fact that many correlation functions have an exponential form, i.e. $\langle \mathcal{X}(t)\mathcal{X}(0) \rangle \propto \exp(-t/\tau_{\mathcal{X}})$, where the exponent of the exponential is the inverse of the correlation time. For short times, $t \simeq \tau_{\mathcal{X}}$, the value of $\mathcal{X}$ is correlated to its initial value and so will not contribute independently to the average or fluctuation. To do so, the simulation needs to be several periods of length $\tau_{\mathcal{X}}$ long so that several independent *blocks* of $\mathcal{X}$ values have been generated. It should be noted that by no means all correlation functions decay exponentially and that some decay very slowly at long times. They are said to have *long-time tails*. The calculation of the correlation functions for such properties can be a computationally demanding task.

To finish the discussion on errors we summarize by emphasizing that it can take data generated from several or even many *long* simulations to obtain results to within a reasonable precision. Even then, however, readers should be warned that great care must always be exercised when interpreting the significance of the results of any numerical simulation!

Two new modules that can help in performing some of the analyses described above are introduced in this section. The first is called TRAJECTORY_ANALYSIS and it contains three subroutines that calculate specific quantities from a trajectory. Its definition is

```
MODULE TRAJECTORY_ANALYSIS

CONTAINS

   SUBROUTINE TRAJECTORY_DSELF ( FILE, TSTOP, SELECTION )
      CHARACTER ( LEN = * ), INTENT(IN) :: FILE
      INTEGER,                INTENT(IN) :: TSTOP
      LOGICAL, DIMENSION(:), INTENT(IN), OPTIONAL :: SELECTION
   END SUBROUTINE TRAJECTORY_DSELF

   SUBROUTINE TRAJECTORY_RDF ( FILE, NBINS, UPPER, SELECTION )
      CHARACTER ( LEN = * ), INTENT(IN) :: FILE
      INTEGER,                INTENT(IN) :: NBINS
      REAL,                   INTENT(IN) :: UPPER
      LOGICAL, DIMENSION(:), INTENT(IN), OPTIONAL :: SELECTION
   END SUBROUTINE TRAJECTORY_RDF
```

```
     SUBROUTINE TRAJECTORY_RDF_CROSS ( FILE, NBINS, UPPER, SELECTION1, &
                                                        SELECTION2 )
        CHARACTER ( LEN = * ), INTENT(IN) :: FILE
        INTEGER,               INTENT(IN) :: NBINS
        REAL,                  INTENT(IN) :: UPPER
        LOGICAL, DIMENSION(:), INTENT(IN) :: SELECTION1, SELECTION2
     END SUBROUTINE TRAJECTORY_RDF_CROSS

  END MODULE TRAJECTORY_ANALYSIS
```

The subroutines are `TRAJECTORY_DSELF`, which calculates the function in the Einstein relation of equation (10.9), `TRAJECTORY_RDF`, which calculates the radial distribution function of equation (10.10) for a single set of atoms, and `TRAJECTORY_RDF_CROSS`, which calculates a radial distribution function for two different sets of atoms.

The subroutine `TRAJECTORY_DSELF` requires two input arguments. The first is a character variable, `FILE`, that gives the name of the trajectory file containing the data to be analysed. The second argument is an integer variable, `TSTOP`, which specifies the maximum time difference for which the Einstein relation function is to be calculated. This time difference is an integer because it is expressed in terms of frames in the trajectory rather than in picoseconds. The subroutine converts the times internally to picoseconds by using the time interval between frames that is stored in the trajectory file. Because the Einstein relation function is undefined for values of `TSTOP` that are greater than or equal to the number of frames in the trajectory, an error will occur if this is the case.

The subroutine `TRAJECTORY_RDF` requires three input arguments. The first argument is the trajectory file name. The second argument, `NBINS`, is an integer that gives the number of distance intervals or *bins* for which the radial distribution function is to be calculated. The third argument, `UPPER`, is a real that gives the maximum distance to consider for the calculation of the function. All interparticle distances greater than this are discarded. The width of each bin (equivalent to δr in equation (10.10)) is `UPPER/NBINS`.

The subroutines `TRAJECTORY_DSELF` and `TRAJECTORY_RDF` can also take an optional argument that is a logical array having the length of the number of atoms, i.e. the array must be defined as `SELECTION(1:NATOMS)`. If this array is present, only the data for the atoms for which the corresponding element in the array `SELECTION` is `.TRUE.` will be analysed and the data for the remaining atoms will be ignored. If `SELECTION` is not present, the default is to analyse the data for all the atoms. This means, in particular, that the subroutine `TRAJECTORY_DSELF` calculates the functions of equation (10.9) for each selected atom and then prints their average.

The third subroutine, TRAJECTORY_RDF_CROSS, has five arguments, all of which are essential. The first three correspond exactly to the essential arguments of the subroutine TRAJECTORY_RDF. The other two arguments, SELECTION1 and SELECTION2, are logical arrays that define the sets of atoms for which the distribution function is to be calculated. They must both have dimensions (1:NATOMS).

The second module is more general and is designed to help users with their own analyses. It is called STATISTICS and contains two subroutines. Its definition is

```
MODULE STATISTICS

CONTAINS

    SUBROUTINE STATISTICS_ACF_DIRECT ( X, ACF, NORMALIZED )
        REAL, DIMENSION(:),  INTENT(IN)  :: X
        REAL, DIMENSION(0:), INTENT(OUT) :: ACF
        LOGICAL, INTENT(IN), OPTIONAL :: NORMALIZED
    END SUBROUTINE STATISTICS_ACF_DIRECT

    SUBROUTINE STATISTICS_CCF_DIRECT ( X, Y, CCF, NORMALIZED )
        REAL, DIMENSION(:),  INTENT(IN)  :: X, Y
        REAL, DIMENSION(0:), INTENT(OUT) :: CCF
        LOGICAL, INTENT(IN), OPTIONAL :: NORMALIZED
    END SUBROUTINE STATISTICS_CCF_DIRECT

END MODULE STATISTICS
```

The subroutine STATISTICS_ACF_DIRECT calculates the autocorrelation function of a property. There are two essential arguments. The first, X, is an input real array that contains the time series for which the autocorrelation function is to be calculated. The second is a real array, ACF, which contains the autocorrelation function on output. This array must have as its lower bound the element 0. Note that, if the upper bound of the array ACF is greater than the number of elements in X, then the elements in ACF for times greater than the length of the simulation cannot be calculated and will be returned undefined. The third argument for the subroutine is optional and it determines whether the calculated autocorrelation function is to be normalized. If the argument is not present no normalization is done and the function $\langle \mathcal{X}(t)\ \mathcal{X}(0) \rangle$ is returned, whereas if it is present with the value .TRUE. the function from equation (10.4) is calculated. STATISTICS_ACF_DIRECT would normally be employed for the analysis of trajectory data using a scheme similar to that in Example 13 of section 8.5. The frames of a trajectory would be read one by one and the data of interest extracted and stored in an array. The autocorrelation function for the stored data would then be calculated after the complete trajectory had been read.

The subroutine STATISTICS_CCF_DIRECT is exactly equivalent to STATISTICS_ACF_DIRECT except that it calculates the cross-correlation function of equation (10.5). The time-series data for the two properties are input in the arrays X and Y; the correlation function is output in the array CCF.

The algorithm used by the subroutines STATISTICS_ACF_DIRECT and STATISTICS_CCF_DIRECT is called a *direct* algorithm because it calculates the autocorrelation function using an expression that is based upon equation (10.7). The time required by the subroutine is, therefore, $O(n_c(2n_t - n_c)) \simeq O(n_c n_t)$, where n_t is the number of elements in the time-series array, X, and n_c is the number of elements for which the autocorrelation function is to be calculated. For the full function the time required is $O(n_t^2)$. There is a faster algorithm for the calculation of the full correlation function that employs the FFT method and scales as $O(n_t \ln n_t)$. This will be more efficient when there are large amounts of data, although no subroutine implementing this algorithm is given here.

The subroutine TRAJECTORY_DSELF also employs a direct algorithm for its calculation, which scales as $O(n_f N \text{ TSTOP})$, where n_f is the number of frames in the trajectory file, N is the number of particles for which the calculation is to be performed and TSTOP has the same meaning as above. The algorithm, as it is implemented here, requires that all data from the trajectory that is to be analysed $(\simeq O(n_f N))$ be stored in memory simultaneously. Both these factors will limit the size of the analyses that can be performed by this subroutine. The calculations performed by TRAJECTORY_RDF and TRAJECTORY_RDF_CROSS can also become expensive because the time required is $O(n_f N^2)$ but, unlike TRAJECTORY_DSELF, each frame of the trajectory is processed separately and so large amounts of memory are not needed.

In the current implementations, the subroutines TRAJECTORY_RDF and TRAJECTORY_RDF_CROSS can be used only to analyse trajectory data generated for a periodic system whereas the analysis performed by TRAJECTORY_DSELF is independent of whether the system has periodic boundary conditions. This restriction means that, for the calculation of radial distribution functions, the system must be periodic and the value given by the argument UPPER must be less than half the length of the simulation box.

To conclude this section, we note that the dimensions of a periodic box, if one is present, are written out to the trajectory file at each step, together with the atomic data. This is done by introducing a third, optional argument for

the subroutines TRAJECTORY_READ and TRAJECTORY_WRITE into the module
TRAJECTORY_IO. Their modified definitions are

```
MODULE TRAJECTORY_IO

CONTAINS

   SUBROUTINE TRAJECTORY_READ ( TRAJECTORY, ATMDAT, BOX_SIZE )
      TYPE(TRAJECTORY_TYPE), INTENT(INOUT) :: TRAJECTORY
      REAL, DIMENSION(:,:),  INTENT(OUT)   :: ATMDAT
      REAL, DIMENSION(1:3),  INTENT(OUT), OPTIONAL :: BOX_SIZE
   END SUBROUTINE TRAJECTORY_READ

   SUBROUTINE TRAJECTORY_WRITE ( TRAJECTORY, ATMDAT, BOX_SIZE )
      TYPE(TRAJECTORY_TYPE), INTENT(INOUT) :: TRAJECTORY
      REAL, DIMENSION(:,:),  INTENT(IN)    :: ATMDAT
      REAL, DIMENSION(1:3),  INTENT(IN), OPTIONAL :: BOX_SIZE
   END SUBROUTINE TRAJECTORY_WRITE

END MODULE TRAJECTORY_IO
```

The array BOX_SIZE contains the lengths of the sides of the periodic box
along each of the Cartesian axes.

10.3 Example 17

The example program presented in this section is a simple one that illustrates
the use of the module TRAJECTORY_ANALYSIS by analysing the molecular
dynamics trajectory that was generated in Example 16 of section 9.6. The
program is

```
PROGRAM EXAMPLE17

... Declaration statements ...

! . Local array declarations.
LOGICAL, ALLOCATABLE, DIMENSION(:) :: SELECTION

! . Construct the system file for water.
CALL MM_SYSTEM_CONSTRUCT ( ''solvent.opls_bin'', ''h2o_box729.seq'' )

! . Read in the coordinates.
CALL COORDINATES_READ ( ''h2o_box729.crd'' )

! . Allocate and initialize the selection array.
ALLOCATE ( SELECTION(1:NATOMS) ) ; SELECTION = ( ATMNAM == ''O'' )

! . Calculate the self-diffusion function for the oxygens.
CALL TRAJECTORY_DSELF ( ''h2o_box729.traj'', 100, SELECTION )
```

```
! . Calculate the radial distribution function for the oxygens.
CALL TRAJECTORY_RDF ( ''h2o_box729.traj'', 100, 10.0, SELECTION )

! . Deallocate the selection array.
DEALLOCATE ( SELECTION )


END PROGRAM EXAMPLE17
```

As usual the program starts by defining the system and reading in a set of coordinates. The latter is not strictly necessary because the coordinates in the file will not be used later in the program but the act of reading the coordinate file indicates to the program that the system has cubic symmetry. An alternative, but probably less convenient way, would be to call the subroutine SYMMETRY_CUBIC_BOX from the module SYMMETRY directly with an appropriate value for the box length.

After the the system has been set up, the logical array SELECTION, which is defined at the start of the program, is allocated so that it has a length equal to the number of atoms. Its element are then initialized so that elements corresponding to atoms that have the name O are set to .TRUE. and all others are set to .FALSE..

The program performs two analyses using the trajectory h2o_box729.traj that was created in Example 16. The first is effected by a call to TRAJECTORY_DSELF, which calculates the function from the Einstein relation of equation (10.9). The function is calculated for a maximum time difference of 100 trajectory frames (out of a possible 201), which corresponds to a time of 5 ps because the interval at which frames were stored was 0.05 ps. As explained earlier, the results for the function for time differences of the order of the trajectory length are much less precise than those at shorter times and are usually not worth calculating. The second analysis is carried out by the subroutine TRAJECTORY_RDF, which determines the radial distribution function of equation (10.10). The function is calculated using 100 bins in the range 0–10 Å. Both subroutines print their results directly to the output file. The calculations are performed for the oxygen atoms of the water molecules only, because the array SELECTION is passed as an argument to both analysis subroutines.

Plots of the simulation results are shown in figures 10.1 and 10.2. The value of the diffusion coefficient can be calculated from the slope of the line in figure 10.1 at large times, giving a value of 0.42 $Å^2$ ps^{-1} or 4.2×10^{-9} m^2 s^{-1}, which is large relative to the experimental value of 2.3×10^{-9} m^2 s^{-1} at 25 °C. To verify this result fully, a similar analysis would have to be carried out on a longer trajectory to ensure that the function plotted in figure 10.1 had indeed reached its limiting value at long times.

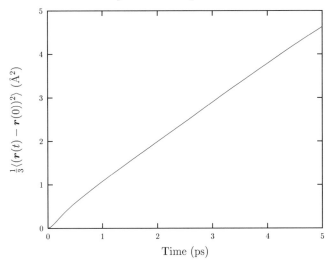

Figure 10.1. The function $\frac{1}{3}\langle(r(t)-r(0))^2\rangle$ calculated for the oxygen atoms of the water molecules using the molecular dynamics trajectory generated in Example 16.

The radial distribution function of figure 10.2 is also found to reproduce reasonably closely the experimental neutron and X-ray results, although the experimental curve has a lower first peak and has more pronounced oscillations at longer range. The fact that $g_{OO}(r) = 0$ for $r \leq 2.4$ Å indicates that the probability of finding two oxygens at these distances apart is negligible whereas the presence of oscillations after this implies that the water molecules are preferentially located in particular regions that correspond to the various 'coordination' shells in the liquid. The structure in the first coordination shell is especially marked.

10.4 Temperature and pressure control in molecular dynamics simulations

Three molecular dynamics simulations have been described up to now. The general procedure when doing a simulation has been to assign velocities to the atoms in the system at a low temperature, heat the system up to the temperature desired during the course of a short simulation and then equilibrate it at this temperature for another short period before starting the simulation for which a trajectory is to be generated. In the heating and equilibration phases of the simulations described previously, the temperature of the system was set to the required value by the simple, *ad hoc* procedure of scaling the velocities of all the atoms every few steps. When no temperature modification is done, we have seen that one measure of the accuracy of the

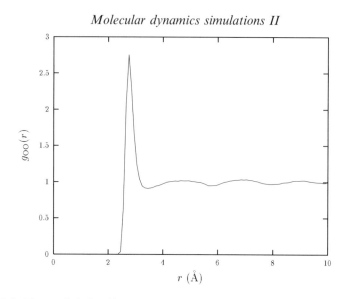

Figure 10.2. The radial distribution function calculated for the oxygen atoms of the water molecules using the molecular dynamics trajectory generated in Example 16.

integration of the equations of motion for the particles is that the total energy for the system is conserved. In thermodynamic terms, the simulations are said to have been performed in the *microcanonical* or *NVE ensemble*. In other words, the number of particles, the volume and the energy of the system are constants.

Molecular dynamics simulations within the microcanonical ensemble are the easiest to perform, but it is evident that, if the aim of our simulations is to mimic the conditions under which systems are investigated experimentally, then the microcanonical ensemble might not be the most appropriate. In particular, it is common to do experiments under conditions in which the ambient temperature and/or pressure are constants. The thermodynamic ensembles that correspond to such conditions are the *canonical* or *NVT ensemble*, the *isothermal–isobaric* or *NPT ensemble* and the *isobaric–isoenthalpic* or *NPH ensemble*, respectively.

Before we go on to discuss methods that allow molecular dynamics simulations to be performed for these ensembles, we shall introduce an algorithm developed by H. J. C. Berendsen, J. P. M. Postma, W. F. van Gunsteren, A. Di Nola and J. R. Haak for the control of the temperature and pressure in a molecular dynamics simulation. It should be emphasized that this algorithm does *not* generate trajectories drawn from the NVT, NPT or NPH ensembles, although the results need not be too different in particular cases. The reason that we use it here is that it is simple and robust and, importantly, entails only slight modifications to the dynamics modules that were presented in chapter 8. It also acts as an introduction to certain concepts that will be needed in the discussion of the more precise algorithms that are to be mentioned later.

An essential idea behind the method is that the system that we want to simulate is, in fact, not isolated but interacts, or is *coupled*, to an *external bath*. Let us consider temperature control first. Remember that, in a simulation in the microcanonical ensemble, the energy remains constant and the temperature fluctuates. A coupling to an external system means that energy can be transferred into and out of the system that we are simulating and so its energy will fluctuate. It is this transfer, properly formulated, which allows the algorithm to control the temperature.

Berendsen *et al.* proposed the following modification of the equation of motion for the velocities of the atoms, V, to accomplish the coupling:

$$\dot{V} = \mathbf{M}^{-1}F + \frac{1}{2\tau_T}\left(\frac{T_B}{T} - 1\right)V \tag{10.12}$$

where T_B is the reference temperature, which is the temperature of the external *thermal bath*, T is the instantaneous temperature defined in equation (8.15) and τ_T is a coupling constant (with units of time). The extra term added to the equations of motion acts like a *frictional force*. When the actual temperature of the system is higher than the desired temperature, the force is negative. This results in the motions of the atoms being damped and the kinetic energy and, hence, the temperature being reduced. If the temperature is too small, the reverse happens because the frictional force is positive and energy is supplied to the system. The coupling constant, τ_T, determines the strength of the coupling to the external bath. For large values of τ_T the coupling is weak and the temperature of the system will be steered only slowly towards the temperature of the bath. For small values the coupling is stronger and the dynamics of the system will be more strongly perturbed.

The principles behind the control of pressure are similar to those for the control of temperature. The major difference, however, is that the control of pressure manifests itself as a modification of the equation of motion for the coordinates *and* the volume of the system, V. Note that, because volume changes are involved, simulations that control the pressure only make sense in the condensed phase. The pressure-control equations of motion proposed by Berendsen *et al.* are

$$\dot{R} = V - \frac{\beta}{3\tau_P}(P_B - \mathcal{P})R \tag{10.13}$$

$$\dot{V} = -\frac{\beta}{\tau_P}(P_B - \mathcal{P})V \tag{10.14}$$

where P_B is the reference pressure and $\mathcal{P}$ is the *instantaneous pressure*. The parameters β and τ_P are the *isothermal compressibility* of the system, which has units of inverse pressure, and the pressure coupling constant, which has units of time. It is the ratio of these two parameters, β/τ_P, that determines the size of the coupling of the system to the external *pressure bath*. Equations (10.13) and (10.14) behave in a similar way to equation (10.12). If the actual pressure is less than the desired pressure the system contracts whereas if the actual pressure is too large the system expands.

An instantaneous pressure can be defined in the same way as an instantaneous temperature. Thus, the thermodynamic pressure, P, is the average of the instantaneous pressure:

$$P = \langle \mathcal{P} \rangle \tag{10.15}$$

The instantaneous pressure can be calculated using the formula

$$\mathcal{P} = \frac{1}{V}\left(\frac{2}{3}\mathcal{K} + \mathcal{W}\right) \tag{10.16}$$

where $\mathcal{K}$ is the kinetic energy and $\mathcal{W}$ is the *instantaneous internal virial* for the system, defined as

$$\mathcal{W} = \frac{1}{3}\mathbf{F}^{\mathrm{T}}\mathbf{R} \tag{10.17}$$

The above formulae are valid only for an isotropic system in which the pressure acts equally no matter what the direction. In the general case of an anisotropic system, the pressure is written as a tensor, . This tensor, which has nine components, is sometimes known as the *stress tensor*. Its form is

$$\mathbf{\Pi} = \begin{pmatrix} \Pi_{xx} & \Pi_{xy} & \Pi_{xz} \\ \Pi_{yx} & \Pi_{yy} & \Pi_{yz} \\ \Pi_{zx} & \Pi_{zy} & \Pi_{zz} \end{pmatrix} \tag{10.18}$$

The pressure in the isotropic case is given by one third of the trace of the stress tensor:

$$P = \frac{1}{3}\left(\Pi_{xx} + \Pi_{yy} + \Pi_{zz}\right) \tag{10.19}$$

It is possible to generalize equations (10.13)–(10.17) so that simulations can be performed with all components of the stress tensor. This results in the shape of the simulation box changing as well as its size. Such techniques are particularly useful for studying crystals and other solids in which there are changes of phase.

The above algorithm has been incorporated into the module `VELOCITY_VERLET_DYNAMICS` that was described in section 8.2. The only difference from the outline of the module given there is the addition of some extra optional arguments to the subroutine DYNAMICS. The new definition is

```
MODULE VELOCITY_VERLET_DYNAMICS

CONTAINS

    SUBROUTINE DYNAMICS ( DELTA, NSTEP, NPRINT, ASSIGN_VELOCITIES,         &
                          INITIAL_TEMPERATURE, SCALE_FREQUENCY, SCALE_OPTION, &
                          TARGET_TEMPERATURE, TEMPERATURE_COUPLING,        &
                          TARGET_PRESSURE,    PRESSURE_COUPLING,           &
                          SAVE_FREQUENCY, COORDINATE_FILE, VELOCITY_FILE )
        INTEGER, INTENT(IN) :: NPRINT, NSTEP
        REAL,    INTENT(IN) :: DELTA
        CHARACTER ( LEN = * ), INTENT(IN), OPTIONAL :: COORDINATE_FILE, &
                                                       SCALE_OPTION,    &
                                                       VELOCITY_FILE
        INTEGER, INTENT(IN), OPTIONAL :: SCALE_FREQUENCY, SAVE_FREQUENCY
        LOGICAL, INTENT(IN), OPTIONAL :: ASSIGN_VELOCITIES
        REAL,    INTENT(IN), OPTIONAL :: INITIAL_TEMPERATURE, PRESSURE_COUPLING, &
                                         TARGET_PRESSURE, TARGET_TEMPERATURE,    &
                                         TEMPERATURE_COUPLING
    END SUBROUTINE DYNAMICS

END MODULE VELOCITY_VERLET_DYNAMICS
```

The new arguments to the subroutine are PRESSURE_COUPLING, TARGET_PRESSURE and TEMPERATURE_COUPLING. PRESSURE_COUPLING gives the value of the ratio τ_P/β in equations (10.13) and (10.14) in units of ps atm, TARGET_PRESSURE gives the value of the reference pressure, P_B, in atmospheres and TEMPERATURE_COUPLING the value of the temperature coupling constant, τ_T, from equation (10.12) in picoseconds. The reference temperature is specified using the existing argument, TARGET_TEMPERATURE. All of these quantities are given nonsensical values by default, so they must be explicitly defined if they are to be used in a calculation.

To specify whether the algorithm developed by Berendsen *et al.* is to be employed, the character argument SCALE_OPTION must be specified with one of the following values: ''BERENDSEN_P'', ''BERENDSEN_PT'' and ''BERENDSEN_T''. These ask for pressure, pressure and temperature and temperature control, respectively, and must be accompanied by the appropriate coupling and target arguments described above. Note that pressure control is allowed only when periodic boundary conditions are being employed and that all changes to the size of the simulation box are made isotropically.

The implementation of the algorithm is straightforward. The essential difference from a standard dynamics algorithm is that, at appropriate points in the integration, the velocities, coordinates and box size are scaled by factors determined by equations (10.12)–(10.14). For temperature control the velocity scale factor, ζ_T, is

$$\zeta_T = \sqrt{1 + \frac{\Delta}{\tau_T}\left(\frac{T_B}{T} - 1\right)} \qquad (10.20)$$

while for pressure control the coordinate scale factor, ζ_P, is

$$\zeta_P = \sqrt[3]{1 - \frac{\Delta\beta}{\tau_P}(P_B - \mathcal{P})} \qquad (10.21)$$

The volume is scaled by the factor ζ_P^3. In both equations, Δ is the timestep for the integration of the equations of motion.

The calculation of the pressure requires the calculation of the internal virial (equation (10.17)). It is not difficult to show, although it is a little tedious, that this depends upon the pairwise bonding and non-bonding interactions only and the contributions from the angle and dihedral force-field terms cancel out. Note that this is true only for the trace of the stress tensor – its individual components will depend upon the angle and dihedral energies. For pairwise interactions the internal virial can be expressed as

$$\mathcal{W} = \frac{1}{3}\sum_{i=2}^{N}\sum_{j=1}^{i-1} r_{ij}^T f_{ij} \qquad (10.22)$$

where f_{ij} is the force acting on particle i due to particle j and the vector r_{ij} is the minimum image vector between the particles.

The algorithm described above is the one that we shall use in this book. As already remarked, though, it does not generate trajectories in the NVT, NPT or NPH ensembles (depending upon whether temperature and/or pressure control is being used). There are methods that do this but before we describe some of them it is important to define exactly what is meant by generating trajectories in an appropriate ensemble. To do this we need to define the *probability density distribution functions* for the various ensembles. These are crucial to the theory of statistical thermodynamics because they can be used to calculate the properties of any equilibrium system. Thus, if the probability density distribution function in a particular ensemble is ρ, the average of a property, $\mathcal{X}$, can be written as an integral:

$$\langle\mathcal{X}\rangle = \int d\Gamma\, \mathcal{X}(\Gamma)\rho(\Gamma) \qquad (10.23)$$

where Γ are the ensemble variables which will include the coordinates and momenta of the particles and dΓ indicates the volume element for a multi-dimensional integration over these variables.

The probability density is easiest to define for the microcanonical ensemble. If we suppose that the energy in the ensemble has a value E, then the probability of a configuration that does not have this energy will be zero for, by definition, the energy is a constant in the ensemble. Hence, the probability of a configuration that has this energy will be simply the reciprocal of the total number of states with an energy E. Mathematically, the number of states with an energy E must be written as an integral over the *phase space* of the system because the system's coordinates, R, and momenta, P, are continuous variables. Taking this into account, the probability density for the microcanonical ensemble, ρ_{NVE}, is

$$\rho_{NVE} = \frac{\delta(\mathcal{H}(P, R) - E)}{\int dP \, dR \, \delta(\mathcal{H}(P, R) - E)} \tag{10.24}$$

where $\mathcal{H}$ is the Hamiltonian for the system and the Dirac delta functions have the effect of selecting only those configurations with total energy E.

In the canonical ensemble, it is the thermodynamical temperature that is constant and the energy fluctuates. The probability density is written as

$$\rho_{NVT} = \frac{\exp[-\mathcal{H}(P, R)/(k_{\mathrm{B}}T)]}{\int dP \, dR \, \exp[-\mathcal{H}(P, R)/(k_{\mathrm{B}}T)]} \tag{10.25}$$

where k_{B} is Boltzmann's constant and T is the temperature. Note that this equation is a statement of the familiar *Boltzmann distribution law* which says that the probability of a configuration is proportional to its *Boltzmann factor*, $\exp[-\mathcal{H}/(k_{\mathrm{B}}T)]$.

In the isobaric–isothermal ensemble the density function is similar to the canonical function but the volume of the system is also a variable:

$$\rho_{NPT} = \frac{\exp[-(\mathcal{H}(P, R) + PV)/(k_{\mathrm{B}}T)]}{\int dP \, dR \, dV \, \exp[-(\mathcal{H}(P, R) + PV)/(k_{\mathrm{B}}T)]} \tag{10.26}$$

What we would like is a molecular dynamics method that generates configurations that are representative either of the canonical or of the isobaric–isothermal ensembles, i.e. one that generates states with a probability distribution appropriate for the ensemble. Many methods that do this have been proposed and we shall only briefly mention a few of them here. As was the case with the Berendsen algorithm, it is usual to divide the methods into those which keep the temperature constant and those that maintain the pressure. Simulations in the NPT ensemble are then performed by combining algorithms for each type of control separately.

We consider constant-temperature algorithms for molecular dynamics simulations first. One technique was originally proposed by S. Nosé and later extended by W. G. Hoover and by G. J. Martyna, M. L. Klein and M. E. Tuckerman. The basis of the method is to introduce an extra, *thermostating* degree of freedom that represents the external thermal bath to which the system is coupled. In the original Nosé–Hoover algorithm, there is a single bath coordinate, η, and an associated momentum, p_η, in addition to the atomic coordinates and momenta. The modified equations of motion for the combined or *extended system* are

$$\dot{R} = \mathbf{M}^{-1} P \tag{10.27}$$

$$\dot{P} = F(R) - \frac{p_\eta}{Q} P \tag{10.28}$$

$$\dot{\eta} = \frac{p_\eta}{Q} \tag{10.29}$$

$$\dot{p}_\eta = P^{\mathrm{T}} \mathbf{M}^{-1} P - N_{\mathrm{df}} k_{\mathrm{B}} T \tag{10.30}$$

where N_{df} is the number of coordinate degrees of freedom. The parameter Q is the 'mass' of the thermostat (with units of mass times length squared) which determines the size of the coupling. A good choice of Q is crucial. Large values result in equations that approximate Newton's equations (and, hence, a constant-energy simulation), whereas small values produce large couplings and, as Nosé showed, lead to dynamics with poor equilibration. Physically, the momentum variable, p_η, in equation (10.28) acts like a friction coefficient. When its value is positive, the kinetic energy of the system is damped, and, when the value is negative, it is increased. The value of p_η is determined by equation (10.30) whose right-hand side is proportional to the difference between the actual temperature and the desired temperature. If the current temperature is too big, energy is removed from the system because the friction coefficient increases, whereas the reverse happens if the temperature is too small.

To perform a constant-temperature simulation, these equations are integrated in the normal way (with the extra two degrees of freedom). Like calculations in the microcanonical ensemble, it is important to be able to verify the precision of the simulation by monitoring conserved quantities. In the microcanonical ensemble it is the energy defined by the classical Hamiltonian (equation (8.1)) which is perhaps the most important. The equivalent in the Nosé–Hoover algorithm is defined by the Hamiltonian, $\mathcal{H}_{\mathrm{NH}}$:

$$\mathcal{H}_{\text{NH}} = \frac{1}{2}\boldsymbol{P}^{\text{T}}\mathbf{M}^{-1}\boldsymbol{P} + \mathcal{V}(\boldsymbol{R}) + \frac{p_\eta^2}{2Q} + N_{\text{df}}k_{\text{B}}T\eta \tag{10.31}$$

Nosé and Hoover showed that the system of equations (10.27)–(10.30) produced trajectories of atomic coordinates and momenta drawn from a canonical distribution. It was observed, however, that in some cases the control of temperature was inadequate or poor. This led Martyna *et al.* to propose adding more thermostating degrees of freedom by introducing a 'chain' of thermostats. Their equations were

$$\dot{\boldsymbol{R}} = \mathbf{M}^{-1}\boldsymbol{P} \tag{10.32}$$

$$\dot{\boldsymbol{P}} = \mathbf{F}(\boldsymbol{R}) - \frac{p_{\eta_1}}{Q_1}\boldsymbol{P} \tag{10.33}$$

$$\dot{\eta}_i = \frac{p_{\eta_i}}{Q_i} \; \forall \, i = 1, \ldots, M \tag{10.34}$$

$$\dot{p}_{\eta_1} = \boldsymbol{P}^{\text{T}}\mathbf{M}^{-1}\boldsymbol{P} - N_{\text{df}}k_{\text{B}}T - p_{\eta_1}\frac{p_{\eta_2}}{Q_2} \tag{10.35}$$

$$\dot{p}_{\eta_j} = \frac{p_{\eta_{j-1}}^2}{Q_{j-1}} - k_{\text{B}}T - p_{\eta_j}\frac{p_{\eta_{j+1}}}{Q_{j+1}} \; \forall \, j = 2, \ldots, M-1 \tag{10.36}$$

$$\dot{p}_{\eta_M} = \frac{p_{\eta_{M-1}}^2}{Q_{M-1}} - k_{\text{B}}T \tag{10.37}$$

where M is the total number of thermostats. Note that it is possible to formulate equivalent equations for the case in which there are several different chains, each of which is coupled to a different part of the system.

Nosé–Hoover thermostating is not the only way in which the temperature can be controlled. An early algorithm was developed by H. C. Andersen, who suggested that, at intervals during a normal simulation, the velocities of a randomly chosen particle or molecule could be reassigned from a Maxwell–Boltzmann distribution. This is equivalent to the particle 'colliding' with one of the particles in a heat bath. The algorithm produces trajectories in the canonical ensemble but, because of the reassignment of velocities, they are discontinuous. A related approach, which is both elegant and widely used, is one in which a stochastic analogue of Newton's equation of motion, the *Langevin equation*, is employed to describe the dynamics of a particle interacting with a thermal bath. The Langevin equation has two extra force terms arising from this interaction – a *random force* that buffets the particle about and a frictional force, proportional to the particle's velocity, which dissipates excess kinetic energy. A third method that is based upon a different concept is one in which the kinetic energy and, hence, the temperature are constrained to be constant at each step. The modification of Newton's equations which achieves this is straightforward and uses *Gauss's principle of least constraint*. The problem with this technique is that the coordinate configurations produced are drawn from a canonical ensemble, but the momentum configurations are not.

There is a smaller variety of algorithms for performing constant-pressure molecular dynamics simulations. The most common type is extended system algorithms, although algorithms based upon Gauss's principle of least constraint have also been developed. All methods change the volume of the simulation box.

One of the first extended system algorithms to be proposed was that of Andersen, who introduced the volume of the simulation box, V, as an additional dynamical variable. His equations are

$$\dot{R} = \mathbf{M}^{-1}P + \frac{1}{3}\frac{\dot{V}}{V}R \tag{10.38}$$

$$\dot{P} = F(R) - \frac{1}{3}\frac{\dot{V}}{V}P \tag{10.39}$$

$$\ddot{V} = \frac{1}{W}(\mathcal{P} - P_\mathrm{B}) \tag{10.40}$$

where W is the 'mass' of the volume or *barostat* degree of freedom with units of mass times length to the fourth power. The Hamiltonian corresponding to this system of equations, $\mathcal{H}_\mathrm{A}$, is conserved and is

$$\mathcal{H}_\mathrm{A} = \mathcal{H} + \frac{1}{2}W\dot{V}^2 + P_\mathrm{B}V \tag{10.41}$$

Andersen showed that these equations generate trajectories consistent with the isobaric–isoenthalpic (NPH) ensemble. Hoover later modified them by adding thermostat variables of the types found in equations (10.27)–(10.30) so that the NPT ensemble could be sampled. More recent improvements to these algorithms have been suggested by Martyna and co-workers.

All the constant-pressure algorithms described above can be generalized to allow the shape as well as the size of the simulation box to change during the course of a simulation. In these cases extra degrees of freedom must be introduced, which correspond to the position vectors of the sides of the box and add considerably to the complexity of the equations. M. Parrinello and A. Rahman and Nosé and Klein did early work to adapt Andersen's equations for simulations of this sort.

10.5 Example 18

To illustrate the pressure and temperature coupling algorithm implemented in the module VELOCITY_VERLET_DYNAMICS, we perform a simulation of water with a program very similar to Example 16 of section 9.6. The program is

```
PROGRAM EXAMPLE18

... Declaration statements ...

! . Read in the system file.
CALL MM_SYSTEM_READ ( ''h2o_box729.sys_bin'' )

! . Read in the coordinates.
CALL COORDINATES_READ ( ''h2o_box729.crd'' )

! . Initialize the non-bonding list cutoff options.
CALL ENERGY_NON_BONDING_OPTIONS ( LIST_CUTOFF  = 13.5, OUTER_CUTOFF = 12.0, &
                                  INNER_CUTOFF = 8.0, MINIMUM_IMAGE = .TRUE. )

! . Calculate an energy.
CALL ENERGY

! . Initialize the random number seed.
CALL RANDOM_INITIALIZE ( 314159 )

! . Equilibrate the system at 300K and 1 atm. for 1ps.
CALL DYNAMICS ( 0.001, 1000, 10, INITIAL_TEMPERATURE  = 300.0,          &
                                 SCALE_OPTION         = ''BERENDSEN_PT'', &
                                 TARGET_TEMPERATURE   = 300.0,          &
                                 TEMPERATURE_COUPLING = 0.1,            &
                                 TARGET_PRESSURE      = 1.0,            &
                                 PRESSURE_COUPLING    = 2.0E+3 )

! . Run dynamics for 10ps using the old velocities.
CALL DYNAMICS ( 0.001, 10000, 10, ASSIGN_VELOCITIES   = .FALSE.,        &
                                  SCALE_OPTION         = ''BERENDSEN_PT'', &
                                  TARGET_TEMPERATURE   = 300.0,          &
                                  TEMPERATURE_COUPLING = 0.1,            &
                                  TARGET_PRESSURE      = 1.0,            &
                                  PRESSURE_COUPLING    = 2.0E+3,         &
                                  SAVE_FREQUENCY       = 50,            &
                                  COORDINATE_FILE      = ''h2o_box729_cpt.traj'' )

! . Print some statistics about the non-bonding energy calculation.
CALL ENERGY_NON_BONDING_STATISTICS

END PROGRAM EXAMPLE18
```

The program starts by reading in the system file for the box of 729 water molecules and an appropriate set of coordinates. Next the non-bonding options for the calculation are set and an energy is calculated. The dynamics run on the water box consists of two simulations, an equilibration of 1 ps and a data-collection phase of 10 ps. Both are performed with pressure and temperature control at a reference pressure of 1 atm and a reference temperature of 300 K. The temperature coupling constant is 0.1 ps while the pressure coupling constant is 2000 ps atm, which corresponds to having values of β and τ_P from equations (10.13) and (10.14) of 5×10^{-5} atm^{-1} and 0.1 ps, respectively. The values of 0.1 ps for the parameters τ_P and τ_T are the minimum values recommended by Berendsen *et al.*

The simulation above produces results for the static and dynamical properties of the water which do not differ markedly from the simulation performed within the microcanonical ensemble in Example 16. Likewise, the values for the averages of the temperature are very similar for the two simulations, although the average from the simulation with the Berendsen algorithm is almost exactly 300 K (as it should be) whereas it is about 302 K for the simulation of Example 16. In both cases the RMS deviations of the temperature are small, with values of 3–4 K.

In contrast, the pressure exhibits much greater fluctuations. The instantaneous pressure as a function of time from the simulation is plotted in figure 10.3. Although the average at the end of the simulation is of the order of 1

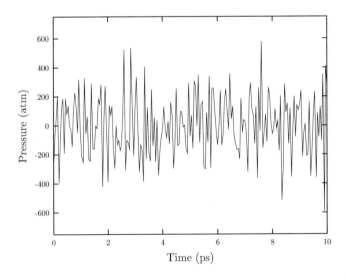

Figure 10.3. The instantaneous pressure as a function of time from the simulation of Example 18.

atm, the instantaneous pressure can deviate by as much as 600 atm from this. The volume as a function of time is shown in figure 10.4. The changes here are also quite marked. There is an initial drop in the value of the volume, which indicates that a longer period of equilibration is probably necessary, before the value gradually increases and appears to stabilize. To verify this, though, the simulation would need to be continued for longer.

10.6 Calculating free energies: umbrella sampling

In section 7.6 we saw how it was possible to estimate various thermodynamic quantities for a gas-phase system within the rigid-rotor, harmonic oscillator approximation. This approximation, although it can give useful results, is limited in that it relies on data from a very limited part of the potential energy surface, namely a stationary point. Molecular dynamics techniques, because they explore the phase space more fully, can be used to determine thermodynamic quantities more rigorously. In this section, we introduce this topic by considering a method that can be employed to calculate the free energies of certain types of processes either in the gas or in condensed phases. Additional techniques for the calculation of free energies will be left to section 11.5.

In what follows we shall limit the discussion to the canonical ensemble, although the arguments generalize to the isobaric–isothermal case. In classical statistical thermodynamics the partition function, Z_{NVT}, for a system of

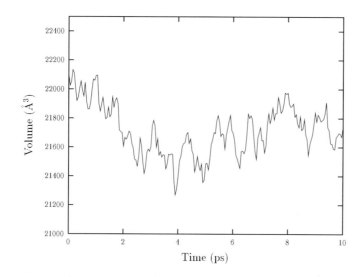

Figure 10.4. The volume as a function of time from the simulation of Example 18.

N indistinguishable particles at constant temperature and volume is

$$Z_{NVT} = \frac{1}{h^{3N} N!} \int d\mathbf{P} \int d\mathbf{R} \exp[-\mathcal{H}(\mathbf{P}, \mathbf{R})/(k_B T)] \qquad (10.42)$$

For Hamiltonians of the form given in equation (8.1), it is possible to perform the integration over the momentum variables, leaving an integral, called the *configuration integral*, of the position coordinates only. The probability density distribution function of equation (10.25) can be expressed in terms of the partition function as

$$\rho_{NVT} = \frac{1}{h^{3N} N!} \frac{\exp[-\mathcal{H}(\mathbf{P}, \mathbf{R})/(k_B T)]}{Z_{NVT}} \qquad (10.43)$$

Once the partition function and the density distribution function are known, the thermodynamic quantities for the system can be determined. Thus, for example, the Helmholtz free energy, A, is given by

$$A = -k_B T \ln Z_{NVT} \qquad (10.44)$$

It is probably not evident from this expression how the free energy for a system can be calculated from a simulation but we can, with a little manipulation, rewrite the partition function of equation (10.42) as an ensemble average of the form found in equation (10.23). The argument is as follows:

$$Z_{NVT} \propto \int d\mathbf{R} \exp[-\mathcal{V}/(k_B T)]$$

$$\propto \frac{\int d\mathbf{R} \exp[-\mathcal{V}/(k_B T)]}{\int d\mathbf{R} \exp[-\mathcal{V}/(k_B T)] \exp[\mathcal{V}/(k_B T)]}$$

$$\times \int d\mathbf{R} \exp[-\mathcal{V}/(k_B T)] \exp[\mathcal{V}/(k_B T)]$$

$$\propto \frac{1}{\langle \exp[\mathcal{V}/(k_B T)] \rangle} \qquad (10.45)$$

In this derivation we have neglected the integrals over the atomic momenta, which can be treated separately. In the second step we note that the integral introduced in the numerator and denominator reduces to $\int d\mathbf{R}$ which evaluates to a constant equal to V^N, where V is the volume of the system and N is the number of particles.

It might be thought that a possible way to determine the free energy is to perform a molecular dynamics simulation, evaluate the average $\langle \exp[\mathcal{V}/(k_B T)] \rangle$ along the trajectory and thus calculate the partition function and hence the free energy. This approach turns out, however, to be impractical because it is extremely difficult to get reliable values for the average. The reason is that the simulation will preferentially sample configurations with large negative potential energies because these are the configurations which have a higher probability. The configurations which contribute significantly to the average, though, will be those with large potential energies because their factors, $\exp[\mathcal{V}/(k_{\mathrm{B}}T)]$, will be large.

Although the problem of adequate sampling may be especially acute when we are trying to calculate free energies and related thermodynamic properties, it must be borne in mind generally whenever any average is being calculated from a simulation. Several strategies are employed for tackling this problem. One is to use techniques that *enhance* sampling either for the phase space as a whole or in certain regions of it during a simulation. It is one of these methods that we shall discuss in more detail in the remainder of the section. Another approach is to be less ambitious and define quantities that can be calculated without encountering the same convergence problems. These techniques will be left to section 11.5.

One way to enhance sampling in a particular region of phase space is the method of *umbrella sampling*, which was suggested by J. P. Valleau and G. M. Torrie. In this technique a positive biasing function, $\mathcal{B}(\boldsymbol{R})$, is introduced and the ensemble average for a property, $\mathcal{X}$, becomes

$$
\begin{aligned}
\langle \mathcal{X} \rangle &= \frac{\int \mathrm{d}\boldsymbol{R}\, \mathcal{X}(\boldsymbol{R}) \exp[-\mathcal{V}(\boldsymbol{R})/(k_{\mathrm{B}}T)]}{\int \mathrm{d}\boldsymbol{R} \exp[-\mathcal{V}(\boldsymbol{R})/(k_{\mathrm{B}}T)]} \\
&= \frac{\int \mathrm{d}\boldsymbol{R}\, (\mathcal{X}(\boldsymbol{R})/\mathcal{B}(\boldsymbol{R}))\mathcal{B}(\boldsymbol{R}) \exp[-\mathcal{V}(\boldsymbol{R})/(k_{\mathrm{B}}T)]}{\int \mathrm{d}\boldsymbol{R}\, \mathcal{B}(\boldsymbol{R}) \exp[-\mathcal{V}(\boldsymbol{R})/(k_{\mathrm{B}}T)]} \\
&\quad \times \frac{\int \mathrm{d}\boldsymbol{R}\, \mathcal{B}(\boldsymbol{R}) \exp[-\mathcal{V}(\boldsymbol{R})/(k_{\mathrm{B}}T)]}{\int \mathrm{d}\boldsymbol{R}\, (1/\mathcal{B}(\boldsymbol{R}))\mathcal{B}(\boldsymbol{R}) \exp[-\mathcal{V}(\boldsymbol{R})/(k_{\mathrm{B}}T)]} \\
&= \frac{\langle \mathcal{X}(\boldsymbol{R})/\mathcal{B}(\boldsymbol{R}) \rangle_{\text{biased}}}{\langle 1/\mathcal{B}(\boldsymbol{R}) \rangle_{\text{biased}}}
\end{aligned}
\tag{10.46}
$$

In the derivation of this equation, it has been assumed that the property is independent of the atomic momenta and so the integrals concerning them cancel out. The notation $\langle \cdots \rangle_{\text{biased}}$ indicates an ensemble average determined with the biased distribution function, ρ_{biased}:

$$
\rho_{\text{biased}} = \frac{\mathcal{B}(\boldsymbol{R}) \exp[-\mathcal{V}(\boldsymbol{R})/(k_{\mathrm{B}}T)]}{\int \mathrm{d}\boldsymbol{R}\, \mathcal{B}(\boldsymbol{R}) \exp[-\mathcal{V}(\boldsymbol{R})/(k_{\mathrm{B}}T)]}
\tag{10.47}
$$

Equation (10.46) states that the ensemble average for a property can be rewritten as the ratio of the averages of the properties $\mathcal{X}(\boldsymbol{R})/\mathcal{B}(\boldsymbol{R})$ and $1/\mathcal{B}(\boldsymbol{R})$ calculated within the biased ensemble. It is easier to see what this means if we write an expression for the biasing function, $\mathcal{B}(\boldsymbol{R})$, in terms of an *umbrella potential*, $\mathcal{V}_{\text{umb}}$:

$$\mathcal{B}(\boldsymbol{R}) = \exp[-\mathcal{V}_{\text{umb}}(\boldsymbol{R})/(k_{\text{B}}T)] \tag{10.48}$$

With this definition, the distribution function of equation (10.47) corresponds to that of a system whose Hamiltonian has been modified by adding to it an extra term, $\mathcal{V}_{\text{umb}}$. Ensemble averages in the biased ensemble are calculated by performing molecular dynamics simulations for the system with the modified Hamiltonian and then the results for the normal, unbiased ensemble are obtained by applying the formula of equation (10.46).

In this section we shall apply the technique of umbrella sampling to the calculation of a particular type of free energy, the *potential of mean force* (PMF), which is central to many statistical thermodynamical theories. It is, for example, required in some versions of transition-state theory to calculate the rate of reaction between two different states of a system.

The PMF can be obtained as a function of one or more of the system's degrees of freedom, although for clarity we shall restrict our attention to the unidimensional case. This degree of freedom, ξ, can be a simple function of the Cartesian coordinates of the atoms, such as a distance or an angle, or it can take a more complicated form depending upon the process being studied. The expression for the PMF is the same as that for a free energy (equation (10.44)) except that the averaging is done over all degrees of freedom apart from the one corresponding to the variable, ξ. Let us denote ξ_0 as the value of the degree of freedom, ξ, for which the PMF is being calculated and $\xi(\boldsymbol{R})$ as the function which relates ξ to the atomic coordinates, $\boldsymbol{R}$. The PMF, $\mathcal{U}(\xi_0)$, can then be written as

$$\mathcal{U}(\xi_0) = c - k_{\text{B}}T \ln\left(\int d\boldsymbol{R}\, \delta(\xi(\boldsymbol{R}) - \xi_0)\exp[-\mathcal{V}(\boldsymbol{R})/(k_{\text{B}}T)]\right) \tag{10.49}$$

In this equation, the parts of the partition function, such as the prefactor and the integrals over the atomic momenta, which are independent of ξ have been separated off into the arbitrary constant, c. The removal of the degree of freedom, ξ, from the averaging procedure is ensured by the Dirac delta function, which selects only those combinations of the atomic coordinates, $\boldsymbol{R}$, which give the reference value of the PMF coordinate, ξ_0.

Although equation (10.49) illustrates the connection between the PMF and equation (10.44), it is more usual to write the PMF in terms of the ensemble average of the probability distribution function of the coordinate, $\langle\rho(\xi_0)\rangle$, which has the expression

$$\langle \rho(\xi_0) \rangle = \frac{\int d\boldsymbol{R}\, \delta(\xi(\boldsymbol{R}) - \xi_0) \exp[-\mathcal{V}(\boldsymbol{R})/(k_B T)]}{\int d\boldsymbol{R} \exp[-\mathcal{V}(\boldsymbol{R})/(k_B T)]} \tag{10.50}$$

The PMF is then

$$\mathcal{U}(\xi_0) = c' - k_B T \ln\langle \rho(\xi_0) \rangle \tag{10.51}$$

where c' is another arbitrary constant.

In principle, the average, $\langle \rho(\xi_0) \rangle$, can be calculated by performing a simulation in the canonical ensemble and then constructing a histogram of the frequencies of occurrence of the different values of the variable ξ_0 along the trajectory, using a similar technique to that for the calculation of the radial distribution function of section 10.2. In practice, for the reasons discussed at the beginning of this section, this will normally not be a feasible approach because of the difficulties of accurately sampling the various configurations that are accessible to the system. One way to tackle this problem is to use the technique of umbrella sampling. The difficulty with this method, however, is that we need to choose a form for the umbrella potential, $\mathcal{V}_{umb}$, that allows efficient sampling throughout the range of the variable, ξ_0, that we are studying. The optimum choice of this potential, or the biasing function, requires a knowledge of the distribution function that we are trying to find which is, of course, unknown beforehand. The solution is to perform a series of calculations, instead of one, with umbrella potentials that concentrate the sampling in different, but overlapping, regions of phase space. The trajectories for each simulation or *window* are then used to calculate a series of biased distribution functions, in the form of histograms, which are pieced together to obtain a distribution function that is valid for the complete range of the coordinate, ξ_0.

The form of the umbrella potential which restricts sampling to a limited range of values of ξ_0 is arbitrary. A common choice, however, and the one that we shall make is a harmonic form, i.e.

$$\mathcal{V}_{umb}(\xi_0) = \frac{1}{2} k_{umb}(\xi_0 - \xi_{ref})^2 \tag{10.52}$$

where k_{umb} is the force constant for the potential and ξ_{ref} is the reference value of the coordinate whose value is changed at each window.

The reconstruction of the full distribution function from the separate distributions for each window is the crucial step in an umbrella sampling calculation. An efficient procedure for doing this is the *weighted histogram analysis method* (WHAM) which was developed by S. Kumar and co-workers from a technique originally due to A. M. Ferrenberg and R. H. Swendsen. The WHAM method aims to construct an optimal estimate for the average distribution function in the unbiased ensemble, $\langle \rho(\xi_0) \rangle$, from the biased distribution functions for each window. Suppose that there are N_w windows, each of which has an umbrella potential, V_{umb}^α, and an associated biased distribution function, $\langle \rho(\xi_0) \rangle_{biased}^\alpha$. The unbiased distribution functions for each window are determined by applying the arguments used in the derivation of equation (10.46) to the expression for the distribution function in equation (10.50). The result is

$$\langle \rho(\xi_0) \rangle_{biased}^\alpha = \frac{\exp[-V_{umb}^\alpha(\xi_0)/(k_B T)]\langle \rho(\xi_0) \rangle_{unbiased}^\alpha}{\langle \exp[-V_{umb}^\alpha(\xi_0)/(k_B T)] \rangle_{unbiased}} \tag{10.53}$$

Note that the unbiased distribution functions would be equivalent to the full distribution function if the form of the umbrella potential allowed a complete sampling of the range of the coordinate, ξ_0. Because this is not the case, the unbiased distributions are likely to provide useful information only for values of ξ_0 around the reference value of the coordinate, ξ_{ref}^α, for each window.

Kumar *et al.* derived an equation for the full distribution function as a weighted sum of the unbiased window distribution functions. The 'best' estimate of the full function was determined by choosing weights that minimized the statistical error associated with the data used to construct the distribution function histograms for each window. The equation is

$$\langle \rho(\xi_0) \rangle = \sum_{\alpha=1}^{N_w} \langle \rho(\xi_0) \rangle_{unbiased}^\alpha \left(\frac{n_\alpha \exp[-(V_{umb}^\alpha(\xi_0) - \mathcal{F}_\alpha)/(k_B T)]}{\sum_{\beta=1}^{N_w} n_\beta \exp[-(V_{umb}^\beta(\xi_0) - \mathcal{F}_\beta)/(k_B T)]} \right) \tag{10.54}$$

where n_α is the number of independent data points employed for the generation of the distribution function for a window and $\mathcal{F}_\alpha$ is a window free energy that is related to the denominator of equation (10.53) by the following expression:

$$\exp[-\mathcal{F}_\alpha/(k_B T)] = \langle \exp[-V_{umb}^\alpha(\xi_0)/(k_B T)] \rangle_{unbiased} \tag{10.55}$$

Equation (10.54) can be rewritten in terms of the biased distribution functions using equation (10.53) as

$$\langle \rho(\xi_0) \rangle = \frac{\sum_{\alpha=1}^{N_w} n_\alpha \langle \rho(\xi_0) \rangle_{biased}^\alpha}{\sum_{\beta=1}^{N_w} n_\beta \exp[-(V_{umb}^\beta(\xi_0) - \mathcal{F}_\beta)/(k_B T)]} \tag{10.56}$$

To complete the derivation an estimate of the constants, $\mathcal{F}_\alpha$, is needed. These can be determined from equation (10.55) using the expression for the full distribution function given in equation (10.56), i.e.

$$\exp[-\mathcal{F}_\alpha/(k_\text{B} T)] = \int \text{d}\xi_0 \, \langle \rho(\xi_0) \rangle \exp[-\mathcal{V}_\text{umb}^\alpha(\xi_0)/(k_\text{B} T)] \qquad (10.57)$$

Equations (10.56) and (10.57) provide the means of calculating the average distribution function, $\langle \rho(\xi_0) \rangle$, and, hence, the PMF from a set of window distribution functions. The equations must be solved iteratively because both the distribution function, $\langle \rho(\xi_0) \rangle$, and the N_w free energies, $\mathcal{F}_\alpha$, are unknown initially. The procedure is to start by guessing a set of values for the free energies of each window (usually zero) and, with these, calculate $\langle \rho(\xi_0) \rangle$ for the complete range of ξ_0 from equation (10.56). This estimate of the distribution function is then used to determine the window free energies from equation (10.57) and the process is repeated until the values both of $\langle \rho(\xi_0) \rangle$ and of the set of $\mathcal{F}_\alpha$ no longer change. Experience has shown that this procedure is stable but that accurate results will be obtained only if the histograms corresponding to neighbouring distribution functions overlap to a reasonable extent.

A single module that allows umbrella sampling calculations to be performed is introduced in this section. The module is called ENERGY_EXTRA and it has four subroutines that we shall need later. Its definition is

```
MODULE ENERGY_EXTRA

CONTAINS

    SUBROUTINE UMBRELLA_DEFINE ( TYPE, ATOM1, ATOM2, ATOM3, ATOM4, &
                                                  FORCE, REFERENCE )
        CHARACTER ( LEN = * ), INTENT(IN) :: TYPE
        INTEGER, INTENT(IN), OPTIONAL :: ATOM1, ATOM2, ATOM3, ATOM4
        REAL,    INTENT(IN), OPTIONAL :: FORCE, REFERENCE
    END SUBROUTINE UMBRELLA_DEFINE

    SUBROUTINE UMBRELLA_START ( FILE, PRINT_FREQUENCY )
        CHARACTER ( LEN = * ), INTENT(IN) :: FILE
        INTEGER, INTENT(IN), OPTIONAL     :: PRINT_FREQUENCY
    END SUBROUTINE UMBRELLA_START

    SUBROUTINE UMBRELLA_STOP
    END SUBROUTINE UMBRELLA_STOP

    SUBROUTINE UMBRELLA_WHAM ( DATA, NBINS, TEMPERATURE )
        INTEGER, INTENT(IN) :: NBINS
        REAL,    INTENT(IN) :: TEMPERATURE
        CHARACTER ( LEN = * ), DIMENSION(:), INTENT(IN) :: DATA
    END SUBROUTINE UMBRELLA_WHAM

END MODULE ENERGY_EXTRA
```

The subroutine UMBRELLA_DEFINE defines the umbrella potential that is to be employed. There is a single essential argument, TYPE, which is an input character variable that can take the values ''DISTANCE'', ''ANGLE'', ''DIHEDRAL'' and ''NONE''. The first three values define umbrella potentials constraining the distance between two atoms, an angle formed by three atoms and a dihedral involving four atoms, respectively, whereas the value ''NONE'' has the effect of removing any existing definition of an umbrella potential in the module. If a potential is being specified then arguments giving the numbers of the atoms involved in the potential must be specified too. Thus, ATOM1 and ATOM2 must be given for a distance constraint, ATOM1, ATOM2 and ATOM3 for an angle constraint and ATOM1, ATOM2, ATOM3 and ATOM4 for a dihedral constraint. For all three types, the arguments FORCE and REFERENCE must be present because these define the force constant for the umbrella potential and the reference value about which the geometrical parameter is to be constrained, respectively. Note that the module allows only a single umbrella potential to be defined at a time and a call to UMBRELLA_DEFINE will erase any memory that the module has of previous umbrella potentials. All the umbrella potentials are harmonic and have the same form as in equation (10.52).

In an umbrella sampling calculation, the values of the variable being constrained by the umbrella potential must be analysed. A call to the subroutine UMBRELLA_START initiates writing of the value of this variable to an external file. The subroutine has two arguments. The first, FILE, is an essential argument that specifies the name of the file and the second, PRINT_FREQUENCY, is optional and gives the frequency with which the variable is to be written out. The default value of this argument is 1, which means that the value of the umbrella potential variable will be written *each* time an energy is calculated. Normally it is only useful to write during a dynamics simulation and not in other circumstances. The first line in each umbrella potential file lists the force constant for the potential and its reference value.

The subroutine UMBRELLA_STOP terminates writing of the umbrella potential variable by closing the external file specified in the call to UMBRELLA_START. It takes no arguments.

The fourth subroutine, UMBRELLA_WHAM, solves the WHAM equations (10.56) and (10.57) for data generated in a series of previous umbrella sampling simulations and then calculates and prints the PMF. There are three essential input arguments. The first, DATA, is a character array that gives the names of the files that contain the umbrella sampling data. The dimension of the array gives the number of windows that are to be analysed. The second argument, NBINS, is an integer that gives the number of bins to use in the histogram analysis of the data. The subroutine works by first scanning all the data in the input data files to find the maximum and minimum of the range of the umbrella potential variable before dividing this range into NBINS divisions of equal width for construction of the data histograms. The third argument, TEMPERATURE, gives the temperature at which the PMF is to be determined. This will usually be the temperature at which the dynamics simulations that generated the data were performed.

It should be noted that there are also other procedures in the module that are concerned, among other things, with the calculation of the energy and derivatives of the umbrella potential energy term. These subroutines are automatically called at the appropriate times if an umbrella potential has been defined.

10.7 Examples 19 and 20

In this section, the potential of mean force between two conformations of the bALA molecule in vacuum is computed. The reaction coordinate is chosen as the distance between the carbonyl oxygen of the N-terminal acetyl group and the amide hydrogen of the C-terminal *N*-methyl group. At short distances there is an intramolecular hydrogen bond but this is broken as the distance increases.

There are two example programs. The first performs the molecular dynamics simulations to generate the umbrella sampling data for a series of windows and the second uses these data to calculate the PMF by solving the WHAM equations.

The first program is

```
PROGRAM EXAMPLE19

... Declaration statements ...

! . Local parameters.
INTEGER, PARAMETER :: NWINDOWS = 5
REAL,    PARAMETER :: INCREMENT = 1.0, MINIMUM = 1.5

! . Local scalars.
INTEGER :: I
REAL    :: DISTANCE

! . Read in the system file.
CALL MM_SYSTEM_READ ( ''bALA.sys_bin'' )

! . Read the coordinates.
CALL COORDINATES_READ ( ''bALA.crd_1pt5'' )
```

```
! . Loop over the values of the distance.
DO I = 1,NWINDOWS

   ! . Calculate the distance.
   DISTANCE = INCREMENT * REAL ( I-1 ) + MINIMUM

   ! . Define the potential.
   CALL UMBRELLA_DEFINE ( ''DISTANCE'', ATOM1 = 6, ATOM2 = 18, &
                          FORCE = 20.0, REFERENCE = DISTANCE )

   ! . Initialize the random number seed.
   CALL RANDOM_INITIALIZE ( 314159 + I )

   ! . Heat the system.
   CALL DYNAMICS ( 0.001, 10000, 1000, INITIAL_TEMPERATURE = 1.0,        &
                                       SCALE_OPTION        = ''LINEAR'', &
                                       SCALE_FREQUENCY     = 500,        &
                                       TARGET_TEMPERATURE  = 300.0 )

   ! . Equilibrate the system.
   CALL DYNAMICS ( 0.001, 10000, 1000, ASSIGN_VELOCITIES   = .FALSE.,        &
                                       SCALE_OPTION        = ''BERENDSEN_T'', &
                                       TARGET_TEMPERATURE  = 300.0,           &
                                       TEMPERATURE_COUPLING = 1.0 )

   ! . Start writing.
   CALL UMBRELLA_START ( ''bALA.window''//TRIM ( ENCODE_INTEGER ( I ) ), &
                                       PRINT_FREQUENCY = 5 )

   ! . Collect data.
   CALL DYNAMICS ( 0.001, 100000, 1000, ASSIGN_VELOCITIES   = .FALSE.,        &
                                        SCALE_OPTION        = ''BERENDSEN_T'', &
                                        TARGET_TEMPERATURE  = 300.0,           &
                                        TEMPERATURE_COUPLING = 1.0 )

   ! . Stop writing.
   CALL UMBRELLA_STOP

END DO

END PROGRAM EXAMPLE19
```

The central portion of this program is a loop, each iteration of which
performs a molecular dynamics simulation for a different window of the
umbrella sampling calculation. The number of windows in this example is
five and is specified by the parameter NWINDOWS which is defined at the start
of the program. The constraint distance in the umbrella potential varies from
1.5 Å up to 5.5 Å in 1 Å increments. The first two statements in the loop
define the constraint distance between the oxygen (ATOM1) and the hydrogen

(ATOM2) and then set up the umbrella potential with a force constant of 20 kJ mol^{-1}. The next statements perform the molecular dynamics simulation. There is an initial heating phase of 10 ps, followed by an equilibration phase of 10 ps and then a longer simulation of 100 ps during which the values of the constrained distance are written out at five-step intervals. The equilibration and data-collection simulations are both performed with the temperature control option ''BERENDSEN_T'' and with a temperature coupling constant of 1 ps.

The umbrella sampling data for each window are written out to a different file. The names of the files are all of the form bALA.window followed by an integer that specifies the window. To convert the loop index, I, into a character string suitable for use as part of the file name, the function ENCODE_INTEGER from the utility module STRING (see section 1.3) is employed. The character string returned by this function is then treated with the intrinsic FORTRAN 90 function TRIM which removes any trailing blanks.

The input coordinate file in the program, ''bALA.crd_1pt5'', corresponds to a minimum energy structure for the bALA molecule in which the distance between the N-terminal carbonyl oxygen and the C-terminal amide hydrogen is constrained to be 1.5 Å. This is done so that the application of the first umbrella potential, which has a constraint distance of 1.5 Å, does not lead to a structure of such a high potential energy that the subsequent dynamics could be significantly perturbed. The constrained minimum energy structure was prepared using the technique described at the end of this section. For subsequent windows in the calculation the last structure produced by the data-collection dynamics simulation of the previous window is taken as the starting configuration.

As a final remark, we note that the system is small enough that all the non-bonding interactions can be calculated during the simulation. Thus, either the version of the module ENERGY_NON_BONDING described in section 4.6 or the version of section 9.2 with the default, very large values for the non-bonding interaction cutoffs can be used. Either way there is no need to call the subroutine ENERGY_NON_BONDING_OPTIONS explicitly.

The second program is much more straightforward and is

```
PROGRAM EXAMPLE20

... Declaration statements ...

! . Local parameters.
INTEGER, PARAMETER :: NBINS = 100, NWINDOWS = 5
REAL,    PARAMETER :: TEMPERATURE = 300.0
```

```
! . Local scalars.
INTEGER :: I

! . Local arrays.
CHARACTER ( LEN = 32 ), DIMENSION(1:NWINDOWS) :: DATA

! . Set up the file names.
DO I = 1,NWINDOWS
   DATA(I) = ''bALA.window''//TRIM ( ENCODE_INTEGER ( I ) )
END DO

! . Calculate the PMF.
CALL UMBRELLA_WHAM ( DATA, NBINS, TEMPERATURE )

END PROGRAM EXAMPLE20
```

The first executable part of the program sets up the names of the window data files in the character array, DATA. The syntax used is similar to that in the previous program. In the last part of the program the WHAM equations are solved using the window data with a call to the subroutine UMBRELLA_WHAM. The number of bins to use in the analysis and the temperature at which the PMF is to be calculated are defined as parameters at the top of the program.

The results of these calculations are shown in figures 10.5 and 10.6. Figure 10.5 shows a histogram analysis of the O—H distance data generated for the windows of the umbrella sampling calculation. Each distribution is relatively smooth and there is a large overlap between the adjacent curves. The corre-

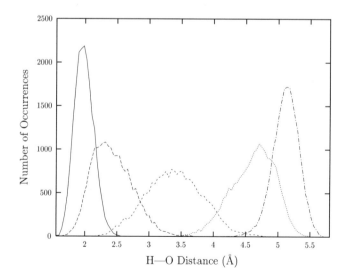

Figure 10.5. The distribution of O—H distances for each window of the umbrella sampling simulation of Example 19.

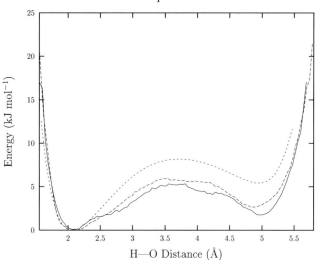

Figure 10.6. The energy profiles for the bALA molecule as a function of the O—H distance. The PMF calculated with Examples 19 and 20 is shown as a solid line. The profile calculated using energy minimization is given by the dotted line and the PMF determined using 17 windows, instead of five, is the dashed line.

sponding PMF is shown in figure 10.6 as the solid line. The more stable minimum is at an O—H distance of about 2.1 Å whereas the second minimum has an energy 2 kJ mol^{-1} higher and is at a distance of 5 Å. This means that the hydrogen-bonded structure is the more stable. The barrier to the interconversion of the two forms is about 6 kJ mol^{-1} starting from the hydrogen-bonded form. Note that the minima in the PMF correspond closely to the configurations of highest probability in figure 10.5.

Also plotted in figure 10.6 are two more curves. There is another PMF calculated (the dashed line) using the same basic procedure as above but with 17 windows at intervals of 0.25 Å and a larger force constant for the umbrella potential of 50 kJ mol^{-1} Å^{-2}. Each window simulation was run for 50 ps. It can be seen that the curves for the PMF are in reasonable agreement. The second curve (the dotted line) plots the energy profile that results if the O—H distance is fixed at certain values and then the remaining degrees of freedom are optimized. The differences between the PMFs and the profile obtained from energy minimization are quite large. The minima are in roughly the same place but the energy of the more unstable minima is over twice as large at about 5 kJ mol^{-1}.

In this book there are no methods to fix particular degrees of freedom at given values during minimization calculations but similar effects can be produced by using umbrella potentials. The curve in figure 10.6 was generated with a program similar to Example 19 but using a loop over distances of the form

```
! . Loop over the values of the distance.
DO I = 1,NVALUES

    ! . Calculate the distance.
    DISTANCE = INCREMENT * REAL ( I-1 ) + MINIMUM

    ! . Define the potential.
    CALL UMBRELLA_DEFINE ( ''DISTANCE'', ATOM1 = 6, ATOM2 = 18, &
                           FORCE = 1000.0, REFERENCE = DISTANCE )

    ! . Perform a coordinate minimization.
    CALL OPTIMIZE_CONJUGATE_GRADIENT ( PRINT_FREQUENCY    = 100, &
                                       STEP_NUMBER        = 2000, &
                                       GRADIENT_TOLERANCE = 1.0E-3 )

    ! . Remove the potential.
    CALL UMBRELLA_DEFINE ( ''NONE'' )

    ! . Calculate the energy without the umbrella potential.
    CALL ENERGY ; E(I) = TOTAL_PE

    ! . Store the actual value of the distance.
    D(I) = GEOMETRY_DISTANCE ( ATMCRD, 6, 18 )

END DO
```

There are several points to note. First, the value of the umbrella potential force constant employed is very large because we want to have a final optimized O—H distance that is as close to the reference value as possible. Second, the umbrella potential is removed after the minimization has been performed and the energy of the structure and the O—H distance are recalculated and stored in the arrays E and D, respectively. It is these values that were plotted in figure 10.6. Finally, note that no window data are written because they are not of much use if produced in a minimization calculation.

Exercises

10.1 The diffusion coefficient for an atom or molecule can also be calculated from the integral of the velocity autocorrelation function (equation (10.8)). Do this calculation for the water box system used in Examples 16 and 17. Note that it will be necessary to repeat the molecular dynamics simulation so that a trajectory containing the velocities is generated and then calculate the average of the correlation functions for each of the oxygens in the system. How do the results for the diffusion coefficients compare? By looking at the form of the velocity autocorrelation function estimate the length of a simulation that is needed to calculate this property – is 10 ps reasonable, is a longer simulation needed or would a shorter simulation be adequate?

10.2 Repeat the constant-pressure and -temperature calculations for water using different sets of coupling parameters, both smaller and larger than 0.1 ps. How do the static and dynamic quantities calculated from the simulation differ from each other and from those calculated in the microcanonical ensemble and at constant volume and constant temperature?

10.3 Calculate a PMF for a problem of your own choosing. Possible examples include the determination of a PMF as a function of an internal torsion angle of a molecule (such as butane) or a PMF as a function of the distance between two molecules (to simulate an association or dissociation). Do the calculation in vacuum and, once the results are satisfactory, repeat it, if possible, in solution. Note that these calculations will be much more expensive!

11

Monte Carlo simulations

11.1 Introduction

In the previous chapters a variety of techniques for the simulation of molecular systems has been covered. These have included methods such as energy minimization and reaction-path-finding algorithms, which explore a relatively limited portion of the potential energy surface of a system, and the molecular dynamics method, which makes accessible a much larger region of the potential energy surface and with which time-dependent events and properties can be studied. The ability of molecular dynamics simulations to sample a large region of the phase space of the system is important, as we have seen, for locating global potential energy minima and for calculating thermodynamic quantities.

There is an alternative technique, the *Monte Carlo method*, that is distinct from the molecular dynamics method but can also sample the phase space of the system and, hence, is appropriate for calculating thermodynamic quantities or for performing simulated annealing calculations. Unlike the molecular dynamics method it cannot be used to study time-dependent properties but it does have other features that are advantageous in some circumstances.

11.2 The Metropolis Monte Carlo method

Consider the integral, $\mathcal{I}$, of a function, $f(x)$, over a region $[a, b]$:

$$\mathcal{I} = \int_a^b f(x)\,dx \tag{11.1}$$

A normal way to estimate the integral, for well-behaved functions, would be to divide the region $[a, b]$ into n equally spaced slices, each of width $\Delta = (b - a)/n$ and then use a standard integration formula of the type

$$\mathcal{I} \simeq \Delta \sum_{i=0}^{n} w_i f(a + i\Delta) \tag{11.2}$$

where the w_i are weights whose values depend upon the formula being used. For the well-known Euler formula these would be 1 except at the end points, where they would be $\frac{1}{2}$.

The basis of the Monte Carlo approach is to realize that, instead of using a regular discretization of the integration variable, as in equation (11.2), it is possible to use a stochastic method in which the values of the integration variable are chosen randomly. Let n denote the *number of trials* and x_i (with $i = 1, \ldots, n$) the values of the integration variable that are chosen at random from a uniform distribution with values between a and b. The integral can be evaluated as

$$\mathcal{I} \simeq \Delta \sum_{i=1}^{n} f(x_i) \tag{11.3}$$

where, as before, $\Delta = (b - a)/n$, but this time it represents the average distance between integration points rather than the exact distance.

This formula works reasonably well for functions whose values do not change too much from one place to another in the integration range. For functions whose values vary greatly or are peaked in certain areas, the formula in equation (11.3) will be inefficient because the values of the function at many of the randomly chosen integration points will contribute negligibly to the integral. In these cases, it is more useful to be able to choose values of x that are concentrated in areas in which the function will be large. To do this the integral of equation (11.1) can be rewritten as

$$\mathcal{I} = \int_{a}^{b} \left(\frac{f(x)}{\rho(x)} \right) \rho(x) \, dx \tag{11.4}$$

where $\rho(x)$ is a probability density function that is large where it is thought that the function will be large. The integral can now be approximated by choosing values of the integration variable randomly from the function $\rho(x)$ in the range $[a, b]$, instead of from the uniform distribution, and averaging over the values of $f(x_i)/\rho(x_i)$ that are obtained, i.e.

$$\mathcal{I} \simeq \frac{1}{n} \sum_{i=1}^{n} \frac{f(x_i)}{\rho(x_i)} \tag{11.5}$$

That this formula is the same as equation (11.3) in the case of a uniform distribution follows because the probability distribution function for the uniform distribution is $1/(b - a)$. The use of a function ρ in this way to enhance sampling in certain regions of space is known as *importance sampling*.

The stochastic method outlined above cannot usually compete with numerical methods of the type given in equation (11.2) if there is a small number of integration variables. However, the number of function evaluations required by simple discretization schemes for the estimation of an integral becomes prohibitively large as the number of dimensions, $N_{\dim}$, increases. To see this, suppose that n points are chosen for the discretization in each direction, then the number of function evaluations required is $n^{N_{\dim}}$. It is in these cases that stochastic methods are often the only realistic approaches for tackling the problem.

As we have already seen in section 10.6, the integrals that are of interest in thermodynamics are almost always multidimensional. As an example, consider a property, $\mathcal{X}$, of the system that is a function of the $3N$ coordinates of the atoms, $\boldsymbol{R}$, only. The ensemble average of the property in the canonical ensemble is then the ratio of two multidimensional integrals:

$$\langle \mathcal{X} \rangle = \frac{\int \mathrm{d}\boldsymbol{R}\, \mathcal{X}(\boldsymbol{R}) \exp[-\mathcal{V}(\boldsymbol{R})/(k_{\mathrm{B}}T)]}{\int \mathrm{d}\boldsymbol{R} \exp[-\mathcal{V}(\boldsymbol{R})/(k_{\mathrm{B}}T)]} \tag{11.6}$$

where $\mathcal{V}$ is the potential energy of the system. This equation can be rewritten in a form reminiscent of equation (11.4) by employing the probability density distribution function for the canonical ensemble, ρ_{NVT}. Thus

$$\langle \mathcal{X} \rangle = \int \mathrm{d}\boldsymbol{R}\, \mathcal{X}(\boldsymbol{R})\rho_{NVT}(\boldsymbol{R}) \tag{11.7}$$

If, somehow (and this is the difficult part!), it is possible to choose configurations for the system drawn from the function, ρ_{NVT}, then the average, $\langle \mathcal{X} \rangle$, can be calculated using a formula analogous to equation (11.5), i.e.

$$\langle \mathcal{X} \rangle \simeq \frac{1}{n} \sum_{I=1}^{n} \mathcal{X}(\boldsymbol{R}_I) \tag{11.8}$$

where n is the number of configurations generated in the simulation and $\boldsymbol{R}_I$ is a vector of the coordinates of the atoms at each configuration. Owing to the presence of the exponential in the Boltzmann factor, it is crucial to employ importance sampling for these integrals because, for most properties, only the configurations of lowest potential energy will contribute significantly.

The Monte Carlo method for integration was formalized in the late 1940s by N. Metropolis, J. von Neumann and S. Ulam, but a Monte Carlo method for generating configurations drawn from a canonical distribution was introduced by Metropolis and co-workers in the early 1950s to study atomic systems. In outline, it is as follows.

1. Choose an initial configuration for the system, R_0, and calculate its potential energy, V_0. Set $I = 0$.
2. Generate, at random, a new configuration for the system, R_J, from the current configuration. How to generate the new configuration is unimportant for the moment and will be discussed in detail later. Metropolis *et al.* used a recipe in which the probability, p_{IJ}, of generating a state J from a state I was equal to the probability, p_{JI}, of generating the state I from state J. They also insisted that the method should allow, in principle, every state to be accessible from all other possible states, if not as a result of a single change, then as a result of a sequence of changes.
3. Calculate the potential energy of the new state, V_J.
4. If the difference in the potential energies of the two states, $V_J - V_I$, is less than zero, choose state J as the new configuration, i.e. set R_J to R_{I+1}.
5. If $V_J - V_I > 0$ fetch a random number in the range [0, 1]. If the number is less than $\exp[-(V_J - V_I)/(k_B T)]$ *accept* the new configuration, otherwise *reject* it and keep the old one.
6. Accumulate any averages that are required using equation (11.8) and the new configuration, R_{I+1}. Note that, even if the 'new' configuration is the same as the old one, it still must be re-used if proper averages are to be obtained.
7. Increment I to $I + 1$ and return to step 2 for as many steps as are desired in the simulation.

The above scheme is an ingenious one and it avoids any explicit reference to the configuration integral or the partition function for the system. It is possible to prove rigorously that the scheme generates configurations drawn from a canonical ensemble using the theory of *Markov chains* because technically what the Metropolis algorithm does is to produce a Markov chain of configurations with the limiting distribution of the canonical ensemble. However, in their paper, Metropolis *et al.* argued as follows. Suppose that there exists an ensemble of identical systems in various states and that n_I is the number of systems in the ensemble in state I. Consider two states, I and J with $V_I > V_J$. During a simulation of the *entire* ensemble using the Metropolis algorithm, the nett transformation of states I to states J will be $p_{IJ} n_I$ and of states J to I, $p_{JI} n_J \exp[-(V_I - V_J)/(k_B T)]$. Thus, the nett flow from states J to states I will be

$$p_{JI} n_J \exp[-(V_I - V_J)/(k_B T)] - p_{IJ} n_I = p_{IJ} \left\{ n_J \exp[-(V_I - V_J)/(k_B T)] - n_I \right\}$$
$$(11.9)$$

because it has been assumed that $p_{IJ} = p_{JI}$. After sufficiently many configurations the nett flow and, hence, the term in brackets will tend to zero, so

$$\frac{n_I}{n_J} \sim \frac{\exp[-V_I/(k_B T)]}{\exp[-V_J/(k_B T)]} \tag{11.10}$$

This is exactly what we are seeking because equation (11.10) gives the ratios
of the populations of two states in the canonical ensemble. As an aside, it
should now be clear why the calculation of correct averages requires that
'old' configurations be re-used. If this were not the case, it would mean that
every time a configuration was rejected and a system was left in its original
state it would be eliminated from the ensemble.

There is a further point worth discussing about the Metropolis scheme. The
condition on the method of generation of new configurations that all states
be accessible from all others (see step 2) means that the full phase space of the
system can, in principle, be explored. This property, which is called *ergodicity*, is important because the values of any properties calculated from a
simulation are likely to be significantly in error if the method becomes
trapped in a small region of phase space. Of course, even if a method is
theoretically ergodic, little is implied about how long a simulation needs to
be or how many configurations need to be sampled to obtain averages to
within a given precision.

The Metropolis Monte Carlo method has the great advantage that it can be
easily extended to generate chains of configurations from other ensembles.
W. W. Wood first showed how this could be done in the isothermal–isobaric
(NPT) ensemble. In this ensemble the average of a property, analagous to the
average in the NVT ensemble of equation (11.6), is

$$\langle \mathcal{X} \rangle = \frac{\int d\mathbf{R} \, dV \, \mathcal{X}(\mathbf{R}, V) \exp[-(\mathcal{V}(\mathbf{R}) + PV)/(k_B T)]}{\int d\mathbf{R} \, dV \, \exp[-(\mathcal{V}(\mathbf{R}) + PV)/(k_B T)]} \tag{11.11}$$

where V is the volume of the system and P is the pressure. It is to be noted
that in these integrals the coordinates of the atoms, $\mathbf{R}$, and the volume are not
independent variables. To ensure a correct derivation, as Wood showed, it is
necessary to transform the absolute coordinates of the atoms, $\mathbf{R}$, to fractional
atomic coordinates, $\mathbf{S}$. For a cubic box of side L, $\mathbf{R} = L\mathbf{S}$ and the volume
element for the integral $dV \, d\mathbf{R}$ becomes $dV \, d\mathbf{S} \, V^N$, where $V = L^3$. Putting
the V^N factor into the exponential gives a probability density distribution
function, ρ_{NPT}, for the ensemble which is proportional to
$\exp[-(\mathcal{V} + PV)/(k_B T) + N \ln V]$. The procedure for performing a Monte
Carlo simulation in the NPT ensemble can now be formulated and it turns

out to be identical to that for simulations in the NVT ensemble except for the following differences.

1. When generating new configurations, the fractional coordinates of the atoms and the volume of the system are changed instead of just the atomic coordinates.
2. The criterion for accepting a configuration is no longer based on the difference between the potential energies of the old and new configurations but on the difference between the the quantities in the exponential of ρ_{NPT}, i.e. on $V_J - V_I + P(V_J - V_I) - Nk_BT \ln(V_J/V_I)$, which reduces to the difference of potential energies when $V_I = V_J$. With this new quantity the Metropolis tests are applied in exactly the same way as before.

11.3 Monte Carlo simulations of molecules

The Metropolis algorithm, although it was the first of a number of Monte Carlo algorithms to have been developed, is still the most successful and widely used Monte Carlo method for the simulation of molecular systems. The Monte Carlo technique itself has a number of features that can make it preferable to the molecular dynamics method in some circumstances. These include the following.

1. Only the energy is required whereas molecular dynamics methods require the forces.
2. It is straightforward to perform simulations in the NVT and NPT ensembles whereas to do so with molecular dynamics simulations requires more complicated techniques.
3. It is easy to 'constrain' various degrees of freedom in the system. As we shall see below, this is done by keeping these degrees of freedom fixed when new configurations for the system are generated.
4. In principle, very different configurations of a system can be sampled during a Monte Carlo simulation if efficient schemes for the construction of new states can be devised. In the molecular dynamics method the prescription for the generation of new states is inherent to the formulation and is determined by the integration of the equations of motion of the atoms. The difference between successive structures, which is governed by the timestep, is small and so it can take very long simulations to probe very different regions of the system's phase space. Likewise, it may be difficult or even impossible to explore certain regions of phase space if the intervening energy barriers are large.

The last advantage of the Monte Carlo technique is also its disadvantage because recipes to generate new configurations for the system must be conceived. It happens that this is relatively simple for systems comprised of atoms or of small molecules. For large, flexible molecules it has proved more difficult to come up with an efficient method and so the application of the Monte Carlo technique in these cases has been relatively limited (with the caveat that an atomic model of the molecule is being used – the Monte Carlo method has been employed widely for polymer studies with simplified molecular models).

Because of the added complexity in dealing with large molecules, the applications of the Monte Carlo method in this book will be more limited than those of the molecular dynamics methods that have already been described. The latter can be used to study almost any molecular system with certain limitations, such as those that are due to the applicability of the force field and the nature of the symmetry conditions that can be imposed. For the Monte Carlo modules, however, we shall restrict ourselves to studying systems comprised of atoms or of small, *rigid* molecules.

With these restrictions, there are three types of change that have to be considered in order to generate new configurations for a system. These are translations and rotations of the molecules and changes in the volume of the simulation box. How we deal with each of these types of moves is arbitrary but the ways described below are commonly found and are the ones implemented in the modules to be introduced later.

Translations are the easiest to deal with. Figure 11.1 shows schematically how this is done. Suppose that there is a cube, of side $2\,\delta t$, centered at the center of mass of the molecule. To translate the molecule, choose a vector of three random numbers, $\boldsymbol{u}$, uniformly in the range $[-1, 1]$ and then translate the molecule by $\delta t\,\boldsymbol{u}$. This type of move satisfies both of the conditions for the

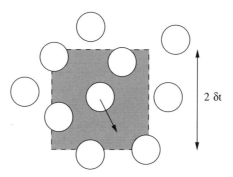

Figure 11.1. A schematic diagram of how a molecule is translated in a Monte Carlo simulation. The two-dimensional representation is easily generalized to three dimensions.

generation of new configurations outlined in step 2 of the Metropolis scheme. First of all, the translation of a molecule from one center ('state'), I, to another, J, will have the same probability as the translation from J to I because the selection of the translation is independent of direction and is uniform within the cube. Secondly, all states (or molecule centers) will be accessible from all others because a molecule can be translated in a single move to 'any' position within its cube (limited by the precision of the computer) and, hence, after a succession of moves, to any position in the simulation box.

The length of a possible translation is governed by the parameter δt. It is obvious that, if δt is large, there will be a high probability that the atoms of two molecules will overlap, making the molecules possess a large, positive interaction energy due to the repulsive part of the Lennard-Jones potential. Such configurations are likely to be rejected by the Metropolis criterion for deciding whether to accept new states. If the parameter δt is small then the probability of accepting the state will be high but it will take a long time for the configuration space of the system to be sampled effectively. The aim, therefore, is to choose a value of δt that is as large as possible while giving a reasonable *acceptance ratio* for new configurations. No comprehensive study seems to have been done to decide what the best value of this ratio is – indeed, it is likely to depend on the nature of the system being studied – but most workers appear to prefer values of the order of 50%. Although the exact value of δt required to attain a certain acceptance ratio might not be known in advance it is straightforward to modify the value of δt in the course of a simulation so that the desired acceptance ratio is approached. This can be done by scaling δt by a small amount every few steps, up if the acceptance ratio is too high and down if it is too small.

Until now no mention of whether all or, if not, which molecules are to be moved during the generation of new states has been made. Because the Metropolis scheme imposes no constraints of its own *per se*, we are free to move one molecule at a time, several molecules or all of them at once, whichever gives the most efficient scheme for sampling the phase space of the system. Once again, no comprehensive study of which method is best appears to have been done, but the preferred choice in the literature and, incidentally, the one that Metropolis *et al.* used in their original work, is the one in which only a *single* molecule is moved at a time. The molecule to be moved can either be chosen at random or by cycling through the molecules in a given order. Why single-molecule moves should be more efficient is not clear. They certainly allow a larger value of δt to be used on average than do multiple-molecule moves but, of course, more moves will be necessary in order to access states that have a substantially different configuration. It should be noted that it is not necessary to recalculate the complete energy of a system that is described with a pairwise additive potential if only a single molecule is moved – it is necessary to recalculate only the *energy of interaction* between the moved molecule and the rest of the system. The difference in potential energies between the old and new states required for the Metropolis test is then simply the difference between the interaction energies of the old and new states. This makes the cost of n single-molecule moves of the same order as the cost of a move in which n molecules are moved at the same time.

Single-molecule moves or moves in which all the molecules are translated at once are appropriate for homogeneous systems but it may be preferable to use other schemes for other systems. Consider, for example, the simulation of a solute molecule in a solvent. In this case, it may be more efficient to move the solute molecule more often than the solvent molecules and to move those solvent molecules which are closer to the solute more frequently than solvent molecules that are further away. Several such *preferential sampling* schemes have been developed but they require a modification of the normal Metropolis procedure and will not be discussed further here.

It is common to rotate a molecule at the same time as it is translated when a new configuration is generated, although some subtleties arise in effecting the rotation. This is because the argument presented in section 11.2 to justify the Metropolis scheme stated that it led to the correct ratio of the populations of two states (equation (11.10)). This is true for Cartesian coordinates and for other coordinate systems whose volume elements are independent of the values of the current coordinates. To see this notice that the probability of a state I is proportional to $\exp[-V_I/(k_B T)]\,d\mathbf{R}$ and the volume elements of the two states I and J will be the same and, hence, cancel out. For a general set of coordinates, the cancellation does not occur and the implementation of the Metropolis algorithm must be modified.

Many sets of coordinates that specify orientation, such as the *Euler angles*, do not have volume elements that will be the same for two states. It is feasible to alter the Metropolis scheme to use such coordinates but it is also possible to select an appropriate set of coordinates that leaves the Metropolis scheme unchanged. One such set, the set that will be used below, involves choosing one of the Cartesian axes, x, y or z, at random and then rotating the molecule about this axis by a random angle chosen in the range, $[-\delta\eta, \delta\eta]$, where the parameter $\delta\eta$ plays exactly the same role for the rotation as δt does when translating a molecule. In the same way as for the translations, the center of mass of the molecule is often chosen as the origin of the rotation. Thus, the new coordinates of the atoms, r_i', in a molecule will be generated from the old ones, r_i, using the following transformation:

$$r'_i = \mathbf{U}(r_i - R_c) + R_c \qquad (11.12)$$

where R_c is the center of mass of the molecule and $\mathbf{U}$ is the rotation matrix, which for a rotation about the x axis by an angle, η, will have the form

$$\mathbf{U}_x = \begin{pmatrix} 1 & 0 & 0 \\ 0 & \cos\eta & \sin\eta \\ 0 & -\sin\eta & \cos\eta \end{pmatrix} \qquad (11.13)$$

Because rotational and translational moves are often performed together, it is best if the values of the parameters $\delta\eta$ and δt are compatible so that the rotational space and the translational space available to the molecules are sampled with roughly the same efficiency.

The final type of move is a volume move that is required for Monte Carlo simulations in the NPT ensemble. It is possible to combine moves in which molecules are translated and rotated with those in which the volume is changed simultaneously. A commoner strategy, however, is to intersperse moves in which only the volume is changed with moves in which only molecule rotations and translations are performed. A volume change can be done in exactly the same way as a rotation or a translation. A random number, υ, in the range $[-1, 1]$ is generated and the volume of the system is changed by $\delta V \upsilon$, where δV is the maximum change in the volume that is allowed in any single move. Once a new value for the volume has been selected, it is necessary to change the coordinates of all the atoms in the simulation box in an appropriate manner. Note that, if this were not done, an increase in volume would lead to cavities forming at the boundaries of the simulation box, whereas a decrease in volume would lead to overlap of molecules in the same regions. For a system comprised of atoms, the coordinates of the atoms are changed such that their fractional coordinates, S, remain the same. Thus, for a cubic box, the coordinates of the atoms in the new configuration, R', are generated by scaling the coordinates of the atoms in the old configuration, R, by the ratio of the old to the new box lengths, L/L'. Such a scaling for a molecular system, in which molecules are supposed to be rigid, will not work because it will lead to changes in their internal geometries. In this case, the same scaling factor is applied instead to the coordinates of the centers of mass of each molecule. Thus, the molecule as a whole is moved but its internal geometry is left unchanged. The transformation relating the new, r'_i, and old, r_i, coordinates for an atom is then

$$r'_i = r_i + \left(\frac{L}{L'} - 1\right)R_c \qquad (11.14)$$

The value of the parameter determining the size of the volume moves, δV, can be adjusted during the simulation in exactly the same way as for the rotational and translational move parameters so that the desired acceptance ratio is obtained.

It is feasible to vary the internal geometry of a molecule in a Monte Carlo simulation. The simplest way, which leaves the Metropolis scheme described above unchanged, is to move the atoms by a small amount. This is easy to do but it suffers from the same disadvantage as the molecular dynamics technique in that the geometry of the molecule will probably change by only a small amount, so it may take many moves, or may even be impossible, for the system to exit from its local region of phase space and explore neighbouring configurations.

A more attractive method, in principle, which allows larger conformational changes and, thus, a more efficient exploration of the phase space of a system, is to alter the internal coordinates and, in particular, the dihedral angles of the molecule directly. The problem with this approach is the one alluded to earlier in the discussion of the rotational moves of molecules, namely that *generalized coordinates*, $\boldsymbol{Q}$, must now be used instead of Cartesian coordinates. It is possible to express integrals of the type given in equation (11.6) in terms of the generalized coordinates, but the equation that results is the following:

$$\langle \mathcal{X} \rangle = \frac{\int d\boldsymbol{Q}\, \mathcal{X}(\boldsymbol{Q}) \sqrt{\|\mathbf{J}\|} \exp[-\mathcal{V}(\boldsymbol{Q})/(k_{\mathrm{B}}T)]}{\int d\boldsymbol{Q} \sqrt{\|\mathbf{J}\|} \exp[-\mathcal{V}(\boldsymbol{Q})/(k_{\mathrm{B}}T)]} \tag{11.15}$$

where $\mathbf{J}$ is a matrix whose elements are related to the transformation between the two sets of coordinates:

$$J_{\alpha\beta} = \sum_{i=1}^{N} m_i \left(\frac{\partial \boldsymbol{r}_i}{\partial Q_\alpha} \right)^{\mathrm{T}} \frac{\partial \boldsymbol{r}_i}{\partial Q_\beta} \tag{11.16}$$

and Q_α and Q_β are components of the generalized coordinate vector, $\boldsymbol{Q}$. It is the square root of the determinant of the matrix, $\sqrt{\|\mathbf{J}\|}$, that causes the problem because its value, in general, will be dependent upon the conformation and so must be correctly accounted for in the Metropolis procedure. For simple molecules, the determinant is relatively straightforward to evaluate but for larger molecules it is much more complicated and it is this that has limited the application of the Monte Carlo technique to these types of systems. Having said all this, it should be noted that some workers have done Monte Carlo simulations with a 'normal' Metropolis algorithm in which some of the internal degrees of freedom of the molecules are altered. The assumption is that the errors introduced by the neglect of terms like those involving the matrix **J** in equation (11.16) are small, although little work seems to have been done to verify this.

The implementation and the use of the Monte Carlo modules in this chapter owe much to the work of W. Jorgensen and co-workers, who have been some of the principal exponents of the Monte Carlo method for the simulation of condensed phase systems. Indeed, the OPLS-AA force field that was chosen for the examples in this book was developed and tested extensively with the aid of such Monte Carlo simulations.

Two modules are introduced in this chapter to do Monte Carlo simulations. They are MONTE_CARLO_ENERGY, which contains procedures for the calculation of energies for periodic systems within the minimum image convention, and MONTE_CARLO_SIMULATION, which contains a subroutine for performing the simulations. To limit their use to systems comprised of small molecules the following restrictions are made.

1. Each residue in the system's sequence must correspond to a separate molecule, which implies that only non-bonding interactions can exist between the atoms of different residues, not bonding interactions.
2. The geometry of each molecule is rigid. Only the relative orientation and position of the molecules can be changed. No check is done on the geometry of the molecules by the modules so it is necessary to ensure beforehand that all molecules of the same type have the same values for their internal coordinates.

The module MONTE_CARLO_ENERGY has been made completely independent of all the other energy modules described until now. This was done in the interest of simplicity and the fact that the requirements of the Monte Carlo simulation module are slightly different from those of the molecular dynamics modules, although one of the versions of the non-bonding energy module, ENERGY_NON_BONDING, could have been adapted in an appropriate manner. The exact form of the energy calculated by the module MONTE_CARLO_ENERGY, V_{MC}, is a sum of all the non-bonding interaction energies (electrostatic and Lennard-Jones) for molecules. If I and J denote different molecules and i and j the atoms in those molecules, the energy can be written as

$$ V_{MC} = \sum_{I=1}^{N_{mol}} \sum_{J=1}^{I-1} S(R_{IJ}) \left\{ \sum_{i \in I} \sum_{j \in J} \frac{q_i q_j}{4\pi\varepsilon_0 \varepsilon r_{ij}} + 4\epsilon_{ij} \left[\left(\frac{s_{ij}}{r_{ij}}\right)^{12} - \left(\frac{s_{ij}}{r_{ij}}\right)^6 \right] \right\} \qquad (11.17) $$

The function $S(R_{IJ})$ is a non-bonding truncation function like those discussed in section 9.2, except that it is a function of the distance between the centers of mass of each molecule, R_{IJ}. This means that *all* the individual interactions between two molecules will be scaled by the *same* amount. The form of the function that is used is the same as that chosen by Jorgensen and his co-workers in much of their work and is

$$S(r) = \begin{cases} 1 & r < r_L \\ \left(\dfrac{r_U^2 - r^2}{r_U^2 - r_L^2}\right) & r_L \le r < r_U \\ 0 & r \ge r_U \end{cases} \qquad (11.18)$$

where r_L and r_U are upper and lower cutoff distances, respectively. Note that this function, unlike the one used in the atom-based force-switching version of the module ENERGY_NON_BONDING, does not have continuous first derivatives because the Monte Carlo method makes no use of the derivatives. It is, however, a good idea to have some smoothing to avoid artefacts that arise when similar configurations of a system have very different energies because some molecules happen to lie within the truncation distance and some do not. This will be especially important for systems comprised of atoms or of molecules which have a non-zero charge.

The definition of the energy module is

```
MODULE MONTE_CARLO_ENERGY

CONTAINS

    SUBROUTINE MONTE_CARLO_ENERGY_OPTIONS ( CUTOFF, DIELECTRIC, SMOOTH, &
                                                               PRINT )
        LOGICAL, INTENT(IN), OPTIONAL :: PRINT
        REAL,    INTENT(IN), OPTIONAL :: CUTOFF, DIELECTRIC, SMOOTH
    END SUBROUTINE MONTE_CARLO_ENERGY_OPTIONS

    REAL FUNCTION MONTE_CARLO_ENERGY_FULL ( )
    END FUNCTION MONTE_CARLO_ENERGY_FULL

    REAL FUNCTION MONTE_CARLO_ENERGY_ONE ( CHOSEN )
        INTEGER, INTENT(IN) :: CHOSEN
    END FUNCTION MONTE_CARLO_ENERGY_ONE

END MODULE MONTE_CARLO_ENERGY
```

The module MONTE_CARLO_ENERGY contains three procedures. The first is MONTE_CARLO_ENERGY_OPTIONS, which sets the options for the energy calculation in a Monte Carlo simulation. There are four arguments, all of which are optional. CUTOFF is a real argument that gives the cutoff to use for the calculation of the intermolecular interactions. Because the calculation of the energies requires that the minimum image convention be satisfied, the value of CUTOFF should be less than half the length of the simulation box, otherwise an error will occur. The real argument DIELECTRIC specifies the value of the dielectric constant to use in the calculation of the electrostatic interactions. It has a default value of 1. The argument SMOOTH is also real and gives the width of the smoothing region for the calculation of the intermolecular interactions using the truncation function defined in equation (11.18). In this equation, the value of the argument CUTOFF corresponds to r_U and the value of the argument SMOOTH to $r_U - r_L$ so that r_L is given by CUTOFF - SMOOTH. If the value of the argument SMOOTH is zero, negative or greater than the value of the argument CUTOFF, no smoothing is done and the interactions are calculated with simple truncation. The final argument, PRINT, is a logical variable that specifies whether a summary of the options is to be printed by the subroutine MONTE_CARLO_ENERGY_OPTIONS. The default is to print.

The other two procedures in the module are real functions that calculate energies. The function MONTE_CARLO_ENERGY_FULL calculates the complete energy for the system whereas the function MONTE_CARLO_ENERGY_ONE calculates the energy of interaction between the molecule specified by the single-integer argument CHOSEN and the remaining molecules in the system. Both procedures are pure functions in that only the function result is calculated and no other data structures are altered. They require the charges and Lennard-Jones parameters for the atoms from the module MM_TERMS, the atomic coordinates and masses from the module ATOMS, data about the residues/molecules from the module SEQUENCE and the size of the simulation box from the module SYMMETRY.

Of the three procedures in the module MONTE_CARLO_ENERGY, it is only MONTE_CARLO_ENERGY_OPTIONS that will be needed for performing a simulation. The others were described because they will be necessary for some of the analyses that are to be done later in the chapter.

The definition of the simulation module is

```
MODULE MONTE_CARLO_SIMULATION
CONTAINS

   SUBROUTINE MONTE_CARLO ( NBLOCK, NCONF, ADJUST_FREQUENCY, SAVE_FREQUENCY, &
                                   VOLUME_FREQUENCY, ACCEPTANCE, ROTATION, &
                                   TRANSLATION, VOLUME_MOVE, PRESSURE, &
                                                       TEMPERATURE, FILE )
      INTEGER, INTENT(IN) :: NBLOCK, NCONF
      CHARACTER ( LEN = * ), INTENT(IN), OPTIONAL :: FILE
      INTEGER, INTENT(IN), OPTIONAL :: ADJUST_FREQUENCY, SAVE_FREQUENCY, &
                             VOLUME_FREQUENCY
      REAL,    INTENT(IN), OPTIONAL :: ACCEPTANCE, ROTATION, TRANSLATION, &
                             VOLUME_MOVE, PRESSURE, TEMPERATURE
   END SUBROUTINE MONTE_CARLO

END MODULE MONTE_CARLO_SIMULATION
```

There is a single subroutine that carries out the type of Monte Carlo simulation described earlier. There are two essential integer arguments, NBLOCK and NCONF, which are, respectively, the number of blocks into which the simulation is to be divided and the number of configurations to investigate in each block. The total number of moves in each simulation is thus NBLOCK × NCONF. The reason for using blocks of moves in a simulation rather than just specifying a total number of moves is primarily for convenience, for it allows the progress of a simulation to be monitored closely by looking at the evolution of the averages of various properties for each block. Also, as mentioned in section 10.2 when we discussed the analysis of molecular dynamics trajectories, the division of a simulation into statistically independent blocks can provide a more accurate measure of the convergence of the values of any calculated quantities.

The remaining arguments are all optional. They are as follows.

ADJUST_FREQUENCY is an integer argument that specifies the move frequency at which the maximum values for changing the length of the simulation box and for rotating and translating a molecule are to be adjusted. Each time these parameters are checked they are scaled by 0.95 if the current acceptance ratio is less than the desired value or by 1.05 if it is greater than the desired value. Note that the box and molecule move parameters (rotation and translation) are scaled independently. The default value for this argument is 1000.

SAVE_FREQUENCY is an integer argument that specifies the frequency at which the coordinates of the current configuration are to be saved on a trajectory file. The default value for this argument is 0, which means that no trajectory file will be generated.

VOLUME_FREQUENCY is an integer argument giving the frequency at which volume, as opposed to molecule, moves are to be attempted. The default value of the argument is 500, which means that simulations will automatically be performed in the NPT ensemble. To generate results appropriate for the NVT ensemble this parameter must be set to 0.

ACCEPTANCE is a real argument indicating the acceptance ratio that it is desired to achieve in the Monte Carlo simulation. The default value is 0.4 (i.e. 40%).

ROTATION is the maximum allowable rotation of a molecule in degrees about the x, y or z axis that is to be made during a molecule's move. The rotation will be chosen uniformly in the range [−ROTATION, ROTATION]. The default rotation is 15°.

TRANSLATION is the maximum allowable translation of a molecule in ång-
    ström units along each coordinate axis during a molecule's move. It is
    applied in the same way as the ROTATION argument and has a default
    value of 0.15 Å.

VOLUME_MOVE is the maximum allowable value in Å^3 by which the volume of
    the simulation box can be changed during a volume move. The volume
    change   will   be   selected   uniformly   in   the   range
    [−VOLUME_MOVE, VOLUME_MOVE]. The default value is 400 Å^3.

PRESSURE is the argument giving the value of the pressure in atmospheres for
    the simulation. The default value is 1 atm.

TEMPERATURE is the argument specifying the simulation temperature in kel-
    vins and has a default value of 300 K.

FILE is a character argument containing the name of the file on which coord-
    inate data sets produced during the simulation are to be saved. It is
    needed only if the value of the argument SAVE_FREQUENCY is different
    from zero.

It is important to note that the values of all these options are reset every time
the subroutine is called and are not stored between simulations.

11.4 Example 21

The program in this section uses the modules described above to perform a
Monte Carlo simulation of a small solute, methane, in water. The program is
a simple one and is

```
PROGRAM EXAMPLE21

... Declaration statements ...

! . Process the MM file.
CALL MM_FILE_PROCESS ( ''solvent.opls_bin'', ''solvent.opls'' )

! . Construct the system file.
CALL MM_SYSTEM_CONSTRUCT ( ''solvent.opls_bin'', ''ch4h2o_box215.seq'' )

! . Read in the coordinates.
CALL COORDINATES_READ ( ''ch4h2o_box215.crd'' )

! . Initialize the random number seed.
CALL RANDOM_INITIALIZE ( 314159 )

! . Set the Monte Carlo energy options.
CALL MONTE_CARLO_ENERGY_OPTIONS ( CUTOFF = 8.5, SMOOTH = 0.5 )
```

```
! . Perform the Monte Carlo simulation.
CALL MONTE_CARLO ( 20, 100000, SAVE_FREQUENCY = 100, &
                          FILE = ''ch4h2o_box215.mc'' )

! . Save the current configuration.
CALL COORDINATES_WRITE ( ''ch4h2o_box215.crd2'' )

END PROGRAM EXAMPLE21
```

The program starts by processing the file `solvent.opls` that contains the molecular mechanics definitions of the methane and water molecules and by constructing the system file for a sequence comprising a single methane molecule and 215 water molecules.

In the next step, a coordinate file that contains coordinates for an equilibrated methane–water system that were generated with a previous Monte Carlo simulation is read. Because rigid molecules are being used in the simulations, the geometries of all the water molecules are the same with H—O bond distances of 0.9572 Å and H—O—H bond angles of 104.52°. The methane molecule has tetrahedral symmetry, with C—H bond lengths of 1.090 Å and H—C—H bond angles of 109.47°. These geometries were chosen because they are close to those observed experimentally and they also correspond to those that would result by optimizing the geometries of a single methane or water molecule in vacuum with the OPLS-AA force field.

After reading the coordinate file, various options are defined before the Monte Carlo simulation is performed. First, the random number seed is explicitly initialized so that the same run can be reproduced (by using the same seed) or a different set of data can be generated starting from the same coordinates (by using a different seed). Second, the Monte Carlo non-bonding energy options are set. A cutoff length of 8.5 Å is selected and all intermolecular interactions falling between 8.0 Å and 8.5 Å are 'smoothed' using the function defined in equation (11.18). Because the dimension of the simulation box is about 18.8 Å this cutoff value is small enough that the minimum image convention can be applied throughout the simulation.

The Monte Carlo simulation itself is performed for 2×10^6 steps in 20 blocks of 10^5 steps. Configurations are saved every 100 configurations on the trajectory file `ch4h2o_box215.mc`, giving 20 001 configurations in all for later analysis. The default parameters are used for all the remaining options, which means that the simulation is done in the NPT ensemble at a pressure of 1 atm and a temperature of 300 K. Volume moves are performed every 500 moves. The desired acceptance ratio is 40% and the move sizes are adjusted every 1000 moves so that this ratio is approached. After the simulation the program terminates by saving the coordinates of the current configuration for later re-use.

Simulations of methane and other small alkanes in water have proved popular because they provide simple models for investigating effects such as *hydrophobicity*, which are necessary for understanding the solvation of more complicated molecules such as lipids and proteins. Insights into the solvation of these molecules can be obtained by analysing the structure of the solvent around the solute and the energy of interaction between the solute and the solvent.

The determination of radial distribution functions for the solute and solvent atoms gives preliminary information about the structure of the solute around the solvent. This can be done in exactly the same way as for the molecular dynamics simulation of water in section 10.3. Figure 11.2 shows the radial distribution function for the carbon of the methane and the oxygens of the water molecules. The function peaks at 4 Å with a height of about 1.55, then falls to a minimum at 5.6 Å before tending to a value of 1. It is to be noted that the curve is much rougher than the radial distribution function of figure 10.2, which is due to the more limited set of data that is available for its calculation.

The number of neighbours, $n_n(r)$, may be determined by integrating the radial distribution function using the following formula:

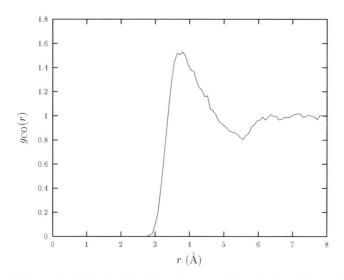

Figure 11.2 The radial distribution function, $g_{CO}(r)$, calculated from the trajectory generated by the program of Example 21.

$$n_n(r) = \frac{4\pi N_O}{V} \int_0^r s^2 g(s)\,ds \qquad (11.19)$$

where N_O is the number of oxygens in the simulation box and V is the average box volume. The values produced are shown in figure 11.3. The number of water molecules in the first solvation shell of the methane molecule may be estimated by taking the value of the neighbour function at 5.6 Å, which is the distance corresponding to the first trough in the radial distribution function. The value is approximately 21.

The trajectory can also be analysed for the energetics of interaction between the solute and solvent. Figure 11.4 shows the average energy of interaction between methane and the water as a function of distance calculated from the same trajectory as the radial distribution function. The energy is zero up to about 2.5 Å and is then positive due to short-range repulsive interactions. At about 3.4 Å it becomes negative as the attractive interactions dominate and the value decreases, rapidly at first and then more slowly. At the cutoff distance, the interaction energy is approximately -12 kJ mol^{-1}. At 5.6 Å, which is the limit of the first solvation shell, the interaction energy is about -10 kJ mol^{-1}, which shows that the water molecules in the first solvation shell contribute the bulk of the interaction energy in the model. Because there are about 21 molecules in the shell the interaction energy per solvent molecule is about -0.47 kJ mol^{-1}.

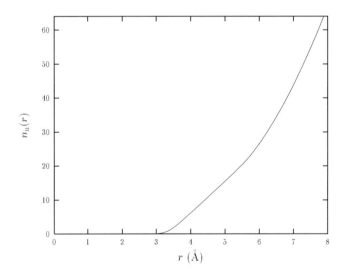

Figure 11.3. The neighbour function, $n_n(r)$, calculated from the radial distribution function $g_{CO}(r)$.

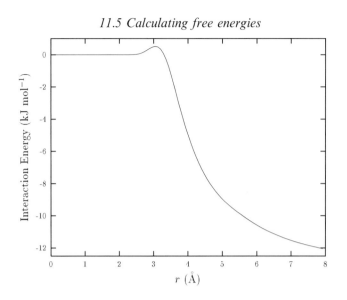

Figure 11.4. The average energy of interaction between methane and the surrounding water molecules as a function of the cutoff distance calculated from the trajectory generated by the program of Example 21.

11.5 Calculating free energies: statistical perturbation theory

In section 10.6, we saw how to calculate the potential of mean force for processes in gas and condensed phase systems using the umbrella sampling technique. This is by no means the only method that is available for calculating free energies and in this section we broach other algorithms for tackling this problem. We shall limit our discussion to the free energies because this is the quantity that is critical for the interpretation of many chemical and physical phenomena. Methods for the calculation of other thermodynamic properties, such as the entropy, exist but they can be more difficult to apply.

Recall that the Helmholtz free energy for a system is written as

$$A = -k_B T \ln Z_{NVT} \tag{11.20}$$

where the partition function, Z_{NVT}, has the form

$$Z_{NVT} = \frac{1}{h^{3N} N!} \int d\boldsymbol{P} \int d\boldsymbol{R} \exp[-\mathcal{H}(\boldsymbol{P}, \boldsymbol{R})/(k_B T)] \tag{11.21}$$

and $\mathcal{H}$ is the Hamiltonian for the system. In principle, it might be possible to calculate this quantity directly from a simulation but, as noted in section 10.6, this proves impossible in practice because extremely long simulation times are required in order to obtain results of acceptable accuracy. This problem was solved with the umbrella sampling method by using a biasing function to restrict a simulation to a certain, smaller region of configuration space that could be sampled adequately. If the property being calculated, which in our case was a PMF, required sampling from a larger region of space, this was achieved by carrying out simulations with slightly different biasing functions and collating the results for each simulation afterwards.

In this section, we consider alternative approaches for overcoming the sampling problem. They are distinct from the umbrella sampling method, but, like it, they work by restricting the region of phase space that need be sampled in a simulation. The way this is done is to compute the *differences* between the free energies of two very similar systems rather than the absolute free energy for a system given by equation (11.20). We introduce two classes of methods to calculate free-energy differences in this section. They are the *thermodynamic integration* and *thermodynamic* or *statistical perturbation* methods. The perturbation methods will be described first.

The free-energy difference between two states, I and J, of a system with partition functions Z_I and Z_J, respectively, is

$$\Delta A_{I \rightarrow J} = A_J - A_I$$

$$= -k_B T \ln\left(\frac{Z_J}{Z_I}\right) \tag{11.22}$$

To simplify the derivation a little we suppose that the kinetic energy parts of the Hamiltonians are the same (i.e. the numbers and the masses of the particles are identical) and that only the potential energy terms, V_I and V_J, differ. The kinetic energy terms in the equation cancel out and we are left with

$$\Delta A_{I \rightarrow J} = -k_B T \ln \frac{\int d\mathbf{R} \exp[-V_J(\mathbf{R})/(k_B T)]}{\int d\mathbf{R} \exp[-V_I(\mathbf{R})/(k_B T)]} \tag{11.23}$$

It is possible to rearrange this equation and express the free-energy difference as an average over an ensemble of configurations for the state, I. To do this, we add and subtract terms involving the potential energy of state I to the exponential in the numerator of equation (11.23) and then notice that the resulting equation resembles equations (11.6)–(11.8). Denoting the probability density function for state I ρ_I and the ensemble average with respect to configurations of the state I $\langle \ldots \rangle_I$ gives

$$\Delta A_{I \rightarrow J} = -k_B T \ln\left(\frac{\int d\mathbf{R} \exp[-(V_J(\mathbf{R}) - V_I(\mathbf{R}))/(k_B T)] \exp[-V_I(\mathbf{R})(k_B T)]}{\int d\mathbf{R} \exp[-V_I(\mathbf{R})/(k_B T)]}\right)$$

$$= -k_B T \ln \int d\mathbf{R} \, \rho_I(\mathbf{R}) \exp[-(V_J(\mathbf{R}) - V_I(\mathbf{R}))/(k_B T)]$$

$$= -k_B T \ln\langle \exp[-(V_J(\mathbf{R}) - V_I(\mathbf{R}))/(k_B T)] \rangle_I \tag{11.24}$$

Equation (11.24) is the main formula for statistical perturbation theory. It states that the difference in free energy between two states, I and J, can be calculated by generating a trajectory for state I (using either a molecular dynamics or a Monte Carlo technique) and calculating the average of the exponential of the difference between the potential energies of states I and J for each configuration divided by $k_B T$.

The average in equation (11.24) will not converge very rapidly unless the differences between the energies of the two states I and J are very small (of the order of $k_B T$). In actual applications this will not normally be the case, so it is usual to break the problem down into smaller pieces. To do this, we introduce a new Hamiltonian that is a function of a *coupling* or *perturbation parameter*, λ. This Hamiltonian, $\mathcal{H}(P, R, \lambda)$, is such that, when $\lambda = 0$ or 1, it is equal to the Hamiltonians of the end states, $\mathcal{H}_I$ and $\mathcal{H}_J$, but at other values it defines a series of *intermediate states* for the transition between I and J.

Many different *coupling schemes* have been used to define the intermediate states. Probably the simplest is a coupling scheme that interpolates linearly between the two states:

$$\mathcal{H}(P, R, \lambda) = (1 - \lambda)\mathcal{H}_I(P, R) - \lambda\mathcal{H}_J(P, R) \qquad (11.25)$$

More complicated coupling schemes have been proposed and may be advantageous in some cases. For example, a straightforward extension of equation (11.25) gives a non-linear scheme:

$$\mathcal{H}(P, R, \lambda) = (1 - \lambda)^n\mathcal{H}_I(P, R) - \lambda^n\mathcal{H}_J(P, R) \qquad (11.26)$$

In other formulations, it is not the Hamiltonians that are scaled directly but the force-field parameters. Thus, for example, a force-field parameter, p, in an intermediate state could be written as a linear function of λ, i.e.

$$p_\lambda = (1 - \lambda)p_I - \lambda p_J \qquad (11.27)$$

Such coupling can, of course, lead to a very complicated functional dependence of the energy of the system on the coupling parameter, λ.

Once the new Hamiltonian has been defined, the free-energy difference between two intermediate states with different values of the perturbation parameter, λ_i and λ_j, can be calculated. Assuming, as before, that the kinetic energy terms cancel out, the difference is

$$\begin{aligned}\Delta A_{i \to j} &= A_j - A_i \\ &= -k_B T \ln\langle\exp[-(V(R, \lambda_j) - V(R, \lambda_i))/(k_B T)]\rangle_{\lambda_i} \qquad (11.28)\end{aligned}$$

The total free-energy difference is, then, the sum of the individual free-energy differences between the intermediate states on going from state I to state J:

$$\Delta A_{I \to J} = \sum_{i=0}^{N_w} \Delta A_{i \to (i+1)} \qquad (11.29)$$

where N_w is the total number of intermediate states or windows and the states $i = 0$ and $i = N_w + 1$ refer to the end states, I and J, respectively.

Up to now, we have focused on the free-energy difference, $A_J - A_I$. An expression for the difference in the reverse direction, $A_I - A_J$, can be obtained from equation (11.24) by interchanging the states I and J. The values of the two differences should be equal in magnitude and opposite in sign but the relationship between the perturbation formulae is not so straightforward. In one case, the ensemble average is for a trajectory generated for state I and in the other it is for state J. The fact that the same quantity can be calculated in two independent ways provides a very useful check on the accuracy of a statistical perturbation theory calculation. If the difference between the changes in free energy for the *forwards* and *backwards* perturbations (or the *hysteresis* of the simulation) is large then the sampling in the simulation has been inadequate and the runs need to be longer or carried out with more intermediate states. It should be noted that it is unnecessary to perform two distinct sets of simulations to obtain the two free-energy differences. This is because the trajectory generated by the simulation of an intermediate state, i, can be used to determine simultaneously the free-energy differences for the windows in the forwards direction, $i \to i + 1$, and in the backwards direction, $i \to i - 1$.

So far the discussion has been a little abstract and it may be unclear to some readers exactly what form the perturbation from one state to another can take in real applications. Perhaps the most important point to note is that one of the major advantages of this technique (as well as the thermodynamic integration method to be discussed below) is that the perturbation does not have to correspond to a physically realizable process. Consider the example studied using the umbrella sampling technique in section 10.7, in which the free-energy profile was calculated for the bALA molecule as a function of one of the intramolecular hydrogen-bonding distances, but this time for the same process in a solvent. If the perturbation technique were applied to this problem, it would be physically most reasonable (and it might also be most efficient) to define the intermediate states as structures having a particular value of the distance. It is also possible, though, to employ one of the coupling formulae, either equation (11.25) or equation (11.26), in which case the intermediate states would correspond to a weighted superposition of the two end states. Suppose that the linear coupling formula were used. This would mean that both structures would be present in the simulation but their interactions with the solvent (and their internal interactions) would be scaled by $1 - \lambda$ and λ, respectively. There would be no interactions between the solute molecules.

Because physically realizable changes do not have to be studied, it is possible to calculate the free-energy differences between 'states' of a system in which the number and the type of atoms are altered. These types of changes are sometimes called *alchemical perturbations*. Thus, for example, the relative hydration free energies of two different solute molecules can be calculated by performing a simulation in which one solute molecule is transformed into another. As we shall see in the next section, we can also obtain absolute free energies of hydration if a solute is made to vanish entirely during a simulation (i.e. the two states of the system correspond to those with the solute molecule present and with it absent). Another common application is the calculation of the relative stability of binding of two different ligand molecules to another molecule, such as a protein. In this case, the transformation would be effected by simulating the transformation between the two ligand molecules as they are bound to the host molecule.

One of the most important aids in the formulation of problems that involve the calculation of free-energy differences is the concept of a *thermodynamic cycle*. Let us take as an example the process of calculating the relative binding affinities of two ligands, A and B, to a host molecule, H. The thermodynamic cycle for this is

$$
\begin{array}{ccccc}
& & \Delta A_{AB} & & \\
& H + A & \longrightarrow & HA & \\
\Delta A_{AB} & \downarrow & & \downarrow & \Delta A_{AHBH} \qquad (11.30) \\
& H + B & \longrightarrow & HB & \\
& & \Delta A_{BH} & &
\end{array}
$$

There are four free-energy contributions to the cycle, the free energies of binding of the ligands to the host in solution, ΔA_{AH} and ΔA_{BH}, the free energy for the conversion of the two ligand molecules in solution, ΔA_{AB}, and that for the conversion between the two ligand complexes, ΔA_{AHBH}. Because the free energy is a thermodynamic state function this means that any free-energy difference depends only upon the nature of the end states and is independent of the path over which the change occurs. Thus, the sum of the individual free energies around the thermodynamic cycle is zero and so

$$
\Delta A_{AH} + \Delta A_{AHBH} - \Delta A_{BH} - \Delta A_{AB} = 0 \qquad (11.31)
$$

The relative binding affinity of the two ligands is determined by the difference $\Delta A_{AH} - \Delta A_{BH}$. For some problems it may be possible to calculate these terms directly, although, according to equation (11.31), the same difference can also be written as $\Delta A_{AB} - \Delta A_{AHBH}$. In many cases, it is easier to compute the free-energy differences for the unphysical conversion of the ligands (in solution and bound to the host) than it is to compute those for the physical process of each ligand binding to the host. Thermodynamic cycles of this sort can be formulated for many other problems of interest.

The other class of methods that we shall mention here is the thermodynamic integration methods. These also calculate the differences between the free energies of two states of a system and employ the same coupling parameter approach as the thermodynamic perturbation theory methods. Instead of equation (11.22), however, they rely on the following identity:

$$\Delta A_{I \to J} = \int_0^1 d\lambda \, \frac{\partial A}{\partial \lambda} \tag{11.32}$$

Determination of the derivatives of the free energy is straightforward. Making use of equations (11.20) and (11.21) and the fact that the Hamiltonian depends upon the coupling parameter, λ, gives

$$\Delta A_{I \to J} = \int_0^1 d\lambda \left\langle \frac{\partial \mathcal{H}}{\partial \lambda} \right\rangle_\lambda \tag{11.33}$$

The derivative of the Hamiltonian with respect to the perturbation parameter is easily evaluated once the coupling scheme has been defined. The integral itself is determined by performing simulations to calculate the ensemble average in the integrand at various values of λ and then applying a standard numerical integration technique to the values that result. With thermodynamic integration methods there is no equivalent of forwards and backwards perturbations because the ensemble averages depend upon only one value of λ. Instead, the precision of the calculations can be judged by estimating the error arising in the calculation of each of the ensemble averages.

We have briefly reviewed some of the principles behind the thermodynamic perturbation and thermodynamic integration methods for the calculation of free-energy differences. It has not been possible in the space available here to describe the many variations of each of these techniques that exist or any alternative methods that may be better for certain problems. It should be borne in mind, though, when calculating free-energy differences that the choice of the most appropriate method and coupling scheme need not be at all obvious and that it will be necessary to experiment to obtain the best approach. Whatever the method, extreme care should be taken to ensure that enough simulations are done and done for long enough that the results are of the precision that is desired. Free-energy calculations of all sorts are notorious for providing pitfalls for the unwary user!

11.6 Example 22

In this book, no specific modules are given for the calculation of free energies using statistical perturbation theory. The reason, as is evident from the previous section and from section 10.6, is that there are numerous ways in which such calculations can be performed and it would be difficult to come up with a module general enough to be applied to a wide range of systems. Instead, in this section, an example program is described that does a statistical perturbation theory calculation but which uses the standard Monte Carlo modules introduced earlier. The program is a complicated one, but it has the advantage that it can be readily adapted to other cases. It can also be used, with suitable modifications, for calculating free energies using the molecular dynamics instead of the Monte Carlo modules.

The problem addressed by the program in this section is the estimation of the free energy of water using statistical perturbation theory. The free energy is determined by taking a box of water molecules and incrementally making one of them vanish. The transformation is effected in two steps. In the first the charges on the atoms in a single water molecule are made to disappear and in the second the Lennard-Jones parameters for the oxygen are gradually reduced to zero (the Lennard-Jones parameters for the hydrogens are already zero). In each case, the parameter values are changed by linearly scaling them with the perturbation parameter, λ, and performing several simulations with values for λ between 0 and 1. In the step in which the charges are made to disappear, 21 simulations are performed, with increments in λ of 0.05. In the second step, in which the Lennard-Jones parameters are removed, 11 simulations are done and the change in λ at each step is 0.1.

The program for the step in which the charges are reduced to zero is

```
PROGRAM EXAMPLE22

... Declaration statements ...

! . Program parameters.
CHARACTER ( LEN = 16 ), PARAMETER :: MCFILE = ''tempfile.mc''
INTEGER,               PARAMETER :: NSIMULATIONS = 21, SOLUTE = 1
REAL,                  PARAMETER :: DLAMBDA = 1.0 / ( NSIMULATIONS - 1.0 ), &
                                    T = 300.0

! . Program scalars.
INTEGER :: ISIM
```

```
REAL     :: LAMBDA

! . Program arrays.
REAL, DIMENSION(1:NSIMULATIONS) :: DGIJ = 0.0, DGJI = 0.0

! . Scalars for the parameter subroutines.
INTEGER :: N, START, STOP

! . Arrays for the parameter subroutines.
REAL, ALLOCATABLE, DIMENSION(:) :: CHARGES

! . Initialize the random number seed.
CALL RANDOM_INITIALIZE ( 181721 )

! . Set the Monte Carlo energy options.
CALL MONTE_CARLO_ENERGY_OPTIONS ( CUTOFF = 8.5, SMOOTH = 0.5 )

! . Read in the system file.
CALL MM_SYSTEM_CONSTRUCT ( ''solvent.opls_bin'', ''h2o_box216.seq'' )

! . Read in the starting equilibrated coordinates.
CALL COORDINATES_READ ( ''h2o_box216.crd'' )

! . Save the molecule parameters.
CALL PARAMETERS_SAVE

! . Loop over the number of simulations.
DO ISIM = 1,NSIMULATIONS

   ! . Set the value of LAMBDA.
   LAMBDA = REAL ( NSIMULATIONS - ISIM ) * DLAMBDA

   ! . Reset the parameter values.
   CALL PARAMETERS_SET ( LAMBDA )

   ! . Equilibrate the system at the new values.
   CALL MONTE_CARLO ( 10, 100000, TEMPERATURE = T )

   ! . Collect data.
   CALL MONTE_CARLO ( 40, 100000, TEMPERATURE = T, SAVE_FREQUENCY = 100, &
                                                   FILE = MCFILE )

   ! . Calculate the averages required from the trajectory.
   CALL CALCULATE_FREE_ENERGIES

END DO

! . Write out a summary of the results.
WRITE ( OUTPUT, ''(/23('-'),A,23('-'))'' ) '' Calculated Free Energies ''
WRITE ( OUTPUT, ''(2(4X,A),2(7X,A),4X,A)'' ) &
                ''Lambda I'', ''Lambda J'', ''dG (I->J)'', ''dG (I<-J)''
```

```
DO ISIM = 1,(NSIMULATIONS-1)
   LAMBDA = REAL ( NSIMULATIONS - ISIM ) * DLAMBDA
   WRITE ( OUTPUT, ''(2F12.4,3G16.4)'' ) LAMBDA, LAMBDA - DLAMBDA, &
                                         DGIJ(ISIM), DGJI(ISIM)
END DO
WRITE ( OUTPUT, ''(A,19X,3G16.4)'' ) ''Total'', SUM ( DGIJ ), SUM ( DGJI )
WRITE ( OUTPUT, ''(72('-'))'' )

! . Free the parameter arrays.
CALL PARAMETERS_FREE

!===============================================================================
CONTAINS
!===============================================================================

   !----------------------------------
   SUBROUTINE CALCULATE_FREE_ENERGIES
   !----------------------------------

   ! . Local parameters.
   REAL, PARAMETER :: RT = R * T

   ! . Local scalars.
   INTEGER :: IFRAME
   REAL    :: EI, EJ, GB, GF

   ! . Type definitions.
   TYPE(TRAJECTORY_TYPE) :: MCTRAJECTORY

   ! . Initialize the trajectory.
   CALL TRAJECTORY_INITIALIZE ( MCTRAJECTORY )

   ! . Activate the trajectory for reading.
   CALL TRAJECTORY_ACTIVATE_READ ( MCFILE, MCTRAJECTORY )

   ! . Initialize the accumulators.
   GB = 0.0
   GF = 0.0

   ! . Loop over the frames.
   DO IFRAME = 1,MCTRAJECTORY%NFRAMES

      ! . Read the next frame.
      CALL TRAJECTORY_READ ( MCTRAJECTORY, ATMCRD, BOXL )

      ! . Calculate the energy at i.
      CALL PARAMETERS_SET ( LAMBDA )
      EI = MONTE_CARLO_ENERGY_ONE ( SOLUTE )

      ! . Calculate the energy at i-1.
      IF ( ISIM > 1 ) THEN
```

```
            CALL PARAMETERS_SET ( LAMBDA + DLAMBDA )
            EJ = MONTE_CARLO_ENERGY_ONE ( SOLUTE )
            GB = GB + EXP ( - ( EJ - EI ) / RT )
        END IF

        ! . Calculate the energy at i+1.
        IF ( ISIM < NSIMULATIONS ) THEN
            CALL PARAMETERS_SET ( LAMBDA - DLAMBDA )
            EJ = MONTE_CARLO_ENERGY_ONE ( SOLUTE )
            GF = GF + EXP ( - ( EJ - EI ) / RT )
        END IF

END DO


! . Scale the values.
GB = GB / REAL ( MCTRAJECTORY%NFRAMES )
GF = GF / REAL ( MCTRAJECTORY%NFRAMES )

! . Calculate the values of the free energies for the window.
IF ( ISIM > 1 ) DGJI(ISIM-1) = - RT * LOG ( GB )
IF ( ISIM < NSIMULATIONS ) DGIJ(ISIM) = - RT * LOG ( GF )

! . Deactivate the trajectory.
CALL TRAJECTORY_DEACTIVATE ( MCTRAJECTORY )

END SUBROUTINE CALCULATE_FREE_ENERGIES

!------------------------
SUBROUTINE PARAMETERS_FREE
!------------------------

! . Deallocate the charge array.
DEALLOCATE ( CHARGES )

END SUBROUTINE PARAMETERS_FREE

!------------------------
SUBROUTINE PARAMETERS_SAVE
!------------------------

! . Get the atom indices for the molecule.
START = RESIND(SOLUTE)+1
STOP  = RESIND(SOLUTE+1)
N     = RESIND(SOLUTE+1) - RESIND(SOLUTE)

! . Allocate the charge array.
ALLOCATE ( CHARGES(1:N) )

! . Save the charges.
CHARGES = ATMCHG(START:STOP)
```

```
END SUBROUTINE PARAMETERS_SAVE


!------------------------------------
SUBROUTINE PARAMETERS_SET ( LAMBDA )
!------------------------------------

! . Scalar arguments.
REAL, INTENT(IN) :: LAMBDA

! . Set the new values of the charges.
ATMCHG(START:STOP) = LAMBDA * CHARGES

END SUBROUTINE PARAMETERS_SET

END PROGRAM EXAMPLE22
```

The program starts off in the usual way by defining the system to be studied and reading in an appropriate set of coordinates for the atoms. After the initialization phase, a local subroutine, PARAMETERS_SAVE, is called, which keeps a copy of the starting values of the parameters that are to be changed in the perturbation procedure. In this case, it is the charges of the atoms of one of the water molecules which are stored in the array, CHARGES. Note that the identity of the molecule being removed is defined by the parameter SOLUTE in the declaration statements at the start of the program. These commands complete the set-up phase of the calculation.

The principal work of the program is done in the loop that follows the initial commands. Each iteration performs a simulation at a different value of the perturbation parameter, λ, which is determined in the first of the loop statements. This value is then used in the call to the subroutine PARAMETERS_SET to fix the values of the parameters with which the Monte Carlo simulation is to be performed. Two simulations are done at a temperature of 300 K and a pressure of 1 atm. First there is an equilibration phase of 10^6 moves and then a simulation of 4×10^6 moves in which the coordinates of the atoms in the water box are saved on a trajectory every 100 configurations (giving 40 001 configurations in all).

The final statement in the loop is a call to the local subroutine CALCULATE_FREE_ENERGIES in which the analysis of the configurations generated in the previous Monte Carlo simulation is carried out. The subroutine loops over each of the configurations stored in the trajectory file in turn and calculates the interaction energy between the water molecule being removed and the remaining molecules in the system. In the most general case (i.e. when λ is not equal to zero or to unity), the interaction energy is calculated for *three* different sets of parameters, which are those of the previous, the current and the subsequent simulations. Denoting these energies $\mathcal{V}_{i-1}$, $\mathcal{V}_i$ and $\mathcal{V}_{i+1}$, respectively, the free energies in the backwards, $\Delta G_{i-1\leftarrow i}$, and forwards, $\Delta G_{i\rightarrow i+1}$, directions are calculated as

$$\Delta G_{i-1\leftarrow i} = -RT \ln\left\langle \exp\left(-\frac{(\mathcal{V}_{i-1} - \mathcal{V}_i)}{RT}\right)\right\rangle \qquad (11.34)$$

$$\Delta G_{i\rightarrow i+1} = -RT \ln\left\langle \exp\left(-\frac{(\mathcal{V}_{i+1} - \mathcal{V}_i)}{RT}\right)\right\rangle \qquad (11.35)$$

where the angle brackets denote the average values of the indicated quantities for each configuration in the trajectory. The program terminates by printing out the calculated free-energy differences both in the forwards and in the backwards direction for each of the simulations and the total values for the complete transformation.

The program for performing the second step in which the Lennard-Jones parameters of the oxygen are reduced to zero is essentially the same as the program given above. Apart from the minor changes that have to be made to such things as the names of files and the random number seed, the major differences between the programs lie in the subroutines PARAMETERS_SAVE and PARAMETERS_SET. The appropriate versions are

```
!-------------------------
SUBROUTINE PARAMETERS_SAVE
!-------------------------

! . Get the atom indices for the molecule.
START = RESIND(SOLUTE)+1
STOP  = RESIND(SOLUTE+1)
N     = RESIND(SOLUTE+1) - RESIND(SOLUTE)

! . Allocate the Lennard-Jones parameter arrays.
ALLOCATE ( EPSILON(1:N), SIGMA(1:N) )

! . Save the Lennard-Jones parameters.
EPSILON = ( ATMEPS(START:STOP) / 2.0 )**2
SIGMA   = ATMSIG(START:STOP)**2

END SUBROUTINE PARAMETERS_SAVE


!------------------------------------
SUBROUTINE PARAMETERS_SET ( LAMBDA )
!------------------------------------

! . Scalar arguments.
REAL, INTENT(IN) :: LAMBDA

! . The charges on the atoms are always zero.
ATMCHG(START:STOP) = 0.0
```

```
!  . Set the new values of the Lennard-Jones parameters.
ATMEPS(START:STOP) = 2.0 * SQRT ( LAMBDA * EPSILON )
ATMSIG(START:STOP) = SQRT ( LAMBDA * SIGMA )


END SUBROUTINE PARAMETERS_SET
```

The main thing to note about these subroutines is that, in contrast to the charges, the Lennard-Jones parameters are stored in the arrays ATMEPS and ATMSIG not as ϵ_{ii} and s_{ii} but rather as $\sqrt{4\epsilon_{ii}}$ and $\sqrt{s_{ii}}$ for each atom. Thus, the values in the arrays ATMEPS and ATMSIG are converted back to the proper ϵ_{ii} and s_{ii} values when filling the parameter arrays EPSILON and SIGMA in the subroutine PARAMETERS_SAVE. In the subroutine PARAMETERS_SET, the charges on the atoms in the water molecule are set to zero for all values of λ while the Lennard-Jones parameters, ϵ_{ii} and s_{ii}, are scaled linearly for all the atoms in the perturbed water molecule. Both the arrays, EPSILON and SIGMA, are defined at the start of the program instead of CHARGES.

Note that, in both these programs, the whole perturbation is carried out at once. In a real study it is more likely that the perturbation would be broken up into several different jobs so that the results of one calculation could be verified before those of the next were begun.

The results of the calculations as functions of the perturbation parameter are shown in tables 11.1 and 11.2. The perturbation in the first step was carried out using 21 simulations (i.e. a change in λ of 0.05) whereas the second step was performed with only 11 simulations. There is reasonable agreement between the free-energy differences calculated in the forwards and backwards directions. The total free energies for the step in which the charges are removed differ by less than 0.2 kJ mol^{-1} while there is a difference of about 1.8 kJ mol^{-1} between the values for the second step. The charges make the biggest contribution to the free-energy change and the charge and Lennard-Jones terms have opposite contributions – the removal of the electrostatic interactions between a water molecule and its neighbours requires energy whereas the removal of the Lennard-Jones interactions is favourable. The average of the forwards and backwards total free-energy changes from the simulations is 25.6 kJ mol^{-1}, which is in good agreement with the experimental value of 26.5 kJ mol^{-1} (at 25 $^\circ$C).

Exercises

11.1 Analyse in more detail the trajectory generated in the program described in section 11.4.

    (a) Write a program to calculate the average of the energies of interaction between methane and the surrounding water molecules. In addition to

Table 11.1 *The free-energy change as a function of the perturbation parameter,* λ, *for the first step of the statistical perturbation calculation of Example 22 in which the electrostatic interactions between a single water molecule and the remainder of the molecules in the system are eliminated. Energies are in* $kJ\,mol^{-1}$.

λ_i	λ_j	$\Delta G_{i \rightarrow j}$	$\Delta G_{i \leftarrow j}$	$\Delta G_{average}$
1.00	0.95	4.73	−4.22	−4.48
0.95	0.90	3.74	−3.99	−3.87
0.90	0.85	3.70	−3.86	−3.78
0.85	0.80	3.57	−3.34	−3.45
0.80	0.75	2.99	−3.17	−3.08
0.75	0.70	2.76	−2.63	−2.69
0.70	0.65	2.40	−2.42	−2.41
0.65	0.60	2.11	−1.93	−2.02
0.60	0.55	1.66	−1.66	−1.66
0.55	0.50	1.42	−1.52	−1.47
0.50	0.45	1.32	−1.47	−1.40
0.45	0.40	1.30	−1.13	−1.22
0.40	0.35	0.96	−1.06	−1.01
0.35	0.30	0.89	−0.58	−0.74
0.30	0.25	0.45	−0.61	−0.53
0.25	0.20	0.46	−0.56	−0.51
0.20	0.15	0.45	−0.43	−0.44
0.15	0.10	0.30	−0.32	−0.31
0.10	0.05	0.18	−0.28	−0.23
0.05	0.00	0.15	−0.10	−0.12
Total		35.55	−35.27	−35.41

reproducing the data in figure 11.4, determine the relative importance of the electrostatic and the Lennard-Jones contributions to the interaction. How does the size of these interactions compare with, for example, the interactions between two water molecules in solution? *Hint: use as a basis for the energy calculation the subroutine* CALCULATE_FREE_ENERGIES *in the program of Example 22.*

(b) What is the structure of the water about the methane molecule? For example, are there hydrogen bonds?

(c) Estimate the size of the effect of the truncation of the intermolecular interactions on the values of the electrostatic and Lennard-Jones energies.

11.2 The programs in section 11.6 calculated the free energy of a water molecule in water. Do similar calculations using the same programs to calculate the free energies of hydration of other small solutes, such as methane (see section 11.4) and the chloride anion. A good strategy in both cases is to try to transform the solute molecules into the Lennard-Jones particle that acted as an intermediate in the

Table 11.2 *The free-energy change as a function of the perturbation parameter,* λ*, for the second step of the statistical perturbation calculation of Example 22 in which the Lennard-Jones interactions between a single, chargeless water molecule and the remainder of the molecules in the system are eliminated. Energies are in* $kJ \, mol^{-1}$.

λ_i	λ_j	$\Delta G_{i \rightarrow j}$	$\Delta G_{i \leftarrow j}$	$\Delta G_{\text{average}}$
1.0	0.9	−0.73	1.25	0.99
0.9	0.8	−1.84	2.21	2.02
0.8	0.7	−1.70	1.31	1.50
0.7	0.6	−1.11	1.21	1.16
0.6	0.5	−1.17	1.18	1.18
0.5	0.4	−0.74	1.43	1.09
0.4	0.3	−0.81	0.95	0.88
0.3	0.2	−0.56	0.63	0.60
0.2	0.1	−0.01	0.34	0.17
0.1	0.0	−0.22	0.16	0.19
Total		−8.90	10.70	9.78

calculations of section 11.6. Try various approaches for changing the parameters. Which are the most effective? Note that, for the negatively charged chloride anion, a correction that accounts for the neglect of the electrostatic interactions beyond the cutoff distance will have to be made in order to obtain reasonable agreement with experimental values. *Hint: use the Born expression of equation (9.10).*

12

Miscellaneous topics

12.1 Introduction

This chapter covers a number of miscellaneous topics that, although they are useful, were not central to the developments in the rest of the book. It starts by describing a widely used method for specifying the structure of a system by means of internal coordinates and continues with some simple techniques for constructing solvent boxes and for solvating molecules. It finishes with general discussions of constraint techniques and strategies for obtaining parameters for empirical force fields.

12.2 Z-matrices

Until now we have assumed that a coordinate file is available for the system that we have been studying. This will often be so because structures can be obtained by means of several different experimental techniques, such as, for example, X-ray crystallography. There will be times, though, when it is necessary to construct a set of coordinates for a system from scratch.

A common way to generate structures is to use a *Z-matrix* that defines the geometry of a molecule in terms of internal coordinates (see section 3.3). The best way to illustrate a Z-matrix is by an example, which here we take as ethane:

```
!===============================================================================
     8   1   1 ! # of atoms, residues and subsystems.
!===============================================================================
Subsystem   1   ETHANE
     1 ! # of residues.
```

```
!============================================================================
  Residue    1 ETHANE
     8 ! # of atoms.
  !        Atom I              Distance      Angle     Dihedral   J   K   L
     1    C1            6        0.0000      0.0000      0.0000   0   0   0
     2    C2            6        1.5400      0.0000      0.0000   1   0   0
     3    H1            1        1.1110    109.5000      0.0000   1   2   0
     4    H2            1        1.1110    109.5000    120.0000   1   2   3
     5    H3            1        1.1110    109.5000   -120.0000   1   2   3
     6    H4            1        1.1110    109.5000    180.0000   2   1   3
     7    H5            1        1.1110    109.5000     60.0000   2   1   3
     8    H6            1        1.1110    109.5000    -60.0000   2   1   3
  !============================================================================
```

It is immediately apparent that the format is similar to that for a coordinate file and exhibits the same partitioning of the system into Subsystems, Residues and Atoms. The only difference is in the specification of the atom data. Instead of the atom line listing the Cartesian coordinates of an atom after its atomic number, it has three real numbers and three integers. The real numbers are, respectively, a distance, an angle and a dihedral angle. The distance is in ångström units and the angles are in degrees. The three integers that follow indicate three other atoms that precede the current atom in the file and define the connectivity for the distance, the angle and the dihedral. If the atom that is defined on the line is denoted I and those defined by the three integers are denoted J, K and L, then the distance in the Z-matrix line is the distance I—J, the angle refers to the angle I—J—K and the dihedral to the dihedral I—J—K—L.

This should become clearer by looking at the fourth atom line of the file which defines the atom H2. This atom is at a distance of 1.111 Å from atom 1 (C1). The angle H2—C1—C2 (atom 2) is 109.5° and the dihedral H2—C1—C2—H1 (atom 3) is 120.0°. Note that, in the definition of the dihedral, the connectivity does not reflect the bonds that are actually present in the molecule. This holds for any of the connectivity definitions in the Z-matrix, for the distance, angle and dihedral specifications are only geometrical constructs, not chemical ones. The remaining atom definitions, atoms 5–8, follow the same pattern as for atom 4.

The Z-matrix definitions for the first three atoms are slightly different because the positions of an atom in a Z-matrix are always defined with respect to the atoms that precede it. Thus, the first atom in a Z-matrix has no antecedents and so its line contains no data. The second atom in the Z-matrix has a single antecedent and is placed so that it is at a certain distance from the first atom (1.54 Å in the example given above). To define the position of the third atom in the Z-matrix it is necessary only to specify a distance (from either the first or the second atom) and an angle. These reduced specifications are manifestations of the fact that only $3N - 6$ degrees of freedom are needed to specify the internal structure of a molecule – the position and orientation of the molecule in space, which are determined by the translational and rotational degrees of freedom, are arbitrary.

It is relatively straightforward to build structures for small systems with Z-matrices but their use quickly becomes cumbersome as the number of atoms gets larger. Many programs that work on small molecules provide Z-matrix facilities and allow the optimization of a structure, for example, directly in the internal coordinates defined by the Z-matrix rather than in the space of Cartesian coordinates (as done in this book). The advantages of using Z-matrices are that they provide an intuitive notion of the structure of the molecule (compared with that provided by a coordinate file!) and that it is possible to *freeze* certain internal coordinates out of the minimization process. There are a number of problems, however, with Z-matrices. First, it is possible to write many Z-matrices that define the same structure and it is often found that the definition is crucial in determining the effectiveness of the optimization procedure. In particular, it sometimes occurs that the Z-matrix becomes ill-defined during the optimization. This will happen, for example, if three consecutive atoms in a Z-matrix line become collinear, in which case the dihedral specified in the line cannot be defined. Care must be taken to avoid such eventualities. A better approach, at least for the optimization of small-to-medium-sized molecules, is to use algorithms that generate sets of internal coordinates automatically without the intervention of the user. Unfortunately, there is not enough space to cover such methods here.

There are two modules in the program library that handle Z-matrices. They are ZMATRIX, which defines the Z-matrix data structure, and ZMATRIX_IO, which deals with Z-matrix input and output. The definition of ZMATRIX_IO is

```
MODULE ZMATRIX_IO

CONTAINS

   SUBROUTINE ZMATRIX_DEFINE ( FILE )
      CHARACTER ( LEN = * ), INTENT(IN), OPTIONAL :: FILE
   END SUBROUTINE ZMATRIX_DEFINE

   SUBROUTINE ZMATRIX_READ ( FILE )
      CHARACTER ( LEN = * ), INTENT(IN), OPTIONAL :: FILE
   END SUBROUTINE ZMATRIX_READ

   SUBROUTINE ZMATRIX_WRITE ( FILE )
      CHARACTER ( LEN = * ), INTENT(IN), OPTIONAL :: FILE
   END SUBROUTINE ZMATRIX_WRITE

END MODULE ZMATRIX_IO
```

This module parallels exactly the module `COORDINATE_IO` of section 2.4. There are three subroutines, `ZMATRIX_DEFINE`, `ZMATRIX_READ` and `ZMATRIX_WRITE`, each of which contains an optional argument that is the name of a file containing a Z-matrix specification. `ZMATRIX_DEFINE`, like `COORDINATES_DEFINE`, not only reads the Z-matrix from a file but also redefines the atom and sequence data structures. `ZMATRIX_READ` reads a Z-matrix from a file in the case in which the atom and sequence data structures already exist, whereas `ZMATRIX_WRITE` writes the Z-matrix to a file or to the default output stream if the file argument is not present. The only difference between these subroutines and those in `COORDINATE_IO` is that there is no equivalent of the remaining optional arguments, `DATA` and `SELECTION`.

The module `ZMATRIX` contains two subroutines that are likely to be of interest. Its definition is

```
MODULE ZMATRIX

CONTAINS

    SUBROUTINE ZMATRIX_BUILD
    END SUBROUTINE ZMATRIX_BUILD

    SUBROUTINE ZMATRIX_FILL
    END SUBROUTINE ZMATRIX_FILL

END MODULE ZMATRIX
```

Neither of the subroutines has arguments. `ZMATRIX_BUILD` constructs a set of Cartesian coordinates from a Z-matrix that has been read using a previous call to `ZMATRIX_DEFINE` or `ZMATRIX_READ`. Both the latter subroutines only read the Z-matrix file. They do not automatically change the Cartesian coordinates of the system to reflect the structure defined in the Z-matrix, which must be done explicitly with a call to `ZMATRIX_BUILD`. This subroutine constructs the coordinates of the system from the Z-matrix definition with an arbitrary orientation and then, if there are no errors, performs a principal axis transformation on the coordinate set (see section 3.5). The subroutine `ZMATRIX_FILL` performs the inverse operation. It takes the Cartesian coordinates defined in the module `ATOMS` and calculates the Z-matrix distance, angle and dihedral variables that are appropriate for the current structure.

Some readers will, no doubt, be familiar with Z-matrices and will have come across the notion of dummy atoms. These are not real atoms but are fictitious atoms that are introduced to facilitate the definition of the Z-matrix. In the Z-matrix modules described above there is no specific dummy-atom option. All the atoms are real in the sense that they must exist in the structure or have been defined in the MM definition file.

12.3 Example 23

A simple example of the use of Z-matrices is given in the following program:

```
PROGRAM EXAMPLE23

... Declaration statements ...

! . Define the atom, sequence and Z-matrix data structures.
CALL ZMATRIX_DEFINE ( ''bALA1.zmatrix'' )

! . Calculate the atomic coordinates.
CALL ZMATRIX_BUILD

! . Print the coordinates.
CALL COORDINATES_WRITE

! . Rebuild the Z-matrix from the coordinates.
CALL ZMATRIX_FILL

! . Print the Z-matrix.
CALL ZMATRIX_WRITE

END PROGRAM EXAMPLE23
```

The program starts by reading a file, using ZMATRIX_DEFINE, that contains a Z-matrix for the blocked alanine molecule, bALA. This command defines the atom and sequence data structures as well as the Z-matrix data structure. The atomic coordinates for the molecule are then built with a call to ZMATRIX_BUILD and written out to the output file with COORDINATES_WRITE. To check that the Z-matrix modules work correctly, the two commands that follow recalculate the Z-matrix variables using the ZMATRIX_FILL subroutine and then write the Z-matrix to the output file. The Z-matrix written out in this way will be equivalent to that in the file bALA1.zmatrix.

12.4 Constructing solvent boxes and Example 24

In the last few chapters of this book, we have concentrated on studying systems in the condensed phase. In each case, it was assumed that a partially or fully equilibrated solvent box was available for the simulations that we performed and we did not mention how to obtain such boxes in the first place. This omission is rectified here.

It should be stated initially that there is no unique way of creating solvent boxes and that many strategies are likely to give satisfactory results. The method adopted here works well for the examples given in this book but it need not be the most efficient method, particularly if the solvent molecules are not approximately spherical in shape.

The program constructs a box of 216 water molecules of the correct density at a temperature of 300 K and at a pressure of 1 atm. With suitable modifications the program can be easily adapted for the construction of different solvent boxes either with different numbers of waters or with other solvent molecules.

The program is

```
PROGRAM EXAMPLE24

... Declaration statements ...

! . Program parameters.
INTEGER, PARAMETER :: NCHANGES = 10
REAL, PARAMETER    :: BIGSIZE  = 10.0, DENSITY = 996.0 ! kg m^-3.

! . Program scalars.
INTEGER :: ICHANGE
REAL    :: NEWSIDE, REDUCE, SIDE, TARGET, VOLUME

! . Program arrays.
REAL, ALLOCATABLE, DIMENSION(:,:) :: COORDINATES

! . Define the atom, sequence and Z-matrix data structures.
CALL ZMATRIX_DEFINE ( ''h2o.zmatrix'' )

! . Calculate the atomic coordinates.
CALL ZMATRIX_BUILD

! . Print the coordinates.
CALL COORDINATES_WRITE

! . Save the coordinates of the molecule.
ALLOCATE ( COORDINATES(1:3,1:NATOMS) ) ; COORDINATES = ATMCRD

! . Read in the system file for the solvent box.
CALL MM_SYSTEM_CONSTRUCT ( ''solvent.opls_bin'', ''h2o_box216.seq'' )

! . Calculate the volume of a box that gives the appropriate density
! . (Angstroms^3).
VOLUME = ( SUM ( ATMMAS(1:NATOMS) ) / DENSITY ) * ( AMU_TO_KG * 1.0E+30 )
```

```
! . Calculate the dimension of a cubic box of the same volume (Angstroms).
TARGET = EXP ( LOG ( VOLUME ) / 3.0 )

! . Start off with a very large box.
SIDE = BIGSIZE * TARGET

! . Calculate the reduction factor.
REDUCE = EXP ( - LOG ( BIGSIZE ) / REAL ( NCHANGES ) )

! . Assign a value to the box size.
CALL SYMMETRY_CUBIC_BOX ( SIDE )

! . Generate coordinates for each molecule.
CALL SET_INITIAL_COORDINATES

! . Define the options for the Monte Carlo modules.
CALL MONTE_CARLO_ENERGY_OPTIONS ( CUTOFF = 8.5, SMOOTH = 0.5 )

! . Do a Monte Carlo simulation (at constant volume).
CALL MONTE_CARLO ( 10, 10 * NRESID, VOLUME_FREQUENCY = 0 )

! . Gradually reduce the size of the box to the desired value.
DO ICHANGE = 1,NCHANGES

   ! . Calculate and set the new box length.
   NEWSIDE = REDUCE * SIDE
   CALL SYMMETRY_CUBIC_BOX ( NEWSIDE )

   ! . Scale the coordinates.
   CALL SET_INTERMEDIATE_COORDINATES

   ! . Do a Monte Carlo simulation (at constant volume).
   CALL MONTE_CARLO ( 10, 10 * NRESID, VOLUME_FREQUENCY = 0 )

   ! . Reset SIDE.
   SIDE = NEWSIDE

END DO

! . Do a final MC calculation for a fuller equilibration.
CALL MONTE_CARLO ( 10, 100000 )

! . Save the coordinates.
CALL COORDINATES_WRITE ( ''water_box216.crd'' )

! . Deallocate the temporary arrays.
DEALLOCATE ( COORDINATES )

!==============================================================================
CONTAINS
!==============================================================================
```

```
!--------------------------------
SUBROUTINE SET_INITIAL_COORDINATES
!--------------------------------

! . Local scalars.
INTEGER :: I, START, STOP

! . Loop over the residues.
DO I = 1,NRESID

   ! . Set the initial values of the coordinates.
   START = RESIND(I)+1
   STOP  = RESIND(I+1)
   ATMCRD(1:3,START:STOP) = COORDINATES

   ! . Randomly translate the molecule within the box.
   CALL TRANSLATE ( ATMCRD(1:3,START:STOP), &
                    SIDE * ( RANDOM_VECTOR ( 3 ) - 0.5 ) )

END DO

END SUBROUTINE SET_INITIAL_COORDINATES

!-------------------------------------
SUBROUTINE SET_INTERMEDIATE_COORDINATES
!-------------------------------------

! . Local scalars.
INTEGER :: I, START, STOP
REAL    :: SCALE

! . Local arrays.
REAL, DIMENSION(1:3) :: DR

! . Calculate the scale factor for changing the coordinates
! . of the molecular centers.
SCALE = NEWSIDE / SIDE

! . Translate the coordinates of the atoms in each molecule.
DO I = 1,NRESID

   ! . Find the atom indices for the molecule.
   START = RESIND(I)+1
   STOP  = RESIND(I+1)

   ! . Find the translation.
   DR = ( SCALE - 1.0 ) * &
        CENTER ( ATMCRD(1:3,START:STOP), ATMMAS(START:STOP) )

   ! . Translate the molecule by the required amount.
   CALL TRANSLATE ( ATMCRD(1:3,START:STOP), DR )

END DO

END SUBROUTINE SET_INTERMEDIATE_COORDINATES

END PROGRAM EXAMPLE24
```

The first part of the program is concerned with generating and saving the coordinates for a *single* water molecule. The calls to the subroutines ZMATRIX_DEFINE and ZMATRIX_BUILD read in a Z-matrix for a water molecule and then construct its atoms' Cartesian coordinates. As noted earlier, ZMATRIX_BUILD automatically performs a principal axis transformation so that the water molecule's center of mass will be at the origin. Finally the coordinates of the molecule are saved in the array COORDINATES, which is allocated to have the correct dimensions.

After the coordinates of the single solvent molecule have been saved, the system file for the solvent box is constructed using the appropriate binary molecular mechanics file and a sequence definition file. At this stage, all the information generated previously for the single water molecule is overwritten except for the coordinates, which were saved in the array COORDINATES.

The next step is to calculate the volume, V, of the solvent box given the solvent's density, ρ. The formula used is

$$V = \frac{\sum_{i=1}^{N} m_i}{\rho} \tag{12.1}$$

where the total mass of all the atoms in the solvent box is used. The desired value for the density is specified by the constant DENSITY, which is defined in a parameter statement at the top of the program. Because the units of the density are kg m^{-3} it is necessary to use a conversion factor of AMU_TO_KG to convert the atomic masses in the array ATMMAS to kilograms from atomic mass units (a.m.u.) and a factor of 10^{30} which converts m^3 to Å^3. The resulting volume has units of Å^3 as required. The constant AMU_TO_KG is defined in the module CONSTANTS. Once the volume is known the length of the side of the cubic solvent box is calculated from $V^{1/3}$ and put into the variable TARGET.

The approach adopted for the determination of the coordinates of the molecules in the box is to start off by placing molecules at random within a box much larger than a box of the target size. The size of the large box is then gradually reduced until the target size is reached. The reason for starting off with a large box size is that the probability of having molecules overlap will be small. In this example, the initial box length is ten times the target length and the reduction procedure is carried out in ten steps. Both these values are specified in parameter statements at the top of the program, the scaling factor for the box length in BIGSIZE and the number of reductions in NCHANGES.

The next statements in the program set the values of the variables SIDE, which holds the current length of the box, and REDUCE, which gives the scaling factor by which to reduce the box length at each reduction step. The latter is calculated from $BIGSIZE^{-1/NCHANGES}$ and is designed so that $REDUCE^{NCHANGES} = BIGSIZE^{-1}$. In other words, NCHANGES applications of REDUCE to the initial, large box length will result in a box length of TARGET, which is the value desired.

Once the details of the reduction procedure have been finalized, the program sets the size of the cubic box using the subroutine SYMMETRY_CUBIC_BOX and chooses initial values for the coordinates of each molecule using the internal subroutine SET_INITIAL_COORDINATES. This subroutine will work correctly only if each residue in the system's sequence is of the same type as the solvent molecule whose coordinates were defined at the beginning of the program. The subroutine takes each residue in turn and copies the coordinates from the array COORDINATES to the part of the array, ATMCRD, that contains the coordinates of the atoms in that residue. The whole residue is then translated randomly by up to a distance of one half of the length of the box, SIDE, in either direction along each of the x, y and z axes. The function, RANDOM_VECTOR, is from the module RANDOM_NUMBERS. It returns a vector of random numbers in the range [0, 1] whose dimension is given by the value of its single integer argument which, in this case, is 3.

Because the random assignment of coordinates could lead to partial overlaps or high-energy contacts between molecules it is best to 'refine' the coordinates in some way. This can be done with a partial geometry optimization, a short molecular dynamics simulation or, as here, a Monte Carlo procedure. To use the Monte Carlo simulation subroutine, it is first necessary to define the non-bonding energy cutoff parameters with the subroutine MONTE_CARLO_ENERGY_OPTIONS. The simulation itself is done at a constant volume – we do not want to change the volume of the box! – and the number of moves is chosen so that each molecule is moved on average 100 times, which should be enough to remove any overlaps between molecules.

The reduction of the box to the required size is performed by the loop that follows the initial Monte Carlo simulation. At each iteration, the reduced value of the box length is calculated and assigned to the variable NEWSIDE. The new value of the box length is then transmitted to the symmetry module using a call to the subroutine SYMMETRY_CUBIC_BOX and new coordinates for the molecules are determined with a call to a second internal subroutine SET_INTERMEDIATE_COORDINATES. These coordinates are refined for the same reasons as given above using a short Monte Carlo simulation procedure and the loop finishes by setting the variable SIDE to the value of the updated box length NEWSIDE.

The subroutine SET_INTERMEDIATE_COORDINATES works in a similar way to the other internal subroutine, SET_INITIAL_COORDINATES, except that it has a different recipe for selecting the amount by which each molecule is to be translated. This method is the same as that used by the Monte Carlo modules described in section 11.3 and it assumes that the fractional coordinates, S_c, of the center of mass of each molecule remain the same even though the box length may have been changed. Thus, if L and L' are the old and new box lengths, respectively, and R_c are the old coordinates of the center of mass of the molecule, the new coordinates of the center of mass are $(L'/L)R_c$ and the translation to move the atoms of the molecule from their old to new positions is $(L'/L - 1)R_c$.

After the reduction procedure the solvent box will have attained the desired volume but the molecules will be by no means fully equilibrated. Thus, the program finishes by performing a long Monte Carlo simulation, this time allowing the volume to change, before writing out the final coordinates to an external file. The coordinates produced by this program will be adequate for use in subsequent Monte Carlo calculations but will need to be equilibrated further if they are to be used in a molecular dynamics simulation.

12.5 Solvating molecules

The last section outlined a method for the construction of solvent boxes. A complementary task is that of how to produce the coordinates of condensed phase systems containing a mixture of molecules, such as, for example, a solute molecule in solvent. Just like for the problem addressed in the last section, there is no unique way of doing this. A possible approach would be to adapt the program of Example 24 to handle molecules of different types. Another method, which is the one we adopt here, is to overlay the solvent box onto the solute molecule and then delete solvent molecules that overlap with any of the solute molecule's atoms.

The program for solvation of a system will be described in the next section but, here, we describe a module, called MM_SYSTEM_EDIT, which is needed for the example. It includes two subroutines that can be used to modify the data in the MM system data structures and its definition is

```
MODULE MM_SYSTEM_EDIT

CONTAINS

   SUBROUTINE MM_SYSTEM_APPEND ( FILE )
      CHARACTER ( LEN = * ), INTENT(IN) :: FILE
   END SUBROUTINE MM_SYSTEM_APPEND
```

```
SUBROUTINE MM_SYSTEM_DELETE ( SELECTION )
    LOGICAL, DIMENSION(1:NATOMS), INTENT(IN) :: SELECTION
END SUBROUTINE MM_SYSTEM_DELETE
```

```
END MODULE MM_SYSTEM_EDIT
```

The subroutine MM_SYSTEM_APPEND appends or joins two systems together to form a single system whereas the subroutine MM_SYSTEM_DELETE removes atoms and their associated data from the MM system data structure to create a system of smaller size.

MM_SYSTEM_APPEND takes a single argument, which is the name of the binary system file that contains the data which are to be added to the system data currently stored by the program. The only restriction in the use of the subroutine is the normal one that all the subsystem names in the combined system must be unique. MM_SYSTEM_DELETE also takes a single argument, which is a logical array with a length equal to the number of atoms that identifies those atoms that are to be deleted from the MM system data structure. An atom is deleted if the corresponding element in the array SELECTION is .TRUE.. The subroutine deletes all reference to the selected atoms including all terms in the molecular mechanics energy function which involve them. The data, including the coordinates, for the atoms that remain are left unchanged.

12.6 Example 25

The example in this section implements the overlay method of solvation mentioned in the previous section and employs both of the subroutines from the module MM_SYSTEM_EDIT. The program is designed to solvate a bALA molecule and is

```
PROGRAM EXAMPLE25

... Declaration statements ...

! . Local parameters.
REAL, PARAMETER :: BUFFER = 2.8

! . Local scalars.
INTEGER :: I, IRES, NSATOMS, NSRESID, O

! . Local arrays.
LOGICAL, ALLOCATABLE, DIMENSION(:) :: FLAG

! . Read in the system file for the solute.
CALL MM_SYSTEM_READ ( ''bALA.sys_bin'' )
```

```
! . Save the number of solute atoms and residues.
NSATOMS = NATOMS
NSRESID = NRESID


! . Append the system file for the box of waters.
CALL MM_SYSTEM_APPEND ( ''h2o_box729.sys_bin'' )


! . Allocate the FLAG array and select the solute atoms.
ALLOCATE ( FLAG(1:NATOMS) ) ; FLAG = .FALSE. ; FLAG(1:NSATOMS) = .TRUE.


! . Read in the coordinates of bALA.
CALL COORDINATES_READ ( ''bALA1.crd'', SELECTION = FLAG )


! . Read in the coordinates of the waters.
CALL COORDINATES_READ ( ''h2o_box729.crd'', SELECTION = .NOT. FLAG )


! . Transform the bALA coordinates to their principal axes.
CALL TO_PRINCIPAL_AXES ( ATMCRD(1:3,1:NSATOMS), ATMMAS(1:NSATOMS) )


! . Translate the water box so that its center of mass is at the origin.
CALL TRANSLATE_TO_CENTER ( ATMCRD(1:3,NSATOMS+1:NATOMS), &
                           ATMMAS(NSATOMS+1:NATOMS) )


! . Re-initialize the FLAG array.
FLAG = .FALSE.


! . Loop over the solvent residues.
DO IRES = (NSRESID+1),NRESID

   ! . Find the oxygen in the residue.
   O = RESIND(IRES)+1

   ! . Find the minimum distance to the non-hydrogen solute atoms.
   DO I = 1,NSATOMS
      IF ( ATMNUM(I) /= 1 ) THEN
         IF ( SUM ( ( ATMCRD(1:3,I) - ATMCRD(1:3,O) )**2 ) <= BUFFER**2 ) THEN
            FLAG(RESIND(IRES)+1:RESIND(IRES+1)) = .TRUE. ; EXIT
         END IF
      END IF
   END DO

END DO


! . Remove the flagged atoms.
CALL MM_SYSTEM_DELETE ( FLAG )


! . Save the system file.
CALL MM_SYSTEM_WRITE ( ''bALAh2o_box729.sys_bin'' )


! . Save the coordinates.
CALL COORDINATES_WRITE ( ''bALAh2o_box729.crd1'' )
```

```
! . Deallocate temporary arrays.
DEALLOCATE ( FLAG )


END PROGRAM EXAMPLE25
```

The first step in the program is to read in, using MM_SYSTEM_READ, the binary system file that defines the solute molecule, bALA. In the next statements, the local variables NSATOMS and NSRESID are used to store the numbers of atoms and residues in the solute molecule, respectively.

After this, the system file for the water box containing 729 water molecules is appended to the system data for the bALA molecule using the subroutine MM_SYSTEM_APPEND. The new, larger system now contains a subsystem corresponding to the solute and a subsystem for the solvent.

Once the system has been defined, it is necessary to read in the coordinates of the atoms. This is done in three steps. First, a logical array of the dimension of the total number of atoms is defined and initialized so that its elements that correspond to solute atoms are .TRUE. and those corresponding to solvent atoms are .FALSE.. Second, the coordinates for the solute molecule are read in from a file using the subroutine COORDINATES_READ and the optional argument SELECTION. Finally, the solvent coordinates are input from a second file, again using COORDINATES_READ, but with the complement of the selection array. The logical array defined by the statement .NOT. FLAG is equivalent to an array in which all the elements for solvent atoms are .TRUE. and all those for solute atoms .FALSE..

Before proceeding to select which solvent molecules are to be deleted, it is necessary to ensure that the coordinates of the solute and solvent molecules are in the same region of space or, in other words, that the solute molecule can be immersed in the solvent. To do this, the subroutine TO_PRINCIPAL_AXES is called to transform the coordinates of the solute atoms and, most importantly, to put the center of mass of the solute molecule at the origin. After this, the coordinates of the solvent molecules are modified using the subroutine TRANSLATE_TO_CENTER so that the center of mass of the solvent box is also at the origin.

We are now ready to select the solvent molecules to delete. The criterion that is adopted, which is appropriate for water, is to say that any water molecule will be deleted if its oxygen atom is less than a certain buffer distance away from *any* non-hydrogen atom of the solute molecule. The buffer distance is contained in the constant BUFFER which is defined in the parameter statement at the top of the program. The value of 2.8 Å is chosen because it corresponds roughly to the 'size' of a water molecule. The loop to determine the minimum distance between the oxygen of the water molecules and the solute atoms is straightforward. The major point to note is the logical array, FLAG, which is being re-used as the array that will flag the atoms to be deleted. Before the loop, all the elements of FLAG are set to .FALSE.. During the loop, all the elements of the array that correspond to the atoms of a particular residue are set to .TRUE. if an oxygen–solute atom distance falls below the minimum distance. The implementation of this search algorithm is specific to water and will need to be modified for other solvents with different numbers of atoms or residues.

The remainder of the program is simpler. The solvent molecules that overlap with the solute molecule are deleted from the system file by passing the array FLAG to the subroutine MM_SYSTEM_DELETE. The new system definition file and the coordinates are then written out in the standard fashion before the temporary array, FLAG, is deallocated.

The coordinates produced using this program, like those resulting from the program in section 12.4, will need to be refined before they can be employed in a simulation. This is because the criterion for deleting molecules is a relatively crude one and may either leave cavities in the solvent or, alternatively, result in large repulsive contacts between some of the solute and solvent atoms. The refinement would optimally be done, depending upon the system, using a constant-pressure Monte Carlo or molecular dynamics technique so that the volume changes. It may also be necessary in certain circumstances to repeat the solvation procedure several times. This is often so for complex solutes, such as proteins, because cavities can appear during the course of a simulation, either in the solvent or between the solute and the solvent, within which it is possible to fit extra water molecules. The strategy in these cases would be to overlay the combined solute–solvent system with another solvent box and then delete all solvent molecules of the second solvent box that overlapped with the atoms of any of the molecules, either solute or solvent, in the first solvated system.

12.7 Constraints

An important topic to which we have only alluded up to now is that of how to constrain various geometrical parameters in a system during a simulation. We have already seen some examples of constraint techniques, notably those which help to constrain a bond, bond angle or dihedral angle about a particular value in umbrella sampling calculations (section 10.6) and those which keep the internal geometry of the molecule fixed during a Monte Carlo simulation (section 11.3). It is not difficult to imagine other cases in which the application of constraints might be helpful. For example, it is often desirable to fix the values of some internal coordinates during a geometry optimization to see how the energy of the optimized structure changes as the values of the fixed coordinates are varied. Another common example is when dealing with structures of macromolecules obtained from X-ray crystallographic or NMR experiments. In these instances, it is often better to start off by minimizing a structure with constraints that keep the atoms close to their original positions and then gradually relax the constraints as the minimization proceeds. A minimization without constraints of any sort can result in structures that are significantly distorted from the experimental ones.

Constraints can be conveniently divided into two types. *Soft constraints* keep the value of a geometrical parameter in the neighbourhood of the target value whereas *hard* or *rigid constraints* keep the value of a geometrical parameter 'exactly' (to within a specified numerical precision) at the target value. Soft constraints, like those used in the umbrella sampling calculations, are relatively simple to apply. They are typically implemented by adding extra terms to the energy function, which give low energies when the constrained parameter is close to the desired value and progressively higher energies as the value deviates further. Examples of such energy terms may be found in equations (6.13) and (10.52).

Hard constraints are, in general, much more complicated and how they are implemented depends a lot on the type of calculations being performed. The most general approach when dealing with hard constraints is to define a new set of geometrical variables that corresponds to the number of degrees of freedom left to the system. This set can be defined either directly by identifying the degrees of freedom themselves or indirectly by specifying the constraints. The first method would be easiest in many cases. For example, when studying polymers it can be illuminating to examine the behaviour of the system if only the torsional degrees of freedom are allowed to change and the bond and angle internal coordinates are fixed. The set of degrees of freedom is then equivalent to the set of torsions that can vary. Another simple case, which is particularly easy to handle in any type of simulation, is that in which only certain atoms are allowed to move and all the others, which act as an environment to the free atoms, are kept fixed at their existing positions.

For other problems, particularly when the constraints involved may be more complicated functions of the coordinates than just bonds, bond angles or dihedral angles, it is necessary to specify the constraints and then generate the degrees of freedom from the set of $3N$ Cartesian coordinates ($3N - 6$ if the rotational and translational degrees of freedom are removed) and the N_c constraints. Two difficulties can arise here. First, the set of N_c constraints need not be independent insofar as some constraints may act on the same degrees of freedom or combinations of them. Second, the generalized coordinates that result from such a procedure are, in general, non-trivial functions of the Cartesian coordinates.

Whatever the approach employed to generate the new set of variables, they are invariably more complicated to use than are Cartesian coordinates. As we have seen in the chapter on Monte Carlo methods (section 11.3), generalized coordinates result in expressions that make statistical properties troublesome to calculate. Equally importantly, the derivatives of the energy with respect to the generalized coordinates are more expensive to compute than are those with respect to the Cartesian coordinates. This is because it is normal to determine the latter first and then transform them into the new set of variables, a step that can become expensive as the size of the system increases. For all these reasons, generalized coordinates have been employed less widely than have Cartesian coordinates in molecular dynamics simulations. In contrast, efficient methods using generalized internal coordinates have been developed for the geometry optimization of small molecules consisting of up to 100 or so atoms.

A second and more widely used approach to the application of constraints for molecular dynamics simulations is the method of *Lagrange multipliers*. Suppose that we have a set of constraints that are functions of the coordinates of the atoms only (and perhaps the time). Such constraints are called *holonomic* and can be written as

$$\Lambda_k(\boldsymbol{R}) = 0 \ \forall \ k = 1, \ldots, N_c \tag{12.2}$$

The dynamics of a system are obviously changed in the presence of the constraints. From classical mechanics, the modified equations of motion are

$$\mathbf{M}\ddot{\boldsymbol{R}} = -\frac{\partial \mathcal{V}}{\partial \boldsymbol{R}} - \sum_{k=1}^{N_c} \lambda_k \frac{\partial \Lambda_k}{\partial \boldsymbol{R}} \tag{12.3}$$

The first term on the right-hand side of the equation corresponds to the forces arising from the potential whereas the second term is the force due to the constraints. The variables λ_k are the Lagrange multipliers which are a function of time and whose values are chosen to ensure that the constraints of equation (12.2) are satisfied at all points along the trajectory.

The way in which these equations are solved varies depending upon the numerical method chosen to integrate the equations of motion. If the standard Verlet algorithm is being used, the first step would be to obtain the unconstrained positions for each atom, $\boldsymbol{R}'(t + \Delta)$, using the formula in equation (8.7) and the forces arising from the potential, $\mathcal{V}$. Note that it is assumed that the previous points along the trajectory satisfy the constraint conditions.

The positions for the atoms which satisfy the constraints, $\boldsymbol{R}(t + \Delta)$, are then written as

$$\boldsymbol{R}(t + \Delta) = \boldsymbol{R}'(t + \Delta) - \Delta^2 \mathbf{M}^{-1} \sum_{k=1}^{N_c} \lambda_k \frac{\partial \Lambda_k}{\partial \boldsymbol{R}} \tag{12.4}$$

and satisfy the conditions

$$\Lambda_k(\boldsymbol{R}(t + \Delta)) = 0 \ \forall \ k = 1, \ldots, N_c \tag{12.5}$$

Substitution of equation (12.4) into equation (12.5) gives a set of N_c equations for the N_c unknown Lagrange multipliers, λ_k, which can be solved fairly straightforwardly by a variety of iterative techniques. These methods are often quite efficient, converging in a small number of steps, but they can have a limited *radius of convergence* for certain types of constraint, which means that they will not work if the unconstrained atom positions deviate too much from their constrained values.

One of the more common applications of the above technique is the *SHAKE* algorithm, which has many variants. J. P. Ryckaert, G. Ciccotti and H. J. C. Berendsen originally developed the algorithm for fixing bond lengths in a simulation, particularly those between hydrogens and other atoms. The reason for doing this is that such bonds have the highest frequency motions in a typical molecular system (3000 cm^{-1} and higher) and it is the highest frequency motions which limit the size of the timestep when integrating the equations of motion in a molecular dynamics simulation. Fixing these bonds can allow a timestep as large as 2 fs to be used in conjunction with Verlet-type algorithms (depending upon the accuracy required). The SHAKE algorithm can also be used for other types of constraints, such as bond angles, although studies have shown that fixing these types of degree of freedom can significantly alter the dynamics of the system.

12.8 Parametrizing force fields

In the last few chapters we have spent much time discussing a wide range of algorithms for performing simulations of molecular systems and how to apply them correctly. No matter how good our simulation techniques are, however, the usefulness of the results produced will be heavily dependent upon how accurately the potential energy function reproduces the potential energy surface of the system being studied. In other words, the quality of parametrization of the force field is crucial to the results that we obtain.

Parametrization is a complex optimization problem with many interdependent parameters for which, unfortunately, there is no 'black-box' procedure.

The parametrization process itself can be a long and time-consuming business. The general approach is to guess values for the parameters that need to be optimized and then use the provisional force field to compute properties for systems that can be directly compared with experimental ones. The force-field parameters are then adjusted and the process repeated until the agreement between the theoretical and observed results is judged to be adequate.

It is impractical to try to optimize all the parameters against all the available experimental data simultaneously so the problem must be broken down into simpler parts. The way in which this is done varies greatly among the various research groups that develop force fields. To some extent these strategies reflect what each group wants their force field to do but the differences are in large part due to the 'parametrization philosophy' that they each adopt. A detailed description of a particular approach will not be given here so readers should consult the references concerning specific force fields for more details.

There are several pieces of common-sense information, though, that can be used in all strategies. First of all, the values of the parameters must be chemically reasonable, which reduces the space within which any optimization procedure must operate. Thus, for example, the radius of an atom is of the order of 1 Å so values for the Lennard-Jones radius parameters should be of roughly the same magnitude. Likewise, the maximum stabilization to be expected from a van der Waals interaction will be a few kJ mol^{-1}, which will limit, in turn, the size of the Lennard-Jones well-depth parameters.

Second, atoms of a given element are well known to have different properties depending upon the chemical environment in which they find themselves. Thus, carbons in alkane molecules can, to a good approximation, be assumed to be of the same type, but they will be different and will need different parameters from carbons in carbonyl or carboxylate groups or in aromatic rings. In general, the aim is to minimize the number of different atom types in the force field in order to keep the number of different parameters as small as possible without sacrificing too much accuracy.

Finally, it is evident that certain properties will be influenced more by the values of particular types of parameters than will others. Thus, it will be possible to split up the optimization process to a certain extent and fit the different types of parameters separately using various subsets of the observational data. Let us consider each type of parameter in turn.

- *Equilibrium bond distances and angles.* The major sources of data for these parameters are structures determined experimentally using techniques such as X-ray crystallography, microwave spectroscopy and electron diffraction. The use of these data is relatively straightforward, although if very high precision is desired care should be taken because the data will include effects such as thermal motion, for which corrections need to be made before they can be used for parametrization. In addition to experimental data, structures resulting from high-quality quantum mechanical calculations are being increasingly employed.

When parametrizing equilibrium bond distances and angles it is best to choose data from relatively small molecules because the observed bond distances and angles are more likely to be those that result from the inherent properties of the bonds and angles. With larger molecules and with molecules between which there are strong interactions, structures can be significantly perturbed by effects such as steric hindrance or by hydrogen bonding. In these cases the observed bond distances and angles arise from a combination of the properties of the bonds and angles themselves, which are accounted for by the equilibrium bond distance and angle parameters, and other interactions, such as the exchange repulsion between atoms, that are accounted for by other parameters in the force field.

- *Bond and angle force constants.* The most rigorous way to derive these parameters is by comparing the results of normal mode calculations on particular molecules versus experimentally or theoretically obtained harmonic frequencies. For very small molecules and for molecules with high symmetry the parametrization process is straightforward because a single parameter can often be correlated directly to a particular vibration. For larger molecules, the process is more complicated because each force constant can influence the values of many different vibrations.

This is one of the situations in which 'black-box' algorithms have been employed fruitfully for parameter fitting. Typically a *non-linear least squares* algorithm would be chosen, in which a function, $\mathcal{F}$, of the following form is minimized with respect to the vector of force-constant variables, k:

$$\mathcal{F}(k) = \sum_{i=1}^{N_o} w_i(\omega_i^{\text{calc}}(k) - \omega_i^{\text{obs}})^2 \qquad (12.6)$$

Here, N_o is the number of *observables* (i.e. the number of vibrational frequencies), w_i are weights, ω_i^{obs} are the experimental values of the frequencies which are constants and ω_i^{calc} are the calculated frequencies which depend implicitly upon the force-constant parameters. Note that, in this equation, it is not sufficient to order the calculated frequencies against the observed ones solely on the basis of the values of the frequencies themselves. This is because two vibrations may have similar frequencies but be completely different in character. To do the alignment properly requires that the forms of the normal mode displacements, both calculated and observed, also be compared. Practically, this is achieved by evaluating the scalar products of the two sets of normal mode vectors and then equating those which have the largest overlap.

- *Dihedral angle parameters.* The dihedral angle parameters are most readily fitted using data about the energy profiles for rotation around particular bonds. These are difficult to measure experimentally but are straightforward to calculate using quantum mechanical methods. This is done by constraining the dihedral angle to have particular values and optimizing the remaining degrees of freedom so that the energy of the fully relaxed molecule as a function of the relevant torsional angle is mapped out. The major complication with the fitting process is that the energies of rotation are determined not only by the dihedral energies about the bond but also by non-bonding energies and, to a smaller extent, by the bond and angle energies. It is easy to see that this is so by considering the case of ethane in which the difference in energy between the most stable 'staggered' and the least stable 'eclipsed' conformations is determined primarily by the balance between the energies of the torsional terms and the 1–4 non-bonding interactions between the methyl hydrogens.

● *Improper angle parameters*. The improper parameters are handled in a similar way to the bond, bond angle and dihedral angle parameters described above.

● *Atom charges*. Information about the charge distribution of molecules can be obtained from experimental and theoretical results, such as the measured multipole moments of molecules (dipole, quadrupole, etc.) and the electron densities determined from X-ray crystallographic experiments or quantum mechanical calculations. The multipole moments are useful sources of data but they must usually be combined with other data. By themselves, they can only be used to parametrize small molecules because each moment provides only a small number of data points (one for the monopole, three for the dipole, five for the quadrupole, etc.) and, in addition, the higher moments are often unavailable because they are difficult to measure experimentally.

Both X-ray crystallography and quantum mechanical calculations give the charge distribution for a molecule directly. There is a problem, though, which is that, in general, there is no unique way of partitioning a charge distribution among atoms. This ambiguity, together with the necessity to have very-high-resolution structures, means that crystallographic data are not often used in parametrization studies. In contrast, there are several convenient recipes for partitioning a quantum mechanical charge distribution, which, even though they give different results, are often used as a rough and ready way to calculate the atomic charges in a force field. Another approach that has been preferred in the parametrization of some force fields is to calculate with a quantum mechanical method the *electrostatic potential* (ESP) at various points around the molecule and then use these data instead for the charge fitting. This circumvents the ambiguity in the partitioning of the charge distribution (although the positions of the points at which the electrostatic potential is to be calculated need to be specified) and the fitting can be efficiently performed using standard linear least squares algorithms.

As mentioned in chapter 5, it is assumed that most parameters in the force field are transferable, which implies that the same parameters can be used for the same types of atoms in different molecules. This is not necessarily the case for the charges. In some force fields, such as AMBER, the charges must be recalculated using ESP data for each different type of molecule. In others, most notably the OPLS-AA force field used in this book, some degree of transferability is assumed and charges are derived, insofar as it is possible, such that the same charges can be used for the same chemical group in different environments. Thus, for example, a methyl group, CH_3, is neutral with a $-0.18e$ charge on the carbon and a $+0.06e$ charge on each of the hydrogens. Incidentally, it should be noted that the atomic charges in the OPLS-AA force field were parametrized primarily by trying to reproduce the properties of organic liquids, not by using quantum mechanical calculations.

- *Lennard-Jones parameters.* These are often the most difficult to determine. In part this is because there is no distinct class of data that can be used to parametrize the Lennard-Jones interactions separately. The sorts of data that are used are those that are determined by intermolecular interactions or by long-range intramolecular interactions and include the crystal structures of small molecules and the properties of organic liquids, such as densities, enthalpies of vaporization and heat capacities. All of these depend heavily upon the electrostatic and sometimes the torsional energies as well as the Lennard-Jones terms.

- *Other parameters.* In the force field used in this book the only other parameters that remain to be discussed are the factors for scaling the 1–4 electrostatic and Lennard-Jones interactions. These can be determined together with the dihedral parameters because their values affect the energetic barriers to rotation about particular bonds.

A last point needs to be emphasized before ending, which is that the final result of any parametrization procedure will inevitably be a compromise. This is both because of the difficulties of the parametrization process itself and because the analytic form of the force field will limit the precision that can be achieved and the type of observational data that can be reproduced.

12.9 Other topics

So, at last, we have come to the end of the book. I very much hope that it has given you, the reader, a flavour of what can be calculated using simulation techniques. It should be emphasized, once again, that the algorithms discussed here represent only a small fraction of the many that are available. The choice of which to include has been highly subjective and, due to space restrictions and the perseverence of the author (!), relatively small. The topics that I most regret omitting are alternative methods for performing molecular dynamics simulations at constant pressure and temperature, multiple-time-step algorithms, surface and volume calculations and continuum methods for calculating solvation effects, not to mention the many quantum mechanical techniques that are available. In any case, readers are encouraged to investigate alternative methods themselves, for many of the techniques presented in the book are the subject of active research and are undergoing continual improvement.

Exercises

12.1 Construct Z-matrices for some reasonably complicated molecules, such as glucose and tryptophan. Try a number of different definitions. Also tackle cases in which the system consists of interacting molecules, such as, for example, a small cluster of water molecules.

12.2 Modify the program of section 12.4 or devise an alternative approach that would allow the construction of simulation boxes containing mixtures of molecules. An example application could be an ionic solution (such as sodium chloride).

12.3 Select a simple molecule and see what effect changing the force-field parameters has on various properties that can be calculated. Appropriate properties to calculate, if ethane were the example, could include its vibrational frequencies and the barrier to rotation about its carbon—carbon bond.

Appendix A

The DYNAMO module library

The modules and the example programs described in the text are available on the World Wide Web at the address `http://www.cup.cam.ac.uk/online-pubs`. Full details about how to use the library and the types of machines upon which it has been tested can be found there.

As mentioned in section 1.3 there are four categories of miscellaneous modules in the DYNAMO program library in addition to the application modules presented in the bulk of the book. Table A.1 lists the miscellaneous

Table A.1 *The miscellaneous modules in the DYNAMO program library.*

Module name	Category
BAKER_OPTIMIZATION	Mathematical
CONJUGATE_GRADIENT	Mathematical
CONSTANTS	Parameter data
DEFINITIONS	Utility
DIAGONALIZATION	Mathematical
DYNAMO	Module declarations
ELEMENTS	Parameter data
FILES	Utility
IO_UNITS	Utility
LINEAR_ALGEBRA	Mathematical
PARSING	Utility
RANDOM_NUMBERS	Mathematical
SORT	Mathematical
SPECIAL_FUNCTIONS	Mathematical
STATISTICS	Mathematical
STATUS	Utility
STRING	Utility
TIME	Utility

modules and their classification while table A.2 gives the remaining modules together with the number of the section in which they were introduced.

There are three different versions of the module Eɴᴇʀɢʏ_Nᴏɴ_Bᴏɴᴅɪɴɢ. The interfaces in each of the versions are such that procedures of the same name can be used interchangeably. Information about how to use a specific version of the module in a program is given on the web pages.

Table A.2 *The application modules in the D*ʏɴᴀᴍᴏ *program library.*

Module name	Section where described
Aᴛᴏᴍꜱ	2.3
Cᴏɴɴᴇᴄᴛɪᴠɪᴛʏ	3.2
Cᴏᴏʀᴅɪɴᴀᴛᴇ_IO	2.4
Eɴᴇʀɢʏ_Cᴏᴠᴀʟᴇɴᴛ	4.6
Eɴᴇʀɢʏ_Exᴛʀᴀ	10.6
Eɴᴇʀɢʏ_Nᴏɴ_Bᴏɴᴅɪɴɢ	4.6, 9.2, 9.5, 9.7
Gᴇᴏᴍᴇᴛʀʏ	3.3
MM_Fɪʟᴇ_Dᴀᴛᴀ	5.3
MM_Fɪʟᴇ_IO	5.3
MM_Sʏꜱᴛᴇᴍ	5.4
MM_Sʏꜱᴛᴇᴍ_Eᴅɪᴛ	12.5
MM_Sʏꜱᴛᴇᴍ_IO	5.4
MM_Tᴇʀᴍꜱ	4.6, 5.4
Mᴏɴᴛᴇ_Cᴀʀʟᴏ_Sɪᴍᴜʟᴀᴛɪᴏɴ	11.3
Mᴏɴᴛᴇ_Cᴀʀʟᴏ_Eɴᴇʀɢʏ	11.3
Mᴜʟᴛɪᴘᴏʟᴇꜱ	7.2
Nᴏʀᴍᴀʟ_Mᴏᴅᴇ	7.2, 7.4
Nᴏʀᴍᴀʟ_Mᴏᴅᴇ_Uᴛɪʟɪᴛɪᴇꜱ	7.2
Nᴜᴍᴇʀɪᴄᴀʟ_Dᴇʀɪᴠᴀᴛɪᴠᴇꜱ	4.6
Oᴘᴛɪᴍɪᴢᴇ_Cᴏᴏʀᴅɪɴᴀᴛᴇꜱ	6.3, 6.5
PDB_IO	Exercise 2.2
Pᴏᴛᴇɴᴛɪᴀʟ_Eɴᴇʀɢʏ	4.6
Rᴇᴀᴄᴛɪᴏɴ_Pᴀᴛʜ	6.7
Sᴇʟꜰ_Aᴠᴏɪᴅɪɴɢ_Wᴀʟᴋ	6.9
SᴇQᴜᴇɴᴄᴇ	2.3
Sᴜᴘᴇʀɪᴍᴘᴏꜱᴇ	3.6
Sʏᴍᴍᴇᴛʀʏ	9.5
Tʜᴇʀᴍᴏᴅʏɴᴀᴍɪᴄꜱ_RRHO	7.6
Tʀᴀᴊᴇᴄᴛᴏʀʏ_Aɴᴀʟʏꜱɪꜱ	10.2
Tʀᴀᴊᴇᴄᴛᴏʀʏ_IO	8.4
Tʀᴀɴꜱꜰᴏʀᴍᴀᴛɪᴏɴ	3.5
Vᴇʟᴏᴄɪᴛʏ	8.2
Vᴇʟᴏᴄɪᴛʏ_Vᴇʀʟᴇᴛ_Dʏɴᴀᴍɪᴄꜱ	8.2, 10.4
Zᴍᴀᴛʀɪx	12.2
Zᴍᴀᴛʀɪx_IO	12.2

Table A.3 *The various types of file used by the* DYNAMO *library.*

File type	Format	Section where described
Coordinate	Text	2.2
MM definition	Text	5.2
MM definition (processed)	Binary	5.3
MM system (processed)	Binary	5.4
Sequence	Text	5.4
Trajectory	Binary	8.4
Umbrella sampling data	Text	10.6
Velocity	Text	8.2
Z-matrix	Text	12.2

The majority of modules were written specifically for the DYNAMO library. The exceptions are the modules CONJUGATE_GRADIENT, DIAGONALIZATION, RANDOM_NUMBERS, SORT and SPECIAL_FUNCTIONS, which contain source code that has been adapted from other sources. Full references may be found on the web pages.

Several types of file are either needed or produced by the DYNAMO library. These are listed together with their format and section of introduction in table A.3.

Bibliography

General references

In keeping with the spirit of the book as a practical introduction, no attempt to give a complete bibliography has been made. Rather the references represent a highly subjective selection and have been chosen because they give overviews of certain topics or because they discuss algorithms or applications that have been mentioned explicitly in the text.

Specific references are given for each chapter. There are, however, a number of general references that readers will probably either need or find useful. Of the many books available in each category, the author's recommendations are listed below.

FORTRAN 90/95

The following provides a nice clear introduction and is the first book that the reader is likely to need!

- M. Metcalf and J. Reid, FORTRAN 90/95 *Explained*. Oxford Science Publications, Oxford, 1996.

Numerical algorithms

Two superb books in the same series that provide a good all round introduction to the subject and are mines of valuable information are

- W. H. Press, S. A. Teukolsky, W. T. Vetterling and B. P. Flannery, *Numerical Recipes in* FORTRAN 77: *The Art of Scientific Computing*. Second Edition, Cambridge University Press, Cambridge, 1993.
- W. H. Press, S. A. Teukolsky, W. T. Vetterling and B. P. Flannery, *Numerical Recipes in* FORTRAN 90: *The Art of Parallel Scientific Computing*. Cambridge University Press, Cambridge, 1996.

Molecular simulations

A very good book that covers a wide range of simulation methodologies for atomic and small-molecule systems is

- M. P. Allen and D. J. Tildesley, *Computer Simulations of Liquids*. Oxford University Press, Oxford, 1987.

Other books covering similar topics are

- J. M. Haile, *Molecular Dynamics Simulation. Elementary Methods*. J. Wiley & Sons, New York, 1992.
- D. Frenkel and B. Smit, *Understanding Molecular Simulation. From Algorithms to Applications*. Academic Press, London, 1996.
- D. C. Rapoport, *The Art of Molecular Dynamics Simulation*. Cambridge University Press, Cambridge, 1997.

A recent book giving an overview of many molecular modeling techniques is

- A. R. Leach, *Molecular Modelling*. Addison-Wesley Longman, London, 1996.

In addition to these general texts, there are volumes that contain compilations of articles about various aspects of molecular simulations. One good series that is published regularly is

- K. B. Lipkowitz and D. B. Boyd (Editors), *Reviews in Computational Chemistry*. VCH, New York.

A very comprehensive set of articles about various methods in computational chemistry may be found in the volumes of the following work:

- P. von Ragué Schleyer (Editor in Chief), *Encyclopedia of Computational Chemistry*. J. Wiley & Sons, Chichester, 1998.

There is an extensive literature about the application of molecular simulation techniques to biomolecular systems. A very nice introduction to this area is

- J. A. McCammon and S. Harvey, *Dynamics of Proteins and Nucleic Acids*. Cambridge University Press, Cambridge, 1987.

A more technical reference is

- C. L. Brooks III, M. Karplus and B. M. Pettitt. 'Proteins: A Theoretical Perspective of Dynamics, Structure and Thermodynamics'. *Adv. Chem. Phys.* **71**, 1–259, 1988.

A good three-volume collection of articles by various authors describing the state of the art in the simulation of biomacromolecular systems is

- W. van Gunsteren, P. Weiner and A. Wilkinson (Editors), *Computer Simulation of Biomolecular Systems: Theoretical and Experimental Applications*. Volumes 1, 2 and 3, ESCOM, Leiden, 1989, 1993 and 1997.

Chapter 2

There are many different coordinate file formats. The PDB (protein data bank) format mentioned in Exercise 2.2 is widespread in X-ray crystallographic and NMR studies of proteins and related molecules. The PDB is described in

- F. C. Bernstein, T. F. Koetzle, G. J. B. Williams, E. F. Mayer, J. M. D. Brice, J. R. Rodgers, O. Kennard, T. Shimanouchi and M. Tasumi. 'The Protein Data Bank: A Computer-based Archival File for Macromolecular Structures'. *J. Molec. Biol.* **112**, 535–42, 1977.

There are a number of tools for converting between coordinate files of different formats. One of the more popular of these is the `Babel` program developed by P. Walters and M. Stahl at the University of Arizona. They can be contacted at the Department of Chemistry, University of Arizona, Tucson, AZ 85721 (electronic mail: `babel@mercury.aichem.arizona.edu`).

Chapter 3

Internal coordinates

The definition for the dihedral angle in the text can be found in

- H. Bekker, H. J. C. Berendsen and W. F. van Gunsteren. 'Force and Virial of Torsional-Angle-Dependent Potentials'. *J. Comput. Chem.* **16**, 527–33, 1995.

Miscellaneous transformations

When dealing with some of the more complicated coordinate transformations it is often handy to have a reference book on classical mechanics. A good one is

- H. Goldstein, *Classical Mechanics*. Second Edition, Addison-Wesley, Reading, MA, 1980.

Superimposing structures

The method of superposition of two sets of coordinates is described in the following articles by Kabsch and by Kneller:

- W. Kabsch. 'A Solution for the Best Rotation to Relate Two Sets of Vectors'. *Acta Cryst.* **A32**, 922–3, 1976.
- W. Kabsch. 'A Discussion of the Solution for the Best Rotation to Relate Two Sets of Vectors'. *Acta Cryst.* **A34**, 827–8, 1978.
- G. R. Kneller. 'Superposition of Molecular Structures using Quaternions'. *Molec. Simul.* **7**, 113–19, 1991.

Chapter 4

Quantum mechanics

The quote from Dirac comes from

- P. A. M. Dirac. 'Quantum Mechanics of Many-Electron Systems'. *Proc. Roy. Soc. (London)* **A123**, 714–33, 1929.

For those wanting to know more about quantum mechanics, a good comprehensive guide is

- A. S. Davydov, *Quantum Mechanics*. Second Edition, Pergamon, Oxford, 1976.

Reviews of various quantum mechanical techniques for the calculation of the energy can be found in

- J. A. Pople and D. L. Beveridge, *Approximate Molecular Orbital Theory*. McGraw-Hill, New York, 1970.
- A. Szabo and N. S. Ostlund, *Modern Quantum Chemistry: Introduction to Advanced Electronic Structure Theory*. First Revised Edition, McGraw-Hill, New York, 1989.
- W. J. Hehre, L. Radom, P. von Ragué Schleyer and J. A. Pople, Ab Initio *Molecular Orbital Theory*. J. Wiley & Sons, New York, 1986.
- R. G. Parr and W. Yang, *Density-Functional Theory of Atoms and Molecules*. Oxford University Press, Oxford, 1989.

A nice overview of potential energy surfaces is given in the article

- B. T. Sutcliffe. 'The Idea of a Potential Energy Surface'. *J. Molec. Struct. (Theochem.)* **341**, 217–35, 1995.

Molecular mechanics

There is an extensive literature on molecular mechanics energy functions. A useful overview by one of its strongest proponents is

- U. Burkert and N. L. Allinger, *Molecular Mechanics*. ACS Monograph 177, American Chemical Society, Washington, 1982.

A shorter reference by another strong proponent of the molecular mechanics method can be found in

- S. Lifson. 'Theoretical Foundations for the Empirical Force Field Method'. *Gazz. Chim. Ital.* **116**, 687–92, 1986.

The various categories of non-bonding energy terms and their origin are discussed in the following classic reference:

- J. O. Hirschfelder, L. Curtiss and R. B. Bird, *Molecular Theory of Gases and Liquids*. J. Wiley & Sons, New York, 1954.

A shorter review is

- A. J. Stone and S. L. Price. 'Some New Ideas in the Theory of Intermolecular Forces: Anisotropic Atom–Atom Potentials'. *J. Phys. Chem.* **92**, 325–35, 1988.

Atomic dipole polarizability models are described in

- A. Warshel and S. T. Russel. 'Calculations of Electrostatic Interactions in Biological Systems and in Solutions'. *Q. Rev. Biophys.* **17**, 283–422, 1984.

Another way of including polarization effects is with *fluctuating charge* or *charge-equilibration* models. An example of an implementation in a molecular mechanics force field is

- A. Rappé and W. A. Goddard III. 'Charge Equilibration for Molecular Dynamics Simulations'. *J. Phys. Chem.* **95**, 3358–63, 1991.

Chapter 5

Force fields

The OPLS-AA force field used in the book is fully described in the following paper

- W. L. Jorgensen, D. S. Maxwell and J. Tirado-Rives. 'Development and Testing of the OPLS All-Atom Force Field on Conformational Energetics and Properties of Organic Liquids'. *J. Am. Chem. Soc.* **118**, 11225–36, 1996.

Jorgensen and his co-workers have published many papers using this force field and its predecessors. A paper giving details of the older united atom OPLS force field for proteins is

- W. L. Jorgensen and J. Tirado-Rives. 'The OPLS Potential Functions for Proteins. Energy Minimizations for Crystals of Cyclic Peptides and Crambin'. *J. Am. Chem. Soc.* **110**, 1657–66, 1988.

There are many other force fields. Not all will be referenced here but some of the more common ones are listed below in alphabetical order. The OPLS-AA force field uses parts of the all-atom AMBER force field described in

- W. D. Cornell, P. Cieplak, C. I. Bayly, I. R. Gould, K. M. Merz, Jr, D. M. Ferguson, D. C. Spellmeyer, T. Fox, J. W. Caldwell and P. A. Kollman. 'A Second Generation Force Field for the Simulation of Proteins, Nucleic Acids, and Organic Molecules'. *J. Am. Chem. Soc.* **117**, 5179–97, 1995.

This force field is an enhancement of previous force fields, details of which can be found in

- S. J. Weiner, P. A. Kollman, D. A. Case, U. C. Singh, C. Ghio, G. Alagona, S. Profeta, Jr and P. Weiner. 'A New Force Field for Molecular Mechanical Simulation of Nucleic Acids and Proteins'. *J. Am. Chem. Soc.* **106**, 765–84, 1984.
- S. J. Weiner, P. A. Kollman, D. T. Nguyen and D. A. Case. 'An All Atom Force Field for Simulations of Proteins and Nucleic Acids'. *J. Comput. Chem.* **7**, 230–52, 1986.

The various incarnations of the CHARMM force field are widely used:

- B. R. Brooks, R. E. Bruccoleri, B. D. Olafson, D. J. States, S. Swaminathan and M. Karplus. 'CHARMM: A Program for Macromolecular Energy, Minimization and Dynamics Calculations'. *J. Comput. Chem.* **4**, 187–217, 1983.
- A. D. MacKerell, J. Wïorkiewicz-Kuczera and M. Karplus. 'An All-Atom Empirical Energy Function for the Simulation of Nucleic Acids'. *J. Am. Chem. Soc.* **117**, 11946–75, 1995.
- A. D. MacKerell, D. Bashford, M. Bellott, R. L. Dunbrack, Jr, J. Evanseck, M. J. Field, S. Fischer, J. Gao, H. Guo, S. Ha, D. Joseph, L. Kuchnir, K. Kuczera, F. T. K. Lau, C. Mattos, S. Michnick, T. Ngo, D. T. Nguyen, B. Prodhom, W. E. Reiher III, B. Roux, M. Schlenkrich, J. C. Smith, R. Stote, J. E. Straub, M. Watanabe, J. Wïorkiewicz-Kuczera, D. Yin and M. Karplus. 'All-Atom Empirical Potential for Molecular Modeling and Dynamics Studies of Proteins'. *J. Phys. Chem.* **B102**, 3586–616, 1998.

Allinger and co-workers have developed a number of force fields over the years. One of the latest ones, MM3, is used extensively, particularly for calculations on smaller molecular systems. There are several papers describing parameters for various types of molecules. Some of these are

- N. L. Allinger, H. J. Geise, W. Pyckhout, L. A. Paquette and J. C. Gallucci. 'Structures of Norbornane and Dodecahedrane by Molecular Mechanics Calculations (MM3), X-ray Crystallography, and Electron Diffraction'. *J. Am. Chem. Soc.* **111**, 1106–14, 1989.

- J. C. Tai, L. R. Yang and N. L. Allinger. 'Molecular Mechanics Calculations (MM3) on Nitrogen-containing Aromatic Heterocycles'. *J. Am. Chem. Soc.* **115**, 11906–17, 1993.
- N. L. Allinger and Y. Fan. 'Molecular Mechanics Calculations (MM3) on Glyoxal, Quinones, and Related Compounds'. *J. Comput. Chem.* **15**, 251–68, 1994.

The MMFF94 force field developed by Halgren is described in a series of five papers in the same issue of the *Journal of Computational Chemistry*. The first is

- T. A. Halgren. 'The Merck Molecular Force Field. I. Basis, Form, Scope, Parameterization and Performance of MMFF94'. *J. Comput. Chem.* **17**, 490–519, 1996.

The QMFF ('Quantum Mechanical Force Field') was developed by Hagler and collaborators using data derived from quantum mechanical calculations:

- J. R. Maple, M. J. Hwang, T. P. Stockfish, U. Dinur, M. Waldman, C. S. Ewig and A. T. Hagler. 'Derivation of Class-II Force Fields. 1. Methodology and Quantum Force Field for the Alkyl Functional Group and Alkane Molecules'. *J. Comput. Chem.* **15**, 162–82, 1994.
- M. J. Hwang, T. P. Stockfish and A. T. Hagler. 'Derivation of Class-II Force Fields. 2. Characterization of a Class-II Force Field, CFF93, for the Alkyl Functional Group and Alkane Molecules'. *J. Am. Chem. Soc.* **116**, 2515–25, 1994.

A novel force field, developed with the aim of covering all elements, is UFF (the 'Universal Force Field'):

- S. L. Mayo, B. D. Olafson and W. A. Goddard III. 'DREIDING: A Generic Force Field for Molecular Simulations'. *J. Phys. Chem.* **94**, 8897–909, 1990.
- A. K. Rappé, C. J. Casewit, K. S. Colwell, W. A. Goddard III and W. M. Skiff. 'UFF, a Full Periodic Table Force Field for Molecular Mechanics and Molecular Dynamics Simulations'. *J. Am. Chem. Soc.* **114**, 10024–35, 1992.

A useful reference for readers interested in the modeling of coordination compounds (which is a separate topic in its own right) is

- B. P. Hay. 'Methods for Molecular Mechanics Modelling of Coordination Compounds'. *Coord. Chem. Rev.* **126**, 177–236, 1993.

Chapter 6

Exploring potential energy surfaces

The analytic form of the Müller–Brown model potential used in figure 6.1 is given in the following reference:

- K. Müller and L. D. Brown. 'Location of Saddle Points and Minimum Energy Paths by a Constrained Simplex Optimization Procedure'. *Theor. Chim. Acta* **53**, 75–93, 1979.

Table 6.1 was adapted, with permission, from

- J. Ma, D. Hsu and J. E. Straub. 'Approximate Solution to the Classical Liouville Equation Using Gaussian Phase Packet Dynamics: Application to Enhanced Equilibrium Averaging and Global Optimization'. *J. Chem. Phys.* **99**, 4024–35, 1993.

Locating minima

Details of optimization algorithms can be found in *Numerical Recipes*. A nice general reference devoted to the mathematical problem of optimization is

- R. Fletcher, *Practical Methods of Optimization*. J. Wiley & Sons, London, 1987.

Two reviews of geometry optimization methods for molecular systems are

- H. B. Schlegel. 'Optimization of Equilibrium Geometries and Transition Structures'. *Adv. Chem. Phys.* **67**, 249–86, 1987.
- J. D. Head and M. C. Zerner. 'Newton Based Optimization Methods for Obtaining Molecular Conformation'. *Adv. Quantum Chem.* **20**, 239–90, 1989.

Locating saddle points

A nice review of transition state location algorithms is

- S. Bell and J. S. Crighton. 'Locating Transition States'. *J. Chem. Phys.* **80**, 2464–75, 1984.

Papers covering the location of transition states using surface walking methods are

- C. J. Cerjan and W. H. Miller. 'On Finding Transition States'. *J. Chem. Phys.* **81**, 2800–6, 1981.
- J. Simons, P. Jorgensen, H. Taylor and J. Ozment. 'Walking on Potential Energy Surfaces'. *J. Phys. Chem.* **87**, 2745–53, 1983.
- J. Baker. 'An Algorithm for the Location of Transition States'. *J. Comput. Chem.* **4**, 385–95, 1986.

Following reaction paths

The formulation by Fukui of the intrinsic reaction coordinate approach is given in

- K. Fukui. 'A Formulation of the Reaction Coordinate'. *J. Phys. Chem.* **74**, 4161–3, 1970.

- K. Fukui. 'The Path of Chemical Reactions – The IRC Approach'. *Acc. Chem. Res.* **14**, 363–8, 1981.

More information on reaction path following may be found in

- K. Ishida, K. Morokuma and A. Komornicki. 'The Intrinsic Reaction Coordinate. An *ab initio* Calculation for HNC $\rightarrow$ HCN and H^- + CH_4 $\rightarrow$ CH_4 + H^-'. *J. Chem. Phys.* **66**, 2153–6, 1977.
- M. W. Schmidt, M. S. Gordon and M. Dupuis. 'The Intrinsic Reaction Coordinate and the Rotational Barrier in Silaethylene'. *J. Am. Chem. Soc.* **107**, 2585–9, 1985.
- C. Gonzalez and H. B. Schlegel. 'An Improved Algorithm for Reaction Path Following'. *J. Chem. Phys.* **90**, 2154–61, 1988.

Determining complete reaction paths

The reaction path methods of Elber and co-workers are described in the following papers:

- R. Elber and M. Karplus. 'A Method for Determining Reaction Paths in Large Molecules: Application to Myoglobin'. *Chem. Phys. Lett.* **139**, 375–80, 1987.
- R. Czermiński and R. Elber. 'Self-Avoiding Walk Between Two Fixed Points as a Tool to Calculate Reactions Paths in Large Molecular Systems'. *Int. J. Quantum Chem.: Quantum Chem. Symp.* **24**, 167–86, 1990.

An application of the method appears in

- R. Czermiński and R. Elber. 'Reaction Path Study of Conformational Transitions in Flexible Systems: Applications to Peptides'. *J. Chem. Phys.* **92**, 5580–601, 1990.

Chapter 7

Calculation of the normal modes

The classic text on the normal mode analysis of molecules is the following book:

- E. B. Wilson, Jr, J. C. Decius and P. C. Cross, *Molecular Vibrations. The Theory of Infrared and Raman Vibrational Spectra*. Dover Publications Inc., New York, 1955.

Rotational and translational modes

Useful discussions of the separation of rotational, translational and vibrational motion may also be found in the book on classical mechanics by Goldstein and in the following paper on reaction path dynamics:

- W. H. Miller, N. C. Handy and J. E. Adams. 'Reaction Path Hamiltonian for Polyatomic Molecules'. *J. Chem. Phys.* **72**, 99–112, 1980.

Calculation of thermodynamic functions

There is a plethora of good books on thermodynamics and statistical thermodynamics. A comprehensive reference is

- D. A. McQuarrie, *Statistical Mechanics*. Harper Collins, New York, 1976.

Two nice introductory texts with very different perspectives are

- E. B. Smith. *Basic Chemical Thermodynamics*. Oxford Chemistry Series 31, Clarendon Press, Oxford, 1982.
- D. Chandler. *Introduction to Modern Statistical Mechanics*. Oxford University Press, Oxford, 1987.

Chapter 8

Molecular dynamics

Good introductions to molecular dynamics calculations can be found in the general books listed at the beginning of the references. Verlet introduced the algorithm that bears his name in

- L. Verlet. 'Computer Experiments on Classical Fluids. I. Thermodynamical Properties of Lennard-Jones Molecules'. *Phys. Rev.* **159**, 98–103, 1967.

The velocity Verlet algorithm is described in

- W. C. Swope, H. C. Andersen, P. H. Berens and K. R. Wilson. 'A Computer Simulation Method for the Calculation of Equilibrium Constants for the Formation of Physical Clusters of Molecules: Application to Small Water Clusters'. *J. Chem. Phys.* **76**, 637–49, 1982.

Increasingly popular are *multiple-timestep integration algorithms* for molecular dynamics simulations. Two papers describing these techniques are

- M. Tuckerman, B. J. Berne and G. J. Martyna. 'Reversible Multiple Time Scale Molecular Dynamics'. *J. Chem. Phys.* **97**, 1990–2001, 1992.
- D. D. Humphreys, R. A. Friesner and B. J. Berne. 'A Multiple Time Step Molecular Dynamics Algorithm for Macromolecules'. *J. Phys. Chem.* **98**, 6885–92, 1994.

Simulated annealing

The classic reference on the method of simulated annealing is

- S. Kirkpatrick, C. D. Gelatt, Jr and M. P. Vecchi. 'Optimization by Simulated Annealing'. *Science* **220**, 671–80, 1983.

A nice review of simulated annealing and related methods and their application to the global optimization of complex molecular systems is

- I. Andricioaei and J. E. Straub. 'Finding the Needle in the Haystack: Algorithms for Conformational Optimization'. *Computers Phys.* **10**, 449–54, 1996.

Another powerful class of techniques for global optimization is the *genetic algorithms*. An accessible review of the principles behind these may be found in

- S. Forrest. 'Genetic Algorithms: Principles of Natural Selection Applied to Computation'. *Science* **261**, 872–8, 1993.

Two examples of applications of genetic algorithms to geometry optimization are

- B. Hartke. 'Global Geometry Optimization of Clusters Using Genetic Algorithms'. *J. Phys. Chem.* **97**, 9973–6, 1993.
- J. Mestres and G. E. Scuseria. 'Genetic Algorithms: A Robust Scheme for Geometry Optimizations and Global Minimum Structure Problems'. *J. Comput. Chem.* **16**, 729–42, 1995.

Chapter 9

Cutoff methods for the calculation of non-bonding interactions

Useful accounts of the calculation of non-bonding interactions can be found in the book by Allen and Tildesley as well as in *Computer Simulation of Biomolecular Systems*, Volume 2, edited by van Gunsteren, Weiner and Wilkinson. Specific references for the atom-based force-switching method are

- R. J. Loncharich and B. R. Brooks. 'The Effects of Truncating Long-Range Forces on Protein Dynamics'. *Proteins: Structure, Function, and Genetics* **6**, 32–45, 1989.
- P. J. Steinbach and B. R. Brooks. 'New Spherical-Cutoff Methods for Long-Range Forces in Macromolecular Simulation'. *J. Comput. Chem.* **15**, 667–83, 1994.

Including an environment

The article by Warshel and Russel cited above contains a description of a wide variety of models for the representation of the environment. Three brief reviews on the calculation of electrostatic interactions in macromolecules, which also discuss methods based upon the Poisson–Boltzmann equation, can be found in

- K. A. Sharp. 'Electrostatic Interactions in Macromolecules'. *Curr. Opin. Struct. Biol.* **4**, 234–9, 1994.
- M. K. Gilson. 'Theory of Electrostatic Interactions in Macromolecules'. *Curr. Opin. Struct. Biol.* **5**, 216–23, 1995.
- B. Honig and A. Nicholls. 'Classical Electrostatics in Biology and Chemistry'. *Science* **268**, 1144–9, 1995.

A solvent-accessible-surface-area model was introduced in

- D. Eisenberg and A. D. McLachlan. 'Solvation Energy in Protein Folding and Binding'. *Nature* **319**, 199–203, 1986.

The generalized Born/surface-area implicit solvation model of Still and co-workers is described in

- W. C. Still, A. Tempczyk, R. C. Hawley and T. Hendrickson. 'Semianalytical Treatment of Solvation for Molecular Mechanics and Dynamics'. *J. Am. Chem. Soc.* **112**, 6127–9, 1990.
- D. Qiu, P. S. Shenklin, F. P. Hollinger and W. C. Still. 'The GB/SA Continuum Model for Solvation. A Fast Analytical Method for the Calculation of Approximate Born Radii'. *J. Phys. Chem.* **A101**, 3005–14, 1997.
- S. R. Edinger, C. Cortis, P. S. Shenklin and R. A. Friesner. 'Solvation Free Energies of Peptides: Comparison of Approximate Continuum Solvation Models with Accurate Solution of the Poisson–Boltzmann Equation'. *J. Phys. Chem.* **B101**, 1190–7, 1997.

Ewald summation techniques

A nice general account of the Ewald technique can be found in the book by Allen and Tildesley. They also give a physical interpretation of the use of convergence functions in the summation of the electrostatic interactions.

Other general accounts may be found in

- D. E. Williams. 'Accelerated Convergence of Crystal-Lattice Potential Sums'. *Acta Cryst.* **A27**, 452–5, 1971.
- N. Karasawa and W. A. Goddard III. 'Acceleration of Convergence for Lattice Sums'. *J. Phys. Chem.* **93**, 7320–7, 1989.

For some original references consult

- P. Ewald. 'Die Berechnung optischer und elektrostatischer Gitterpotentiale'. *Ann. Phys.* **64**, 253–87, 1921.
- B. R. A. Nijboer and F. W. De Wette. 'On the Calculation of Lattice Sums'. *Physica* **23**, 309–21, 1957.

For details about the derivation see

- S. W. de Leeuw, J. W. Perram and E. R. Smith. 'Simulation of Electrostatic Systems in Periodic Boundary Conditions. I. Lattice Sums and Dielectric Constants'. *Proc. Roy. Soc. (London)* **A373**, 27–56, 1980.

Fast methods for the evaluation of non-bonding interactions

The fast multipole methods of Greengard and Rokhlin are described in

- L. F. Greengard and V. Rokhlin. 'A Fast Algorithm for Particle Simulations'. *J. Comput. Phys.* **73**, 325–48, 1987.
- L. F. Greengard, *The Rapid Evaluation of Potential Fields in Particle Systems.* MIT Press, Cambridge, MA, 1987.

Applications of fast multipole methods to molecular systems are described in

- H.-Q. Ding, N. Karasawa and W. A. Goddard III. 'Atomic Level Simulations on a Million Particles: The Cell Multipole Method for Coulomb and London Nonbond Interactions'. *J. Chem. Phys.* **97**, 4309–15, 1992.
- J. A. Board, Jr, J. W. Causey, J. F. Leathrum, Jr, A. Windemuth and K. Schulten. 'Accelerated Molecular Dynamics Simulation with the Parallel Fast Multipole Algorithm'. *Chem. Phys. Lett.* **198**, 89–94, 1992.

The algorithm of Hockney and Eastwood is described in

- R. W. Hockney and J. W. Eastwood, *Computer Simulation Using Particles.* McGraw-Hill, New York, 1981.

The particle mesh Ewald methods are discussed in

- T. Darden, D. York and L. G. Pedersen. 'Particle Mesh Ewald: An $N \ln(N)$ Method for Ewald Sums in Large Systems'. *J. Chem. Phys.* **98**, 10089–92, 1993.
- U. Essmann, L. Perara, M. L. Berkowitz, T. Darden, H. Lee and L. G. Pedersen. 'A Smooth Particle Mesh Ewald Method'. *J. Chem. Phys.* **103**, 8577–93, 1995.

Chapter 10

Analysis of molecular dynamics trajectories

Readers are referred to one of the many standard texts on statistical mechanics, such as the book by McQuarrie, for more details of the types of quantities that can be calculated from simulation data. Another good source of information is the book by Allen and Tildesley which has a background chapter on statistical mechanics as well as sections devoted to the calculation of various properties. This book also discusses some statistical methods for assessing the errors in the averages, fluctuations and time correlation functions calculated from a simulation.

A classic reference that discusses the statistical error arising when calculating averages from trajectories is

- R. Zwanzig and N. K. Ailawadi. 'Statistical Error due to Finite Time Averaging in Computer Experiments'. *Phys. Rev.* **182**, 193–6, 1969.

Another nice, although brief, discussion of statistical errors can be found in

- R. W. Pastor. 'Techniques and Applications of Langevin Dynamics Simulations'. In *The Molecular Dynamics of Liquid Crystals*, G. R. Luckhurst and C. A. Veracini (Editors), 85–138, Kluwer Academic Publishers, The Netherlands, 1994.

Example 17

Examples of molecular dynamics (and Monte Carlo) simulations of various models of water may be found in

- W. L. Jorgensen, J. Chandrasekhar, J. D. Madura, R. W. Impey and M. L. Klein. 'Comparison of Simple Potential Functions for Simulating Liquid Water'. *J. Chem. Phys.* **79**, 926–36, 1983.

Temperature and pressure control in molecular dynamics simulations

A concise description of constant-pressure and -temperature algorithms is given by Allen and Tildesley. The algorithm derived by Berendsen *et al.* that is used in this book may be found in

- H. J. C. Berendsen, J. P. M. Postma, W. F. van Gunsteren, A. Di Nola and J. R. Haak. 'Molecular Dynamics with Coupling to an External Bath'. *J. Chem. Phys.* **81**, 3684–90, 1984.

An early seminal paper on constant-pressure and -temperature algorithms was published by Andersen:

- H. C. Andersen. 'Molecular Dynamics Simulations at Constant Pressure and/or Temperature'. *J. Chem. Phys.* **72**, 2384–93, 1980.

The extension of this method to allowing the simulation box to change shape as well as size is described in

- M. Parrinello and A. Rahman. 'Crystal Structure and Pair Potentials: A Molecular Dynamics Study'. *Phys. Rev. Lett.* **45**, 1196–9, 1980.

The constant-pressure work of Nosé and Klein and Nosé–Hoover thermostating can be found in

- S. Nosé and M. L. Klein. 'Constant Pressure Molecular Dynamics for Molecular Systems'. *Molec. Phys.* **50**, 1055-76, 1983.
- S. Nosé. 'A Unified Formulation of the Constant Temperature Molecular Dynamics Methods'. *J. Chem. Phys.* **81**, 511–19, 1984.
- W. G. Hoover. 'Canonical Dynamics: Equilibrium Phase-Space Distributions'. *Phys. Rev.* A**31**, 1695–7, 1985.

Two more recent papers by Klein and co-workers on the extended system methods are

- G. J. Martyna, M. L. Klein and M. E. Tuckerman. 'Nosé–Hoover Chains: The Canonical Ensemble via Continuous Dynamics'. *J. Chem. Phys.* **97**, 2635–43, 1992.
- G. J. Martyna, D. J. Tobias and M. L. Klein. 'Constant Pressure Molecular Dynamics Algorithms'. *J. Chem. Phys.* **101**, 4177–89, 1994.

A novel recent modification of the Andersen constant pressure technique may be found in

- S. E. Feller, Y. Zhang, R. W. Pastor and B. R. Brooks. 'Constant Pressure Molecular Dynamics Simulation: the Langevin Piston Method'. *J. Chem. Phys.* **103**, 4613–21, 1995.

The constraint techniques are described in

- D. J. Evans and G. P. Morriss. 'Non-Newtonian Molecular Dynamics'. *Computer Phys. Rep.* **1**, 297–343, 1984.

Calculating free energies: umbrella sampling

The method of umbrella sampling is detailed in

- G. M. Torrie and J. P. Valleau. 'Nonphysical Sampling Distributions in Monte Carlo Free Energy Estimation: Umbrella Sampling'. *J. Comput. Phys.* **23**, 187–99, 1977.
- J. P. Valleau and G. M. Torrie. 'A Guide to Monte Carlo for Statistical Mechanics. 2. Byways'. In *Statistical Mechanics A. Modern Theoretical Chemistry*, Volume 5, B. J. Berne (Editor), 169–94, Plenum Press, New York, 1977.

Two references detailing the WHAM method of analysing umbrella sampling data are

- A. M. Ferrenberg and R. H. Swendsen. 'Optimized Monte Carlo Data Analysis'. *Phys. Rev. Lett.* **63**, 1195–8, 1989.

- S. Kumar, D. Bouzida, R. H. Swendsen, P. A. Kollman and J. M. Rosenberg. 'The Weighted Histogram Analysis Method for Free Energy Calculations on Biomolecules. I. The Method'. *J. Comput. Chem.* **13**, 1011–21, 1992.

A nice description of the WHAM methodology and the source for the example used in this book is

- B. Roux. 'The Calculation of the Potential of Mean Force Using Computer Simulations'. *Comput. Phys. Commun.* **91**, 275–82, 1995.

Chapter 11

The Metropolis Monte Carlo method

The book by Allen and Tildesley gives a nice discussion of all aspects of the Monte Carlo technique. The original paper introducing the Metropolis algorithm is

- N. Metropolis, A. W. Rosenbluth, M. N. Rosenbluth, A. H. Teller and E. Teller. 'Equation of State Calculations by Fast Computing Machines'. *J. Chem. Phys.* **21**, 1087–92, 1953.

For more details on the mathematics behind the Metropolis method, its relation to the theory of Markov chains and other Monte Carlo techniques a good review is

- J. P. Valleau and S. G. Whittington. 'A Guide to Monte Carlo for Statistical Mechanics. 1. Highways'. In *Statistical Mechanics A. Modern Theoretical Chemistry*, Volume 5, B. J. Berne (Editor), 137–68, Plenum Press, New York, 1977.

The extension of the Monte Carlo method to the NPT ensemble is discussed in

- W. W. Wood. 'Monte Carlo Calculations for Hard Disks in the Isothermal–Isobaric Ensemble'. *J. Chem. Phys.* **48**, 415–34, 1968.
- W. W. Wood. 'NPT Ensemble Monte Carlo Calculations for the Hard Disk Fluid'. *J. Chem. Phys.* **52**, 729–41, 1970.

Monte Carlo simulations of molecules

Jorgensen and co-workers have done simulation work using the Metropolis Monte Carlo method on a wide variety of systems. References to much of their work can be found in the papers on the development of the OPLS force field (chapter 5) and in the papers listed here. A review giving some idea of their approach, though, is

- W. L. Jorgensen. 'Theoretical Studies of Medium Effects on Conformational Equilibria'. *J. Phys. Chem.* **87**, 5304–14, 1983.

In an interesting paper Jorgensen and Tirado-Rives discussed the relative efficiencies of the Monte Carlo and molecular dynamics methods for sampling conformational space. They estimated that Monte Carlo methods are approximately twice as efficient for the systems they studied.

- W. L. Jorgensen and J. Tirado-Rives. 'Monte Carlo vs. Molecular Dynamics for Conformational Sampling'. *J. Phys. Chem.* **100**, 14508–13, 1996.

The method of preferential sampling is described in

- J. C. Owicki and H. A. Scheraga. 'Preferential Sampling near Solutes in Monte Carlo Calculations on Dilute Solutions'. *Chem. Phys. Lett.* **47**, 600–2, 1977.

More details of generalized coordinates, including how to derive equation (11.15) by integration over the generalized conjugate momenta, can be found in the books by Goldstein and by Wilson, Decius and Cross.

Example 21

Some early Monte Carlo studies on water and methane were performed by Owicki and Scheraga. These are detailed in

- J. C. Owicki and H. A. Scheraga. 'Monte Carlo Calculations in the Isothermal–Isobaric Ensemble. 1. Liquid Water'. *J. Am. Chem. Soc.* **99**, 7403–12, 1977.
- J. C. Owicki and H. A. Scheraga. 'Monte Carlo Calculations in the Isothermal–Isobaric Ensemble. 2. Dilute Aqueous Solution of Methane'. *J. Am. Chem. Soc.* **99**, 7413–18, 1977.

A paper by the Jorgensen group on the same system is

- W. L. Jorgensen, J. Gao and C. Ravimohan. 'Monte Carlo Simulations of Alkanes in Water: Hydration Numbers and the Hydrophobic Effect'. *J. Phys. Chem.* **89**, 3470–3, 1985.

Calculating free energies: statistical perturbation theory

Some articles reviewing the calculation of free energies using molecular dynamics and Monte Carlo simulations are

- W. L. Jorgensen. 'Free Energy Calculations: A Breakthrough for Modeling Organic Chemistry in Solution'. *Acc. Chem. Res.* **22**, 184–9, 1989.
- D. L. Beveridge and F. M. Di Capua. 'Free Energy via Molecular Simulation: Applications to Biomolecular Systems'. *Annu. Rev. Biophys. Biophys. Chem.* **18**, 431–92, 1989.

- T. P. Straatsma and J. A. McCammon. 'Computational Alchemy'. *Annu. Rev. Phys. Chem.* **43**, 407–35, 1992.
- P. A. Kollman. 'Free Energy Calculations: Applications to Chemical and Biochemical Phenomena'. *Chem. Rev.* **93**, 2395–417, 1993.

A recent application of Monte Carlo free-energy perturbation calculations to a complex problem is

- J. W. Essex, D. L. Severance, J. Tirado-Rives and W. L. Jorgensen. 'Monte Carlo Simulations for Proteins: Binding Affinities for Trypsin–Benzamidine Complexes via Free Energy Perturbations'. *J. Phys. Chem.* **B101**, 9663-9, 1997.

Example 22

A paper detailing free-energy simulations of water, methane and the chloride anion is

- W. L. Jorgensen, J. F. Blake and J. K. Buckner. 'Free Energy of TIP4P Water and the Free Energies of Hydration of CH_4 and Cl^- from Statistical Perturbation Theory'. *Chem. Phys.* **129**, 193–200, 1989.

Chapter 12

Constraints

Readers interested in background on the classical mechanical theory of constraints and generalized coordinates are advised to refer to the book by Goldstein. Details on how to calculate forces in internal coordinates may be found in the classic reference of Wilson, Decius and Cross.

For readers interested in geometry optimization methods with automatically generated sets of internal coordinates, the following references will be of value:

- P. Pulay and G. Fogarasi. 'Geometry Optimization in Redundant Internal Coordinates'. *J. Chem. Phys.* **96**, 2856–60, 1992.
- C. Peng, P. Y. Ayala, H. B. Schlegel and M. J. Frisch. 'Using Redundant Internal Coordinates to Optimize Equilibrium Geometries and Transition States'. *J. Comput. Chem.* **17**, 49–56, 1996.
- J. Baker, A. Kessi and B. Delley. 'The Generation and Use of Delocalized Internal Coordinates in Geometry Optimization'. *J. Chem. Phys.* **105**, 192–212, 1996.

One example of an algorithm for performing molecular dynamics simulations with torsional degrees of freedom only is

- L. M. Rice and A. T. Brünger. 'Torsion Angle Dynamics: Reduced Variable Conformational Sampling Enhances Crystallographic Structure Refinement'. *Proteins: Structure, Function, and Genetics* **19**, 277–90, 1994.

The SHAKE algorithm for constraints is described in the following publications:

- J. P. Ryckaert, G. Ciccotti and H. J. C. Berendsen. 'Numerical Integration of the Cartesian Equations of Motion of a System with Constraints: Molecular Dynamics of n-Alkanes'. *J. Comput. Phys.* **23**, 327–41, 1977.
- J. P. Ryckaert. 'Special Geometrical Constraints in the Molecular Dynamics of Chain Molecules'. *Molec. Phys.* **55**, 549–56, 1985.

A version of SHAKE for the velocity Verlet algorithm (called *RATTLE*) is outlined in

- H. C. Andersen. 'RATTLE: a Velocity Version of the SHAKE Algorithm for Molecular Dynamics Calculations'. *J. Comput. Phys.* **52**, 24–34, 1983.

The effect of using SHAKE to constrain various degrees of freedom is discussed in

- W. F. van Gunsteren and H. J. C. Berendsen. 'Algorithms for Macromolecular Dynamics and Constraint Dynamics'. *Molec. Phys.* **34**, 1311–27, 1977.
- W. F. van Gunsteren and M. Karplus. 'Effects of Constraints on the Dynamics of Macromolecules'. *Macromolecules* **15**, 1528–44, 1982.

Parametrization

Much information about parametrization of force fields can be found in the book by Burkert and Allinger. Other useful sources are the papers dealing with the development of particular force fields.

For those interested in non-linear least squares methods, mathematical details are given in the book by Press, Teukolsky, Vetterling and Flannery. Examples of applications to the parametrization process include

- A. T. Hagler and S. Lifson. 'A Procedure for Obtaining Energy Parameters from Crystal Packing.' *Acta Cryst.* **B30**, 1336–41, 1974.
- S. R. Niketic and K. Rasmussen, *The Consistent Force Field: A Documentation.* Lecture Notes in Chemistry 3, Springer, Heidelberg, 1977.
- L. Nilsson and M. Karplus. 'Empirical Energy Functions for Energy Minimization and Dynamics of Nucleic Acids'. *J. Comput. Chem.* **7**, 591–616, 1986.

The methodology for the fitting of charges from electrostatic potential data that is employed by many force fields is described in

- S. R. Cox and D. E. Williams. 'Representation of the Molecular Electrostatic Potential by a Net Atomic Charge Model'. *J. Comput. Chem.* **2**, 304–23, 1981.
- U. C. Singh and P. A. Kollman. 'An Approach to Computing Electrostatic Charges for Molecules'. *J. Comput. Chem.* **5**, 129–45, 1984.
- C. I. Bayly, P. Cieplak, W. D. Cornell and P. A. Kollman. 'Well Behaved Electrostatic Potential Based Method Using Charge Restraints for Deriving Atomic Charges: The RESP Model'. *J. Phys. Chem.* **97**, 10269–80, 1993.

Author index

Adams, J. E. 303
Ailawadi, N. K. 307
Alagona, G. 299
Allen, M. P. 295, 304, 305, 306, 307, 309
Allinger, N. 48, 67, 297, 299, 300, 312
Andersen, H. C. 215, 216, 303, 307, 312
Andricioaei, I. 304
Ayala, P. Y. 311

Baker, J. 95, 301, 311
Bashford, D. 299
Bayly, C. I. 299, 313
Bekker, H. 296
Bell, S. 301
Bellott, M. 299
Berendsen, H. J. C. 208, 209, 218, 285, 296, 307, 312
Berens, P. H. 303
Berkowitz, M. L. 306
Berne, B. J. 303, 308, 309
Bernstein, F. C. 296
Beveridge, D. L. 297, 310
Bird, R. B. 298
Blake, J. F. 311
Board Jr, J. A. 306
Bouzida, D. 309
Boyd, D. B. 295
Brice, J. M. D. 296
Brooks, B. R. 170, 173, 174, 299, 304, 308
Brooks III, C. L. 295
Brown, L. D. 301
Bruccoleri, R. E. 299
Brünger, A. T. 312
Buckner, J. K. 311
Burkert, U. 297, 312
Burlisch, R. 137

Caldwell, J. W. 299
Case, D. A. 299
Casewit, C. J. 300
Causey, J. W. 306
Cerjan, C. 96, 301
Chandler, D. 303

Chandrasekhar, J. 307
Ciccotti, G. 285, 312
Cieplak, P. 299, 313
Colwell, K. S. 300
Cornell, W. D. 299, 313
Cortis, C. 305
Cox, S. R. 313
Crighton, J. S. 301
Cross, P. C. 302, 310, 311
Curtiss, L. 298
Czermiński, R. 302

Darden, T. 194, 306
Davydov, A. S. 297
Decius, J. C. 302, 310, 311
de Leeuw, S. W. 306
Delley, B. 311
De Wette, F. W. 189, 305
Di Capua, F. M. 310
Ding, H.-Q. 306
Di Nola, A. 208, 307
Dinur, U. 300
Dirac, P. A. M. 43, 44, 297
Dunbrack Jr, R. L. 299
Dupuis, M. 302

Eastwood, J. W. 194, 306
Edinger, S. R. 305
Eisenberg, D. 305
Elber, R. 105, 302
Essex, J. W. 311
Essmann, U. 306
Evans, D. J. 308
Evanseck, J. 299
Ewald, P. 305
Ewig, C. S. 300

Fan, Y. 300
Feller, S. E. 308
Ferguson, D. M. 299
Ferrenberg, A. M. 223, 308
Field, M. J. 299
Fischer, S. 299

315

Flannery, B. P. 294, 312
Fletcher, R. 301
Fogarasi, G. 311
Forrest, S. 304
Fox, T. 299
Frenkel, D. 295
Friesner, R. A. 303, 305
Frisch, M. J. 311
Fukui, K. 301, 302

Gallucci, J. C. 299
Gao, J. 299, 310
Geise, H. J. 299
Gelatt Jr, C. D. 155, 304
Ghio, C. 299
Gilson, M. K. 305
Goddard III, W. A. 298, 300, 305, 306
Goldstein, H. 296, 302, 310, 311
Gonzalez, C. 302
Gordon, M. S. 302
Gould, I. R. 299
Greengard, L. F. 194, 306
Guo, H. 299

Ha, S. 299
Haak, J. R. 208, 307
Hagler, A. T. 300, 312
Haile, J. M. 295
Halgren, T. A. 300
Handy, N. C. 303
Hartke, B. 304
Harvey, S. 295
Hawley, R. C. 305
Hay, B. P. 300
Head, J. D. 301
Hehre, W. J. 297
Hendrickson, T. 305
Hirschfelder, J. O. 298
Hockney, R. W. 194, 306
Hollinger, F. P. 305
Honig, B. 305
Hoover, W. G. 214, 215, 308
Hsu, D. 301
Humphreys, D. D. 303
Hwang, M. J. 300

Impey, R. W. 307
Ishida, K. 302

Jorgensen, P. 301
Jorgensen, W. 68, 69, 245, 246, 298, 307, 309, 310, 311
Joseph, D. 299

Kabsch, W. 38, 297
Karasawa, N. 305, 306
Karplus, M. 69, 295, 299, 302, 312
Kennard, O. 296
Kessi, A. 311
Kirkpatrick, S. 155, 304
Klein, M. L. 214, 216, 307, 308
Kneller, G. 38, 297

Koetzle, T. F. 296
Kollman, P. A. 69, 299, 309, 311, 313
Komornicki, A. 302
Kuchnir, L. 299
Kuczera, K. 299
Kumar, S. 223, 224, 309

Lau, F. T. K. 299
Leach, A. R. 295
Leathrum Jr, J. F. 306
Lee, H. 306
Lifson, S. 48, 298, 312
Lipkowitz, K. B. 295
Loncharich, R. J. 304
Luckhurst, G. R. 307

Ma, J. 301
MacKerell, A. D. 299
Madura, J. D. 307
Maple, J. R. 300
Martyna, G. J. 214, 215, 216, 303, 308
Mattos, C. 299
Maxwell, D. S. 298
Mayer, E. F. 296
Mayo, S. L. 300
McCammon, J. A. 295, 311
McLachlan, A. D. 305
McQuarrie, D. A. 303, 306
Merz Jr, K. M. 299
Mestres, J. 304
Metcalf, M. 294
Metropolis, N. 236, 237, 309
Michnick, S. 299
Miller, W. H. 96, 301, 303
Morokuma, K. 302
Morriss, G. P. 308
Müller, K. 301

Ngo, T. 299
Nguyen, D. T. 299
Nicholls, A. 305
Nijboer, B. R. A. 189, 305
Niketic, S. R. 312
Nilsson, L. 312
Nosé, S. 214, 215, 216, 307, 308

Olafson, B. D. 299, 300
Ostlund, N. S. 297
Owicki, J. C. 310
Ozment, J. 301

Paquette, L. A. 299
Parr, R. G. 297
Parrinello, M. 216, 307
Pastor, R. W. 307, 308
Pedersen, L. G. 194, 306
Peng, C. 311
Perara, L. 306
Perram, J. W. 306
Pettitt, B. M. 295
Pople, J. A. 297
Postma, J. P. M. 208, 307

Press, W. H. 294, 312
Price, S. L. 298
Prodhom, B. 299
Profeta Jr, S. 299
Pulay, P. 311
Pyckhout, W. 299

Qiu, D. 305

Radom, L. 297
Rahman, A. 216, 307
Rapoport, D. C. 295
Rappé, A. 298, 300
Rasmussen, K. 312
Ravimohan, C. 310
Reid, J. 294
Reiher III, W. E. 299
Rice, L. M. 312
Rodgers, J. R. 296
Rokhlin, V. 194, 306
Rosenberg, J. M. 309
Rosenbluth, A. W. 309
Rosenbluth, M. N. 309
Roux, B. 299, 309
Russel, S. T. 298, 304
Ryckaert, J. P. 285, 312

Scheraga, H. A. 310
Schlegel, H. B. 301, 302, 311
Schlenkrich, M. 299
Schmidt, M. W. 302
Schulten, K. 306
Scuseria, G. E. 304
Severance, D. L. 311
Sharp, K. A. 305
Shenklin, P. S. 305
Shimanouchi, T. 296
Simons, J. 96, 301
Singh, U. C. 299, 313
Skiff, W. M. 300
Smit, B. 295
Smith, E. B. 303
Smith, E. R. 306
Smith, J. C. 299
Spellmeyer, D. C. 299
Stahl, M. 296
States, D. J. 299
Steinbach, P. J. 170, 173, 174, 304
Still, W. C. 181, 305
Stockfish, T. P. 300
Stoer, J. 137
Stone, A. J. 298
Stote, R. 299
Straatsma, T. P. 311
Straub, J. E. 299, 301, 304

Sutcliffe, B. T. 297
Swaminathan, S. 299
Swendsen, R. H. 223, 308, 309
Swope, W. C. 303
Szabo, A. 297

Tai, J. C. 300
Tasumi, M. 296
Taylor, H. 301
Teller, A. H. 309
Teller, E. 309
Tempczyk, A. 305
Teukolsky, W. T. 294, 312
Tildesley, D. J. 295, 304, 305, 306, 307, 309
Tirado-Rives, J. 298, 310, 311
Tobias, D. J. 308
Torrie, G. M. 221, 308
Tuckerman, M. E. 214, 303, 308

Ulam, S. 236

Valleau, J. P. 221, 308, 309
van Gunsteren, W. F. 208, 296, 304, 307, 312
Vecchi, M. P. 155, 304
Veracini, C. A. 307
Verlet, L. 137, 303
Vetterling, W. T. 294, 312
von Neumann, J. 236
von Ragué Schleyer, P. 295, 297

Waldman, M. 300
Walters, P. 296
Warshel, A. 298, 304
Watanabe, M. 299
Weiner, P. 296, 299, 304
Weiner, S. J. 299
Whittington, S. G. 309
Wilkinson, A. 296, 304
Williams, D. E. 188, 305, 313
Williams, G. J. B. 296
Wilson Jr, E. B. 302, 310, 311
Wilson, K. R. 303
Windemuth, A. 306
Wiórkiewicz-Kuczera, J. 299
Wood, W. W. 238, 309

Yang, L. R. 300
Yang, W. 297
Yin, D. 299
York, D. 194, 306

Zerner, M. C. 301
Zhang, Y. 308
Zwanzig, R. 307

Subject index

ab initio quantum mechanical methods 46
acceptance ratio 241, 248
activated complex 129
alanine dipeptide; *see* N-methyl-alanyl-acetamide
algorithms
    computational effort 26
    design 27
    linear scaling 26, 193–4
    numerical 294
    overhead cost 27
    prescreening step 27
    scaling behaviour 26ff.
amino acids 12, 14, 78, 81
    C- and N-terminal variants 77–8, 81
anharmonic theories 113
arrays
    allocation 3, 18
    syntax 22
asymmetric center 42
atom-based force-switching; *see* non-bonding
    interactions
atom pairs, number of 26
atoms 12ff., 25ff.
    accelerations 113
    bonding distances 26
    bonding radii 26
    charges 35, 53, 72, 259, 263, 288–9
    coordinates 13ff.
    dummy 271–2
    extended 24
    fixing 283
    masses 5, 17, 35, 113, 136
    momenta 136
    names 16, 17
    nuclei 45
    numbers 16, 17
    radii 5
    types 71, 76
    united 24
    velocities 123, 136, 140
    weights 35
average; *see* ensemble, thermodynamic; statistical
    analysis

Baker's algorithm 96–100, 110–11, 122
bALA; *see* N-methyl-alanyl-acetamide
barostat; *see* degrees of freedom
bath, external 209
    coupling to 209ff.
    pressure 210ff.
    thermal 209ff.
Berendsen algorithm 209–10, 213, 218
biasing function; *see* umbrella sampling
Boltzmann distribution law 213
Boltzmann factor 156, 213, 236
bond angles 25, 30ff.
    generation of 28–9, 33
bonding interactions 47–52, 57–8; *see also*
    interactions
bonds 25, 30ff.
    as constraints 161
    generation of 25–9, 32–4
    lengths 30
    pattern of 25
    unchemical 73
Born expression 180, 267
Born model of solvation 180–1
    effective radii 181
    generalized 181, 305
Born–Oppenheimer approximation 43–5, 135
boundary approximations 181–2
boundary potential 181–2
bounding box 27
Buckingham potential 53
buffer distance 26

center of charge 35
center of geometry 35
center of mass 35, 243, 278
    motion 144
central-step finite-difference method 60
characteristic equation; *see* secular equation
charge distribution 52–3, 180, 288
charge-equilibration models; *see* fluctuating charge
    models
charge fitting; *see* electrostatic potential
chirality 34, 42

chloride anion; *see* ions
classical mechanics 43, 113, 127, 135, 296, 302;
  see also Newton's laws
combination rules 54–5
comment statements 8, 15
common block 2
complementary error function 5, 189
computational science 26
condensed phase; *see* systems
conditional convergence, of series 190
configuration integral 220
conjugate gradient algorithm 5, 91ff.
connectivity; *see* system; Z-matrices
conservation conditions 138–9
  total energy 147
constant-pressure simulation; *see* molecular
  dynamics simulations
constants
  chemical 5
  mathematical 5
  physical 5
constant-temperature simulation; *see* molecular
  dynamics simulations
constrained minimization; *see* optimization
constraints 105, 119, 282–5, 311–12
  distance 161
  forces 284
  hard or rigid 283
  holonomic 284
  soft 283
continuum models; *see* reaction field models
convergence criteria geometry optimization 93, 95,
  100, 110
convergence function 188, 305
cooling, simulated annealing 156
coordinate file 12–24
  format 14ff., 296
  input and output 19–23
  protein data bank 24
  structure 14–16
  symmetry in 186
coordinates 13ff.; *see also* transformations
  Cartesian 14, 17, 34, 98, 283–4
  crystallographic 14
  data banks 14
  distance search 25
  fractional 238, 243, 278
  freezing 270
  generalized 244–5, 284, 310, 311
  internal 14, 29–32, 38, 268, 296, 311
  mass-weighted 101, 114, 120
  operations on 25–42
  redundancy in 98, 118
  sets 13
  undefined 17
coordination compounds 300
correlation time 201
coupling; *see also* bath
  constant
    pressure 209, 211, 218, 233
    temperature 210, 211, 218, 229, 233
  parameter 255ff.

scheme 255ff.
crambin
  energy 85
  non-bonding interactions in 175–8
  sequence file for 81
critical system size 28
cubic systems; *see* periodic boundary conditions
curvature; *see* potential energy surface
cutoff, non-bonding; *see* non-bonding interactions
cyclohexane
  geometry optimization 99–100, 111
  normal mode calculation 124–5, 126
  reaction-path following 104–5, 111
  self-avoiding walk calculation 109–10
  structures 100
  thermodynamic quantities 131–4

data collection, molecular dynamics 146
Debye–Hückel parameter 180
default precision 4
degrees of freedom 118
  barostat 216
  constraining 231–2, 239, 312
  freezing 270
  number of 141, 144
  potential of mean force 222ff.
  rotational and translational 106, 118–22, 141,
    270
  thermostat 214–15
  torsional 311
density functional theory 46
derivatives; *see also* differentiation
  first 61, 87, 91–2
  second 61–2, 87, 91–2
dielectric constant 53, 179–80
differential equations, ordinary 114, 136
  integration of 137
differentiation
  analytical 60–1
  numerical 60–1, 65–6
diffusion coefficient 198–9
  water 206, 232
dihedral angles 25, 30ff., 296
  cis 31
  definition in MM file 76–7
  generation of 28–9, 33
  improper 50
  proper 50
  trans 31
dipole moment 53, 116
  derivatives 116
  induced 55–8
  of periodic box 190
  origin-dependence 116
  splitting of; *see* splitting of the dipoles
Dirac delta function 213, 222
distance-dependent dielectric function 180
disulphide bridges 81
DYNAMO module library 1ff., 291–3
  source code distribution 7
  version number 70

Eckart conditions 119–22
effective potential 45
Einstein relation 199, 202
  water 207
electric field 55–6, 118
electron density 25, 288
electron diffraction 286
electrons 45
  π 46
  σ 46
electrostatic potential 179–80, 193
  charge fitting 288, 312–13
elements, chemical 5 radii 26
empirical energy function; *see* potential energy
  functions
empirical potential; *see* potential energy functions
energy; *see also* potential energy functions
  absolute 60
  angle 47–9, 68
  bond 47–8, 68
  bonding 47
  covalent 47
  cross-terms 52
  derivatives 60–2
  dihedral 47, 49–50, 68
  electronic 45, 127
  electrostatic 52–3, 69, 164ff.
  improper dihedral 47, 50, 68
  interaction 242, 251–3, 265
  invariance to rotation and translation 118
  kinetic 44, 122–3, 136, 140, 141, 147
  Lennard-Jones 53–5, 69, 164ff.
  Morse 48–9
  non-bonding 52–9, 164ff., 245–6
  non-polar 181
  nuclear 127
  out-of-plane 51
  polarization 55–8
  potential 43ff., 68ff., 122, 136, 147
  relative 43, 60
  rotational 127
  total 122–3, 136, 147
  translational 127
  vibrational 127
ensemble, thermodynamic
  average 140, 212, 221ff., 236ff.
  biased 221–2
  canonical or NVT 208, 213, 219, 236ff.
  isobaric–isoenthalpic or NPH 208, 216
  isothermal–isobaric or NPT 208, 213, 216, 219,
    238–9, 309
  microcanonical or NVE 208, 213, 214, 233
  probability density distribution 212–13, 220, 236
  unbiased 222, 224
enthalpy 129
  of formation 60
entropy 129
environment; *see also* non-bonding interactions;
  solvent; system
  inclusion of 178–82, 304–5
enzyme active site 47
equilibration

molecular dynamics 146, 214, 218–19, 229
  Monte Carlo 263, 278
equilibrium constants 129, 133–4
ergodicity 238
errors
  in molecular dynamics integration 139
  in theoretical model 200
  run-time 4
  statistical 200–1, 307
ethane
  non-bonding exclusions 66
  Z-matrix 268–9
Euler angles 242
Euler integration formula 235
Ewald summation techniques 183, 187–95, 305–6
  real space term 190
  reciprocal space term 190
  self-energy term 190
  surface correction term 190
  tin-foil boundary conditions 190, 192
exchange repulsion 52, 59, 287
extended system methods 214, 216, 308

fast Fourier transform 194, 204
fast multipole methods; *see* Non-bonding
  interactions
files 4; *see also* FORTRAN DYNAMO library files 293
floating point numbers 3, 4
fluctuating charge models 298
fluctuation; *see* statistical analysis
force, on atoms 113
force field; *see* potential energy function
format
  fixed 4
  free 4, 16
  module 2–4
  program 6–7
FORTRAN
  FORTRAN 77 2
  FORTRAN 90 2, 3, 5, 8, 22, 149, 229, 294
  FORTRAN 95 3, 5, 294
  streams 4
fourier expansion 68
fourier transform 189
fractional coordinates; *see* coordinates
frame, trajectory 124, 125, 148
free energy calculation of 219–33, 253–67, 308,
  310–11
  differences 130, 254ff.
  Gibbs 129
  Helmholtz 129, 220, 253
  hydration 257
  ligand binding 257
  potential of mean force 222ff., 253
  window 224
frequency; *see* Normal modes; Vibrational motion
friction coefficient 214 force 209, 215
function, programming 2ff.

gamma function
  complete 189
  incomplete 189

Gaussian random variables 140
Gauss's principle of least constraint 216
generalized Born model; *see* Born model
generalized coordinates; *see* coordinates
genetic algorithms 304
global minimum; *see* minima
global search algorithms; *see* optimization
gradients 61; *see also* derivatives
    mass-weighted 101
    modified 120
    RMS 64, 93, 100

haemoglobin 12, 13
Hamiltonian 119
    Andersen 216
    classical 136, 138, 214
    in free energy calculations 255, 258
    Nosé–Hoover 214–15
    operator 44
Hamilton's equations 136
harmonic approximation 113, 116, 122–3
harmonic function 47, 223
harmonic oscillators 123
heat capacities
    constant pressure 129, 198
    constant volume 129, 198
heating, molecular dynamics 146, 229
    unwanted 166
Hessian 62; *see also* derivatives
    eigenvalues and eigenvectors 88–9
    mass-weighted 102, 114–15, 120
    modified 98, 122
    projected 122
    storage schemes 63
    updating formulae 98
hydrogen-bonding 59, 227
hydrophobicity 251
hysteresis 256

ideal gas 128
importance sampling 236
infrared intensities 116, 125
initial value algorithms 137
interactions; *see also* bonding interactions; non-
    bonding interactions
    charge–dipole 57, 169–70
    Coulomb 166–9, 172, 188, 190
    dipole–dipole 57, 170
    dispersion 52, 59, 189, 192
    electrostatic 52, 59, 165, 304–5
    excluded 58
    induced 52
    intermolecular, theory of 52
    many-body 57
    non-polar 179, 181
    pairwise additive 57
    polarization 52
integer numbers 3
integration; *see also* differential equations
    ordinary; molecular dynamics simulations
    of a function 234–6
intermediate states 255

internal coordinates; *see* coordinates; Z-matrices
internal energy, thermodynamic 129
intrinsic reaction coordinate; *see* reaction paths
ions
    complexes with water; *see* water
    molecular mechanics definition file 69–73
isothermal compressibility 210, 211, 218
isotope effects 134
isotopes 17, 45, 134

k vectors; *see* vectors
kinetic energy; *see* energy

Lagrange multipliers 39, 284–5
Langevin equation 215
Lennard-Jones clusters
    energies and minima of 90–1
    simulated annealing 157–63
linear algebra 5
line search techniques 102
links, between residues 77, 79, 81
local search algorithms; *see* optimization
logical variables 3
lower triangle storage 63

machine dependence 4, 7
Markov chains 237, 309
matrices
    diagonalization 5, 35
    eigenvalues 5, 35, 88
    eigenvectors 5, 35, 88
    inertia 35
    projection 122
    real symmetric 5, 62
    rotation 34ff.
Maxwell–Boltzmann distribution 140, 144, 215
mechanism, of reaction or transition 101
methane
    Monte Carlo simulation of, in water 249–53,
        310–11
Metropolis Monte Carlo; *see* Monte Carlo
    simulations
microwave spectroscopy 286
minima, on potential energy surfaces 87ff.; *see also*
    stationary point
    definition of 87
    global 90, 155
    location of 91ff., 301
minimization; *see* optimization
minimum image convention 182–7, 192, 246, 250;
    see also periodic boundary conditions
MKSA system of units 53
MM file; *see* molecular mechanics definition file
mode following
    for geometry optimization 95, 98
modes; *see* normal modes
modules 2ff.
    example 2
    initialization 4
    mathematical 5, 291
    miscellaneous 4–6
    module declarations 5ff., 291

parameter data 5, 291
  private items 2
  public items 2
  scientific 4, 291, 292
  utility 4, 291
molecular dynamics simulations 112, 135–63,
    196–233, 303, 306–7, 310, 311
  integration scheme 139
  pressure control 207–19, 307–8
  temperature control 141–2, 207–19, 229, 307–8
  velocity assignment 140–1, 142, 144
molecular mechanics 47, 297–8
molecular mechanics definition file 67–79
    processing of 79–80
molecular orbital theory 46
molecular structure; *see* structures
molecules
  distinguishable 127
  indistinguishable 127
  linear 128, 141
  rigid 240
moments of inertia 35, 128
Monte Carlo simulations 112, 157, 234–67, 277–8,
    307, 309–11
  comparison with molecular dynamics 239, 310
  geometry moves 244–5
  Metropolis algorithm 236–9, 241–2, 244, 309
  molecule moves 241–2
  restrictions on module use 245
  rotational moves 242–3, 249
  translational moves 240–2, 249
  volume moves 243–4, 248–9
Morse potential 48
Müller–Brown potential energy surface 89, 300
multiple-timestep integration algorithms 303
multipole moments 193–4, 288

neighbour function; *see* radial distribution
    function
Newton optimization methods 91–2
  exact 91
  Newton–Raphson 91
  quasi 91
  reduced basis-set 91
  truncated 91
Newton–Raphson step 96
Newton's equations of motion 136, 214ff.
Newton's laws 113
N-methyl-alanyl-acetamide 14, 16
  coordinate input and output 21–3
  coordinate manipulation 40–1
  energy and derivatives 83–5
  internal coordinates 32–4
  molecular dynamics simulations in vacuum
    145–7, 163
  molecular mechanics definition file 73–7
  optimization of 93–5
  ϕ–ψ map 154
  potential energy surface of 110
  potential of mean force calculation 227–33
  sequence file 80–1
  solvating 279–82

system file 83
trajectory analysis 151–5
Z-matrix input and output 272
non-bonding exclusions 58–9, 64–5, 69, 164, 191
  weighting factor 69, 71, 289
non-bonding interactions 47, 52–9, 164–95, 298,
    304; *see also* Interactions
  atom-based force-switching 170–3
  atom-based truncation 169
  cutoff methods 164–78, 245–6, 304
  fast methods 193–5, 306
  in Monte Carlo modules 245–6
  interaction lists 170, 174, 185
  residue-based truncation 169, 173, 245–6
  shift function 166–9
  smoothing function 166
  switch function 166–9
  truncation function 165, 167, 245–6
  truncation methods 165
non-linear least squares algorithm 287, 312
normal modes 115, 287
  amplitude of motion 123
  analysis 112–34, 302
  calculation of 112–18, 302
  degenerate 125
  frequencies 115, 125, 287
    imaginary 115, 123–4, 130
  in geometry optimization 96
  rotational and translational 118–22, 302–3
  soft 98
  trajectories of 122–4
  zero-frequency 120, 123–4
Nosé–Hoover algorithm 214–15, 307
nuclear dynamics 45, 135
nuclear magnetic resonance spectroscopy 14, 283,
    296
nucleic acids 68

observables 60, 287
optical isomers 42
optimization, geometry 5, 90, 301
  constrained 102, 231–2
  global algorithms 90, 155, 304
  in internal coordinates 270, 284, 311
  local algorithms for minima 91–3, 155, 301
  local algorithms for saddle points 95–8, 301
orthorhombic systems; *see* periodic boundary
    conditions

pair distribution function; *see* radial distribution
    function
parameters, empirical 46
parameters, force field; *see also* atoms; non-
    bonding exclusions
  angle 48, 72–3, 286–7
  bond 48, 72–3, 286–7
  definition in MM file 69–79
  dihedral angle 49–50, 72, 76–7, 287–8
  improper dihedral angle 50–1, 72, 77, 288
  Lennard-Jones 53–5, 71, 161, 259, 264–5, 289
  transferability 72, 288
parametrization 46, 59–60, 285–90, 312–13

particle mesh Ewald method 194, 306
particles
  composite 24
  probability density distribution 44
partition function 127–9
  classical 219–20, 253
  electronic 127–8
  nuclear 127–8
  rotational 127–8
  translational 127–8
  vibrational 127–30
peptide bonds 14
periodic boundary conditions 179ff., 200
  cubic 182ff., 238, 243
  dodecahedral 182
  hexagonal 182
  orthorhombic 182, 189
  triclinic 182
  truncated octahedral 182
perturbation
  alchemical 257
  backwards 256, 264
  forwards 256, 264
  parameter; *see* coupling
phase space 213, 219
pointers 3, 29
Poisson–Boltzmann equation 179–80, 305
Poisson equation 180, 194
polarizability 55, 298
  anisotropic dipole 55
  isotropic dipole 55
potential energy; *see* energy
potential energy functions 46ff., 67–9, 297,
    298–300; *see also* parametrization
  AMBER 69, 77, 288, 299
  CHARMM 69, 299
  derivatives 60–2
  hybrid 46
  MM2, MM3 67, 299–300
  MMFF94 300
  OPLS, OPLS-AA 68ff., 77, 191, 245, 250, 288–9,
    298–9, 309
  QMFF 300
  UFF 300
potential energy surface 45, 135, 297
  curvature 96
  exploration of 86–91, 300–1
  model 89
  obtaining energies on 45–7
  walking on 301
potential of mean force; *see* free energy
predictor–corrector integrators 137
preferential sampling 242, 310
pressure 129, 131
  control; *see* molecular dynamics simulations
  fluctuations 218
  instantaneous 210
  isotropic system 210
  reference 209
principal axes 35; *see also* transformations
probability density distribution; *see* ensemble;
    particles

procedure, programming 2ff.
programming languages 2; *see also* FORTRAN
  object-oriented 2
programs, example 6–7
protein data bank (PDB) 24, 296
proteins 12, 14, 68, 78, 296; *see also* crambin;
    haemoglobin

quantum mechanical calculations 46, 60, 288, 297
quantum mechanics 43–5, 113, 127, 135, 297
quantum states 127
quaternions 38–9

radial distribution function 199–200, 202
  methane in water 251
  neighbour function 251–2
  water 207, 208
radius of convergence 285
radius of gyration 41
random force 215
random numbers
  Gaussian distribution 5
  initialization of seed 5, 146, 162, 250
  uniform distribution 5
  vector of 277
rate constants 129–30
RATTLE algorithm 312
reaction field models 179–81
reaction paths 89; *see also* self-avoiding walk
    algorithm
  complete 105, 302
  definition of 100
  following 100–5, 301–2
  intrinsic reaction coordinate 101, 301
  steepest descent 101
real space term; *see* Ewald summation techniques
reciprocal space 189; *see also* vectors
reciprocal space term; *see* Ewald summation
    techniques
reduced units 90, 161
residues 12ff.
  definition in MM file 72, 76
  index array 19
  names 19
  protonation state 13
rigid-rotor, harmonic oscillator approximation
    127–31, 198, 219
root mean square (RMS) coordinate deviation
    38–40, 106, 154–5
root mean square (RMS) deviation; *see* statistical
    analysis
rotation; *see* transformations
rotational motion 36, 98, 118–22; *see also* degrees
    of freedom; normal modes

saddle points 88ff.
  as transition state 130
  definition of 88
  location of 95ff., 301
sampling, of configuration space 220–1, 223
  enhancing 221
scaling; *see* algorithms

Schrödinger equation
  electronic 45, 46
  time-dependent 44, 135
  time-independent 44
screening, non-bonding interactions 180
secular equation 88
self-avoiding walk algorithm 105–8, 119
self-energy term; *see* Ewald summation techniques
semi-empirical quantum mechanical methods 46
sequence
  data 16ff.
  file 80
SHAKE algorithm 285, 312
shift function; *see* non-bonding interactions
simplex method 91
simulated annealing 155–63, 303–4
smoothing function; *see* non-bonding interactions
sodium cation; *see* ions
solvation
  energy 180–2
  of molecules 278–82
  shells 207, 252
solvent bath 12
  boxes, construction of 272–8
  explicit models 179, 181
  implicit models 179–82, 305
solvent-accessible surface area 181, 305
sorting 5
specific heats; *see* heat capacities
spin 44
splitting of the dipoles 169–70, 173
stationary point 5, 87ff.; *see also* minima; saddle
    points
  definition of 87
  dynamics around 115
stationary state 44
statistical analysis 5; *see also* errors
  averages 148, 197–8, 307
  fluctuations 148, 197–8, 307
  root mean square deviation 154, 197
  use of blocks of data 201, 248
statistical mechanics 125–31, 140, 149, 156, 198,
    222, 303, 306
statistical perturbation methods; *see*
    Thermodynamic perturbation methods
statistical thermodynamics; *see* statistical
    mechanics
steepest descent method 91
steepest descent reaction path; *see* reaction paths
Stoermer's rule 137
stress tensor 210
strings, character 4
structures
  chain of 105, 108
  chemical 29
  distance between 106
  molecular 13
  stability of 43
subroutine 2ff.
subsystems 12ff.
  index array 19
  names 19

superposition; *see* transformations
surface correction term; *see* Ewald summation
    techniques
surface free energy 181, 305
switch function; *see* non-bonding interactions
symbols 9–10
symmetry, of a system 184, 186; *see also* periodic
    boundary conditions
symmetry number 128, 133
system
  angular momentum 138
  composition 12ff.
  condensed phase 164ff., 209, 272
  connectivity 25–9
  definition of 80–3
  force on 139
  mass 128
  momentum 138
  size 12, 26, 86
  torque on 139
system file 83 editing 278–9

Taylor expansion 96, 112, 137
temperature
  absolute 123, 129, 140
  control; *see* molecular dynamics simulations
  fluctuations 147–8, 218
  in simulated annealing 156
  instantaneous 140, 209
  reference 209
thermal conductivity 198
thermodynamic cycle 257–8
thermodynamic integration 254, 258
thermodynamic perturbation methods 254ff.,
    310–11
thermodynamic quantities; *see also* ensemble,
    thermodynamic
  calculation of 125–31, 303
thermostat; *see* degrees of freedom
time
  CPU 5, 187
  current 5
time correlation functions 197–9, 203–4, 307
  auto 197
  cross 197
  exponential form 201
  long-time tails 201
  normalization 197
  stationary 198
time series 148, 196
timestep, molecular dynamics 137–40, 142, 163, 285
tin-foil boundary conditions; *see* Ewald summation
    techniques
torsion angles; *see* dihedral angles
trajectory file 103, 107, 124, 143, 250
  analysis 147–8, 151–5, 196–207, 306–7
  input and output 149–51, 204–5
  symmetry in 204–5
transformations 34–42, 296
  improper rotation 34
  principal axis 35ff., 271
  proper rotation 34, 242–3

superposition 37–41, 106, 296
  translation 34, 240–1
transition state structure 129
transition state theory 129, 222
translation; *see* transformations
translational motion 98, 118–22; *see also* degrees of
    freedom; normal modes
transport coefficients 198
truncation, non-bonding; *see* non-bonding
    interactions
tunneling, quantum mechanical 136

umbrella potential 222ff.
  harmonic 223
umbrella sampling 221–32, 253, 282, 308
  biasing function 221–2, 253
units 8, 11, 71, 131
updating formulae; *see* Hessian
upper triangle storage 63

valence bond theory 46
variants, of residues 77–8, 81
vectors
  displacement 96, 113–15
  normalization 5
  normal mode 116, 287
  orthonormal 88
  reciprocal space or *k* vectors 189, 192–3
  translation 34
velocities; *see also* atoms; molecular dynamics
    simulations
  input and output 143–4
velocity autocorrelation function 198–9, 232
Verlet integrators 137–40, 284–5, 303
  leapfrog 138
  velocity Verlet 138, 303, 312

vibrational frequencies; *see* normal modes
vibrational infrared spectroscopy 115
vibrational motion 119, 302
  high frequency 140, 285
virial, instantaneous internal 210
  calculation of 212
viscosity bulk 198 shear 198
volume 128, 129, 189, 209, 216, 220, 238
  fluctuations 219

water
  atom names 12
  box, construction of 272–8
  complexes with ions 85, 169–70
  dimer 85, 170
  free energy of 259–67, 311
  models 307
  molecular dynamics 185–7, 217–19, 307
  molecular mechanics definition file 69–73
  Monte Carlo calculations 307, 310
  size of a molecule 281
  trajectory analysis 205–7
  Z-matrix 276
wavefunction 44
weighted histogram analysis method 223ff., 308–9
window, free-energy calculations 222ff., 256ff.
World Wide Web 291

X-ray crystallography 14, 268, 283, 286, 288, 296

zero-point motion 136
Z-matrices 268–72
  connectivity 269
  input and output 270–1
problems with 270